CORAL REEFS: TOURISM, CONSERVATION AND MANAGEMENT

Coral reefs are an important tourism resource for many coastal and island destinations and generate a range of benefits to their local communities, including as a food source, income from tourism, employment and recreational opportunities. However, coral reefs are under increasing threat from climate change and related impacts such as coral bleaching and ocean acidification. Other anthropogenic stresses include over-fishing, anchor damage, coastal development, agricultural run-off, sedimentation and coral mining.

This book adopts a multidisciplinary approach to review these issues as they relate to the sustainable management of coral reef tourism destinations. It incorporates coral reef science, management, conservation and tourism perspectives and takes a global perspective of coral reef tourism issues covering many of the world's most significant coral reef destinations. These include the Great Barrier Reef and Ningaloo Reef in Australia, the Red Sea, Pacific Islands, South East Asia, the Maldives, the Caribbean islands, Florida Keys and Brazil. Specific issues addressed include climate change, pollution threats, fishing, island tourism, scuba diving, marine wildlife, governance, sustainability, conservation and community resilience. The book also issues a call for more thoughtful development of coral reef experiences where the ecological needs of coral reefs are placed ahead of the economic desires of the tourism industry.

Bruce Prideaux is Professor of Tourism, School of Business and Law, Central Queensland University, Australia. He is the author or editor of 11 books, 120 research articles and 84 book chapters. His interests include sustainability, climate change, marine and rainforest tourism and regional tourism development.

Anja Pabel is a Lecturer in Tourism, School of Business and Law, Central Queensland University, Australia. Her research interests are tourist behaviour, marine tourism, tourism sustainability and humour research.

EARTHSCAN OCEANS

For further details please visit the series page on the Routledge website: http://www.routledge.com/books/series/ECOCE

CORAL REEFS: TOURISM, CONSERVATION AND MANAGEMENT

*Edited by Bruce Prideaux
and Anja Pabel*

LONDON AND NEW YORK

First published 2018
by Routledge
2 Park Square, Milton Park, Abingdon, Oxon OX14 4RN

and by Routledge
711 Third Avenue, New York, NY 10017

Routledge is an imprint of the Taylor & Francis Group, an informa business

British Library Cataloguing-in-Publication Data
A catalogue record for this book is available from the British Library

Library of Congress Cataloging-in-Publication Data
Names: Prideaux, B. (Bruce), editor. | Pabel, Anja, editor.
Title: Coral reefs : tourism, conservation and management / edited by Bruce Prideaux and Anja Pabel.
Description: Abingdon, Oxon ; New York, NY : Routledge, 2019. | Includes bibliographical references and index.
Identifiers: LCCN 2018022726 | ISBN 9781138689831 (hardback) | ISBN 9781138497313 (paperback) | ISBN 9781315537320 (ebook)
Subjects: LCSH: Coral reefs and islands—Recreational use. | Coral reef conservation. | Coral reef management. | Tourism.
Classification: LCC GB461 .C68 2019 | DDC 333.95/53—dc23
LC record available at https://lccn.loc.gov/2018022726

ISBN: 978-1-138-68983-1 (hbk)
ISBN: 978-1-138-49731-3 (pbk)
ISBN: 978-1-315-53732-0 (ebk)

Typeset in Bembo
by Swales & Willis Ltd, Exeter, Devon, UK

This book is dedicated to the many people who have contributed in some way to the efforts to preserve the wonder of coral reefs. The need remains and the fight must be won.

CONTENTS

FIGURES

TABLES

ABBREVIATIONS

CARICOMP	Caribbean Coastal Marine Productivity Program
CBD	Convention on Biological Diversity
CEPF	Critical Ecosystem Partnership Fund
CI	Conservation International
CITES	Convention on International Trade in Endangered Species of Wild Fauna and Flora
COTS	Crown-of-Thorns Starfish
CRISP	Coral Reef Initiatives in the Pacific
DPaW	Department of Parks and Wildlife
GBPMP	Great Barrier Reef Marine Park
GBR	Great Barrier Reef
GBRMPA	Great Barrier Reef Marine Park Authority
GDP	Gross Domestic Product
GoK	Government of Kiribati
ICRC	Indigenous Coral Reef Communities
ICRI	International Coral Reef Initiative
IFAW	International Fund for Animal Welfare
ILUA	Indigenous Land Use Agreements
MPA	Marine Protected Area
NBSAP	National Biodiversity Strategy and Action Plan
NMP	Ningaloo Marine Park
NOAA	National Oceanic and Atmospheric Administration
NZ-DOC	New Zealand Department of Conservation
PIPA	Phoenix Islands Protected Area
QPWS	Queensland Parks and Wildlife Service
SCBD	Secretariat of the Convention on Biological Diversity
SOPAC	Secretariat of the Pacific Islands Applied Geoscience Commission
SPC	Secretariat of the Pacific Community

SPREP	Secretariat of the Pacific Regional Environment Programme
SPTO	South Pacific Tourism Organisation
TUMRA	Traditional Use of Marine Resources Agreements
UNDP	United Nations Development Programme
UNEP	United Nations Environment Programme
UNESCO	United Nations Education, Scientific and Cultural Organization
UNFCC	United Nations Framework on Climate Change
UNWTO	United Nations World Tourism Organization
WHS	World Heritage Site
WTTC	World Travel and Tourism Council
WWF	World Wide Fund for Nature

CONTRIBUTORS

Denis Allemand is the Scientific Director of the Centre Scientifique de Monaco and Professor of Biology at the University of Nice-Sophia Antipolis. His main area of research is related to comparative and conservation physiology of marine organisms such as Cnidarians. He has co-authored over 140 research articles and numerous book chapters.

Roberta Atzori is a Visiting Professor of Sustainable Hospitality Management, College of Business, California State University, Monterey Bay. Her research interests include sustainable tourism and hospitality, and climate change mitigation and adaptation in tourism destinations and hospitality businesses. She has published several peer-reviewed journal articles and book chapters and has spoken at numerous international conferences in Europe and the United States.

Leonie Cassidy is associated with the School of Business and Law at Central Queensland University and is a member of the Centre for Tourism and Regional Opportunities (CTRO), in Cairns, Australia. She graduated with her PhD titled 'Website Benchmarking: A Tropical Tourism Perspective' in 2015 at James Cook University. Her research developed a theory of website benchmarking employing a unique Design Science Research Methodology approach that is programmable, adaptable, and expandable, providing a theoretical framework for comprehensive website benchmarking into the future. Her interest in the online world's application for tourism continues with current research. She also excels in analysis of research data.

Guy Castley is affiliated with Griffith University's Environmental Futures Research Institute, Gold Coast, Queensland, Australia. He is a conservation biologist with experience in landscape and urban ecology, as well as protected area and threatened species management. He has recently extended this background to explore how tourism can contribute to the delivery of net positive contributions for conservation.

Leandra Cho-Ricketts is a marine ecologist and co-founder of the UB Environmental Research Institute. She has a PhD in Marine Sciences from the University of the West Indies, Mona, Jamaica with a focus on coral reef ecology and management. She heads the marine programmes of the UB ERI and the Calabash Caye Field Station in Turneffe. In 2005 she joined the University of Belize as an Assistant Professor and has been training future resource managers of Belize's marine environment for the last 12 years.

Fanny Couture is the coral biologist for the Reefscapers coral restoration programme in Landaa Giraavaru, Maldives. Fanny is originally from the French Basque Country. Her passion for marine life took her to Australia where she completed a Bachelor of Honours in Marine Biology from the University of Queensland in 2010 and a MSc at James Cook University in 2014. Her research initially focused on the response of fish assemblages to Marine Protected Areas in Koh Tao, Thailand. She then moved to the Maldives where she worked as the coral biologist for the Reefscapers coral restoration programme in Landaa Giraavaru, Baa Atoll. She is now pursuing more research endeavours in Vancouver, Canada.

Glen Croy is a senior lecturer at the Department of Management in the Monash Business School, Monash University. His teaching and research interests are in tourism, with special research interests in the role of media in tourism, tourism in natural and protected areas, and higher education. Glen has written on the complex and subtle relationships between film and tourism, visitor management in protected areas particularly for satisfaction of the experience and sustaining the conservation estate, and the explicit development of students' skills.

Sarah Duffy is affiliated with the School of Business, Western Sydney University of Western Sydney. Sarah has worked as a marketing professional, but after studying whale shark tourism at Ningaloo Reef for her PhD she has joined the ranks of academia. This experience has led to research that concerns how marketing practice impacts society, fairness and sustainability.

Larry Dwyer is affiliated with the School of Marketing, University of New South Wales, Sydney. Larry publishes widely in the areas of tourism economics, management and policy, with 200 publications in international journals, government reports, books, book chapters and monographs. Larry has undertaken an extensive number of consultancies for public and private sector tourism organisations within Australia, including Tourism Research Australia, and for international agencies, including the United Nations World Tourism Organisation.

Alan Fyall is Orange County Endowed Professor of Tourism Marketing at the Rosen College of Hospitality Management, University of Central Florida, USA. He has published widely, including 20 books and over 150 journal articles. Alan has examined 27 PhDs in the UK, India, South Africa, Australia, Hong Kong and Malaysia.

Dennis J. Gayle serves as Executive Chancellor at the University of the Commonwealth Caribbean, following senior leadership roles at the American University in the Emirates; Indiana Institute of Technology; the American InterContinental University, London; the University of the West Indies; the University of North Florida; and Florida International University, inter alia. His academic publications include a range of books, book chapters, academic journal articles and book reviews in fields that include development economics, international business, international political economy, sustainable development and university governance. He is a graduate of the University of California at Los Angeles, the London School of Economics, Oxford University and the University of the West Indies.

Margaux Y. Hein is affiliated with the Australian Research Council (ARC) Centre of Excellence for Coral Reef Studies, Townsville, Queensland, Australia. She is currently a PhD candidate at James Cook University, Australia, looking at characterising the potential socio-ecological effectiveness of coral restoration worldwide. Margaux grew up in the Principality of Monaco. She graduated with a Bachelor of Honours from the University of Queensland in 2010. She then volunteered on various marine conservation programmes in South Africa (Oceans Research) and Thailand (NHRCP) before coming back to Australia to study Marine Biology. She completed her MSc at James Cook University in 2014 with a project looking at the "Effects of newly implemented marine protected areas on coral health and diversity in Koh Tao, Thailand". She is now doing a PhD looking at the potential long-term socio-ecological benefits of active coral restoration and applications to the Great Barrier Reef.

Nathalie Hilmi is a specialist in Macroeconomics and International Finance. She joined the Centre Scientifique de Monaco as Section Head of Environmental Economics in 2010 and collaborated with IAEA-Environment Laboratories. Her research looks at the economic impacts of climate change and ocean acidification, the evaluation of ecosystem services and sustainable development policies.

Roger Layton is affiliated with the Australian School of Business, School of Marketing, University of New South Wales. In 1967 Roger was appointed Professor of Marketing and has since had a distinguished academic career. His research interests focus on how marketing practice impacts society, strategic marketing management and complex marketing systems. He has received a number of awards recognising services to marketing including the Order of Australia.

Henrietta L. Marrie is an Elder of the Gimuy Walubara clan of the Yidinji people and Traditional Owner of the land on which the City of Cairns is now located. Over a period spanning nearly four decades, Associate Professor Marrie has conducted research and published in a number of distinct fields, including indigenous visual art, cultural tourism and the advancement of indigenous women. Her research and publication record extends to issues related to the application of the United Nations Millennium Village Development Program to reduce extreme poverty and hunger, and addressing the UN Sustainable Development Goals in how it is relevant to indigenous people, while improving education, health, gender equality and environmental sustainability to indigenous community development.

Kelsey Miller is associated with the International Pole and Line Foundation, London. She is a fisheries scientist whose primary interest is promoting collaboration between scientific researchers and local fishermen to improve sustainability of fisheries. She earned her MSc from James Cook University, Australia. While living in the Maldives, she conducted research at sea on commercial pole-and-line tuna fishing boats.

Putu L. Mustika obtained her PhD on the sustainability of dolphin watching tourism in Bali (Indonesia) from James Cook University in Australia in December 2011. Dr Mustika is an adjunct researcher with the Management, Governance and Tourism academic group, College of Business, Law and Governance, James Cook University. She also actively assists the Indonesian government in fostering the country's marine mammal conservation management, particularly for whales and dolphins. She is currently the database coordinator and trainer for the whale stranding initiative in Indonesia. She also leads a marine mammal-focused NGO in Indonesia ('CETASI', Cetacean Sirenian Indonesia).

Anja Pabel is a lecturer in tourism at Central Queensland University in Cairns, Australia. Her research interests are marine tourism, tourist behaviour and humour research. Anja completed a Bachelor of Business with first class honours in 2009 at James Cook University. Her honours research project examined how certified scuba divers perceive their Great Barrier Reef experience compared to other reef destinations. Her PhD research investigated the role of humour in tourism settings. More specifically she examined in what ways humour influences the tourist experience in making tourists feel comfortable, connected and more mindful.

Fernanda de Vasconcellos Pegas is an adjunct research fellow at Griffith University's Environmental Futures Research Institute, Gold Coast, Queensland, Australia. She is a conservation social scientist with experience in community-based conservation, sea turtle conservation via tourism, private protected areas and cultural uses of wildlife and implications on species conservation and tourism.

Brooke A. Porter is a specialist in the human dimensions of the fisheries and the marine environment. Her work explores tourism as a development and conservation strategy in lesser developed regions, with emphasis on surf and adventure tourism. She is focused on developing simple and effective development and marine conservation strategies for coastal communities. She has worked in various capacities with NGOs, international aid agencies, tour operators and educational institutions in Maui, Hawai'i; New Zealand; Italy; the Philippines and in Africa. She is a postdoctoral research fellow at Auckland University of Technology and serves as a Scientific Adviser to The Coral Triangle Conservancy, an NGO that focuses on reef protection and restoration in the Philippines.

Bruce Prideaux holds the position of Director of the Centre for Tourism and Regional Opportunities at the Cairns campus of Central Queensland University, Australia and is responsible for the Master of Sustainable Tourism Management programme. His research includes issues that affect regional tourism development, sustainable tourism, crisis management, coral reef tourism and destination development issues. He has authored over 300 journal articles, book chapters and conference papers on a range tourism related issues. He has also authored or co-authored eleven books.

Ambrozio Queiroz Neto is a PhD candidate at Griffith University in Australia and a Senior Lecturer at the Department of Tourism, Centro Federal de Educação Tecnológica Celso Suckow da Fonseca – CEFET-RJ in Brazil. His research interests focus on tourism destination management, tourism planning, consumer behaviour and coastal destinations.

Stéphanie Reynaud is a marine biologist whose research focuses on coral calcification to understand how coral skeletons can be used to record environmental conditions and on the effects of climate change (mainly ocean acidification) on coral reefs. She received a PhD in Biogeochemistry in 2000 and a postdoctoral grant from Columbia University.

Alain Safa is the president of Skill-Partners. He teaches at the University of Nice-Sophia-Antipolis, EDHEC Business School and IPAG Business School as visiting professor. He works on the capacity of countries to adapt to economic, environmental and social changes. He focuses his studies on the integration of environmental considerations into national, regional and international governance.

Chad M. Scott is the founder and director of the New Heaven Reef Conservation Program in Koh Tao, Thailand. Chad is originally from Colorado in the United States. Chad first came to Koh Tao with the CPAD foundation while completing his bachelor's degree from the University of Colorado, Boulder, CO. In 2007 he founded the New Heaven Reef Restoration Program with the New Heaven Dive

School in Chalok Bay. Over the last ten years Chad has been the conservation figure of Koh Tao with active involvement in the Save Koh Tao group, advocating for better sustainable practices both in the water and on land.

Naneng Setiasih has been working on marine and coastal resources management since 1995. Her works mainly focus on programme development and management. This includes networking/building collaborative management, running and integrating science for management, developing sustainable financing/community economic development, as well as the policy aspect (national and traditional). She has been working at various NGOs and as consultants for NGOs, university, companies and government. She co-founded Reef Check Indonesia Foundation in 2009. Awarded two scholarships, she finished her master's degrees from Bandung Institute of Technology in 1999 and School of Environmental Science University of East Anglia in 2000.

Semisi Taumoepeau is the Director of Pacific Studies and Relations and is senior lecturer and researcher of the Auckland Institute of Studies as well as an associate director of the New Zealand Tourism Research Institute, Auckland University of Technology. His previous positions included Director of Tourism for the Government of Tonga, CEO of the Tonga national airline and Chairman of the South Pacific Tourism Organisation (SPTO), Suva, Fiji. He has published several publications on Pacific tourism and aviation, and recently initiated, with the Secretariat of the Pacific Regional Environment Programme (SPREP), the first International Whales in a Changing Ocean Conference, April 2017 in Tonga.

Michelle Thompson is a Lecturer in Tourism at Central Queensland University, Cairns, Australia. She has a PhD from James Cook University (2015), where her research examined the development of tourism in agricultural regions. Her research interests focus on regional tourism development, including nature-based (reef and rainforest) tourism and agri-tourism, and extend from her involvement in industry-driven research and government funded projects such as the Marine and Tropical Science Research Facility (MTSRF) and the National Environmental Research Program (NERP). Michelle has published 5 journal articles, 8 book chapters and 11 papers in refereed proceedings.

Bernadette E. Warner serves as Executive Vice President for Academic Affairs, Internationalization and Online Programs at the University of the Commonwealth Caribbean. She previously served as Dean of the Colleges of Business at Al Falah University and the College of Business at the American University in the Emirates, following senior leadership roles at the London College of Commerce, BPP University College (London), and the University of the West Indies, inter alia. Her academic publications include a range of books, book chapters and academic journal articles in fields that include international business, international strategic management, sustainable development and comparative higher education systems. She is a graduate of the University of the West Indies and the University of North Florida.

Riccardo Welters obtained his master's degree at the University of Maastricht, the Netherlands in August 1997. Afterwards, Riccardo became a Research Fellow at the Research Centre for Education and the Labour Market (ROA) at the University of Maastricht. He subsequently received scholarship support to pursue a PhD study at the University of Maastricht, completed in May 2005. He then accepted a Fellowship at the Centre of Full Employment and Equity (CofFEE) at the University of Newcastle in Australia before becoming a lecturer at James Cook University in 2008. Now he is furthering his research agenda in areas relating to socio-economic disadvantage at the Economics and Marketing Academic Group, College of Business, Law and Governance, James Cook University, Australia.

PREFACE

Although coral reefs are important tourism attractions and, for some small tropical islands, are the central focus of their tourist offering, the tourism literature has largely ignored these spectacular natural formations. Unfortunately, coral reefs are now under significant pressure from a range of anthropogenic threats, including overfishing, uncontrolled tourism development, coastal urban development, pollution and, more recently, climate change.

The aim of this book was to bring together a group of authors from the social science and scientific communities and from a diverse range of developed and developing economies to provide a contemporary commentary on coral reef tourism. This book therefore represents both a celebration of the beauty of coral reefs and the immense pleasure they provide to their visitors, as well as a warning on how quickly celebration can turn to tragedy if coral reefs are not adequately protected.

The need for adequate protection of coral reefs is a common theme throughout the various chapters in this book. Many of the chapters point to the damage that has been done, sometimes quite recently, because developers, tourist operators, tourists and governments have failed to understand that coral reefs only thrive if they are cared for and used in a manner that is sustainable. Many chapters report on the work that has yet to be undertaken to save specific coral reef ecosystems and the extent of loss suffered by far too many coral reef systems. Other chapters report on the successes of strategies that have been adopted to care for coral reefs. Overall, the book raises serious questions about the long-term economic sustainability of many coral reef dependent destinations in an era of rapid climate change.

The book also poses the currently unanswerable question of how to define sustainability in an era when the threats from climate change are increasing. It is not possible to predict when and at what level global temperature increase will cease and what this new global temperature normal will mean for all ecosystems, including coral reefs. Our current level of understanding is not adequate to develop an informed view of which coral reefs will survive and in what form.

The book concludes with a warning about what may happen if global temperatures continue to rise but also offers hope that many of the initiatives currently underway to protect coral reefs are successful. Importantly, the final chapter offers a 10 point policy framework (Table 18.1) that, if adopted and enforced, will facilitate ongoing tourism use of coral reef resources but in a form that recognises that further damage to these increasingly fragile ecosystems must be minimised. This framework is one of the major contributions of this book.

Overall, the book is an attempt to redress gaps in the literature and to highlight the successes and failures of tourism use of coral reefs. We hope that the book will stimulate further research into coral reef tourism and provide insights into policy successes and failures that can built on to ensure that future policy and management is more effective than the past.

Bruce Prideaux
Anja Pabel
Editors

1

CORAL REEF SYSTEMS AS A TOURISM RESOURCE

Anja Pabel and Bruce Prideaux

Introduction

2018 was proclaimed as the Third International Year of the Reef by the International Coral Reef Initiative (ICRI), an informal partnership between countries and organisations such as the United Nations that aims to strengthen awareness globally about the value of, and threats to, the world's coral reefs and related ecosystems (ICRI, n.d.). This initiative highlights the interest in coral reefs but also highlights the urgent nature of the problems faced by all coral reef ecosystems. Coral reefs are renowned for their diversity of marine life, colour and beauty. Sadly, however, their use for a range of economic gain, including tourism, has in many areas resulted in their degradation and even loss. While generating an estimated USD$1 trillion per year (Hoegh-Guldberg, 2015) in value, including ecosystem services, academic investigation of coral reefs remains largely the preserve of the sciences. Understanding of the impacts of tourism on coral reef ecosystems remains limited. The aim of this chapter is to introduce a range of issues that have emerged at the intersection of tourism and coral reefs. Later chapters in this book explore many of these issues in greater detail, both thematically and geographically.

Coral reefs provide a range of benefits to their local communities including the provision of food, opportunities for recreation and, in many areas, the generation of economic opportunities through their use as a tourism resource. However, coral reefs are facing an unprecedented range of anthropogenic and natural threats that range from global to local in scale. Global anthropogenic stresses centre on climate change and its related impacts such as coral bleaching and ocean acidification. At a local level, anthropogenic stresses are caused by a range of factors, including overfishing or destructive fishing, anchor damage, coastal development, dredging, pollution and terrestrial run-off, sedimentation, coral mining (Brown & Dunne, 1988; Dubinsky & Stambler, 1996; Hodgson, 1999; Burke et al., 2002; Dinsdale & Harriott, 2004; Fabricius, 2005; Reopanichkul et al., 2009; Burke et al., 2011; Sutherland et al., 2011; McCook et al., 2015) and unsustainable use as a tourism resource. Many of these stresses can be reduced or even eliminated by appropriate controls and by human agents modifying their behaviours. Tourism use of coral reefs, however, need not pose a high level of threat if managed sustainably (Spalding et al., 2017). Unfortunately, this is not always the case as discussed in many of the following chapters in this book.

Coral reefs also face a number of natural threats, including wind storm damage (from cyclones, hurricanes and typhoons), coral diseases, discharges from river catchments and in some areas from crown-of-thorns starfish (COTS) infestations. Despite these threats, coral reefs in many countries continue to be used in an unsustainable manner, either because of ignorance, greed or because of weak governance systems. While anthropogenic related factors pose the greatest threat to coral reefs in the long-term, increasing concerns about the health of coral reefs also offer the greatest hope for their long-term survival. Concerns about coral reefs expressed by individuals and community groups have galvanised a range of national and international strategies to respond to the dangers faced by coral reefs. It is against this background of growing concern about the long-term sustainability of many coral reef systems and hope that impacts can be minimised by strong public advocacy and action that this chapter discusses the role of coral reefs as a tourism resource. This chapter commences with a brief discussion on the value of coral reef systems, followed by an overview of a number of management issues, protection types and threats that effect coral reefs and how these in turn impact on the use of coral reefs as a tourism resource. The chapter concludes with discussion on the importance of community resilience of coral reef dependent destinations.

Coral reefs are unique marine ecosystems that are mainly found in warm tropical and subtropical waters, as illustrated in Figure 1.1. While covering less than 1% of the Earth's surface (Coral Reef Alliance, 2017), they have been described as "rainforests of the sea" due to their high level of biodiversity which supports about 25% of all sea life, including sponges, worms, molluscs, crustaceans, reptiles and fish (Scott, 2014, p. 10). In terms of age, most established coral reef ecosystems are between 5,000 and 10,000 years old (Coral Reef Alliance, 2017); however, geological records indicate that the ancestors of today's coral reefs first emerged about 240 million years ago (Coral Reef Alliance, 2017).

Coral reefs provide a variety of goods and services to humans, including social, cultural and economic benefits. Examples of social and cultural values in relation to indigenous communities are discussed in Chapter 16 in this volume. The economic benefits of coral reefs include a source of food, shoreline protection, tourism, ecosystem services and medicine. Scholars have suggested a number of methodologies that may be used to calculate the economic value of coral reefs.

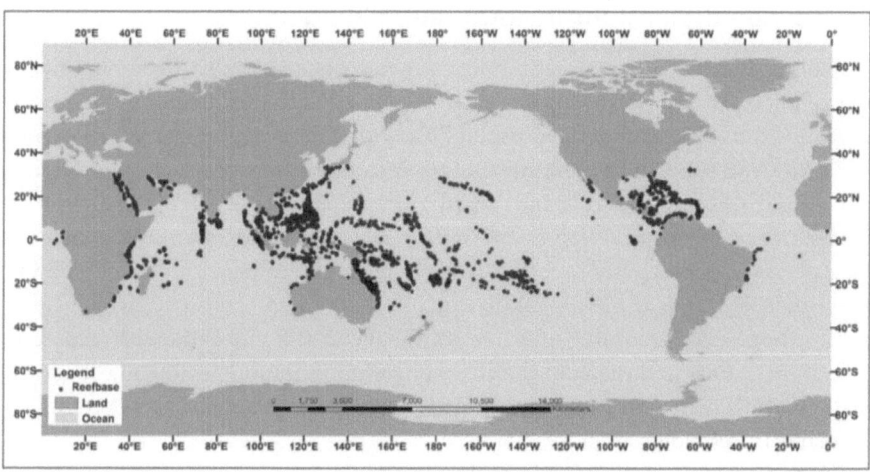

FIGURE 1.1 Distribution of global coral reefs

Source: NOAA National Ocean Service – Communications and Education Division: https://oceanservice.
noaa.gov/education/kits/corals/media/supp_coral05a.html

Brander et al. (2007), for example, suggest that coral reef systems can be valued on
the basis of:

- Direct use values from recreational opportunities for diving, snorkelling and
  viewing.
- Indirect use values from coastal protection and habitat/nursery functions for
  commercial and recreational fisheries.
- Preservation values through the welfare associated with the existence of
  diverse natural ecosystems.

In another approach, Deloitte Access Economics (DAE) (2017) examined the
economic, social and iconic values of Australia's Great Barrier Reef and found
that the value to the community was AUD\$56 billion, which includes the value
of the reef to people who had not yet visited it. On a global basis, the value of
the world's coral reef systems is estimated to be close to USD\$1 trillion per year
(Hoegh-Guldberg, 2015). This figure is based on the economic value of goods and
services provided by coral reefs to the 850 million people who live within 100 km
of coral reefs and derive benefits from ecosystem services such as food and livelihoods, tourism, treatment for disease and shoreline protection (Burke et al., 2011).
The value of coral reef ecosystems includes *residential values* through recreational
activities such as fishing and boating, as well as *tourist values* through destination
competitiveness achieved by promoting coral reefs (Stoeckl et al., 2014). On an
annualised basis, Spalding et al. (2017) estimated that global coral reef tourism is
worth USD\$35.8 billion based on the total within-country expenditure by inter-
national and domestic visitors that can be assigned to the presence of coral reefs.

The human dimensions in coral reef management

Globally, multiple regional economies depend on coral reefs for a significant part of their livelihood through coral reef related activities, including tourism, recreational fishing, commercial fishing, traditional hunting by traditional owner groups and, in some areas, mining and other industrial use. As well as providing habitat and nurseries for fish and aquatic invertebrates, coral reefs also provide barriers from storms and waves to protect sea coasts and importantly help to regulate atmospheric gases (Scott, 2014). Coral reefs are estimated to contribute between 7 and 15% of the global production of calcium carbonate, which helps with carbon sequestration (Suzuki & Kawahata, 2004) and influences the concentration of carbon dioxide in the atmosphere. Continuing decline in the health of the world's reefs will have significant economic as well as social repercussions for many stakeholders.

There is evidence of a shift in societal attitudes from exploitation to recognition of the need to protect natural areas in general and coral reefs in particular. Examples of this shift are discussed in Chapters 4, 9 and 10 in this volume. Societal attitudes are affected by cultural and social norms (Knez & Thorsson, 2006), economic imperatives and politics (Norris, 2000), and by external sources such as mass media (Zillmann & Vorderer, 2008). Societal attitudes influence both community and individual responses to the protection of nature, including protecting coral reef ecosystems. Local residents and tourists are fascinated by certain landscapes through a sense of place (Moskwa, 2012). These place attachments are based on emotional bonds to certain natural sites and can lead to a sense of stewardship and behavioural actions that may improve environmental practices (Moskwa, 2012).

Increasing interest in stewardship actions by many local communities indicates that a paradigm shift is taking place with increased advocatory for ecologically sustainable development, at least in some areas. Stewardship actions in this context can be described as creating awareness through environmental workshops or projects, undertaking environmental monitoring or research, and helping to restore the natural environment, i.e. through tree planting or coral restoration attempts (see Chapter 18 in this volume for examples of coral reef stewardship).

In Australia concern for the long-term well-being of the Great Barrier Reef has resulted in the growth of a strong pro-Great Barrier Reef lobby that has been successful in influencing governments to ban oil drilling, stop overfishing and reduce or prohibit large-scale dredging. Similar examples of stewardship by local communities can be found in many coral reef communities (see Chapter 18 in this volume for examples). However, in many reef-dependent communities the voice of citizens calling for an end to over-exploitation of fishing in particular, or other unsustainable uses of coral reefs including coral mining, is overshadowed by power elites concerned with short-term economic gains (Acemoglu & Robinson, 2008). Examples include destructive fishing practices (using dynamite or cyanide); unplanned coastal and tourism development where some tourist resorts have been built directly on top of reefs, with some resorts emptying their sewage or other wastes into water surrounding coral reefs; and coral mining where coral is removed from reefs to be used as bricks, road-fill or cement for new buildings.

Policymakers often find that it is challenging to integrate human dimensions into decision-making processes particularly in circumstances where power elites have yet to recognise the need for and wisdom of long-term sustainability verses short-term gains (see Chapter 9 in this volume for a discussion on long-term verses short-term use from a common pool resource perspective). This situation will only change when there is an understanding by government that citizens have genuine concerns for iconic national resources that can "promote pride, and instil a sense of individual identity and collective responsibility to protect it" (Goldberg et al., 2016). In relation to the protection of coral reefs, Goldberg et al. (2016, p. 1) stated that these iconic ecosystems play an important role in their potential to "unify seemingly disparate factions of a population around a common goal, for example, the long-term management and preservation of an internationally significant natural resource." The premise is that understanding the underlying perceptions that people have towards specific natural environments is important for their management, including any conservation activities (McCook et al., 2010).

Goldberg et al. (2016) also made the observation that emotional connections to natural areas could be used to foster political support for resource protection. As Prideaux and Pabel (this volume) point out in Chapter 6, coral reef scientists have been very effective in tapping the high level of emotional engagement expressed by Australians towards natural areas, including the Great Barrier Reef, to gain very large ongoing funding for science related research. Prideaux and Pabel (Chapter 6) also point out that while this level of emotional attachment has resulted in substantial investment in scientific research, there has been little social science research undertaken in areas such as tourism, community resilience and social capital. Ultimately, the failure to invest in social science research is counterproductive because it is human actions that are driving the anthropogenic factors leading to coral (and other ecosystem) decline. It seems rather pointless to spend large amounts of money studying how a system is declining without devoting equal resources to help ameliorate that decline.

Research by Prideaux et al. (2018) found that in a survey of a coral reef dependent community (Cairns, Australia), the majority of respondents (75%) supported the current level of funding of coral reef management and protection, and two-thirds of respondents thought that coral bleaching will affect them on a personal level and realised that their own and others' actions individually and nationally can affect the health of the Great Barrier Reef. However, responses to two questions that asked about personal involvement in protecting the Great Barrier Reef, indicated that respondents had yet to move from the position that it was someone else's role to protect the reef to accepting personal responsibility for actions to save the Great Barrier Reef. This finding is similar to research by other scholars, including McKercher et al. (2013).

Ensuring that coral reefs are managed effectively and sustainably is a role that falls on government. Allocation of sufficient resources for research, management and to fund adaption is important, as is enforcing policies that relate to coral reef use and coastal development. Where conflicting demands emerge, governments need to mediate, leaning towards ensuring long-term natural sustainability rather than favouring short-term economic benefits (such as fishing) that often result in long-term environmental costs. Good governance lies at the heart of effective public administration and protection of natural resources such as coral reefs. To be effective, good governance needs to be supported by an informed and engaged public that see it as their personal responsibility to support environmental protection. In many cases, the public still appear to believe that environmental protection is essential but the responsibility of others (McKercher & Prideaux, 2014).

The protection status of coral reefs

The protection status of global coral reefs varies greatly from unprotected coral reefs systems (i.e. some coral reefs around the Philippines, Kenya, Tanzania and the Mexican Caribbean), to coral reef systems that have been declared as marine protected areas (MPAs). The forms of protection include marine parks (i.e. Bonaire Marine Park, Red Sea Marine Peace Park, Pitcairn Islands Marine Reserve, Papahānaumokuākea Marine National Monument) and coral reef systems located in World Heritage sites (i.e. Great Barrier Reef, Belize Barrier Reef Reserve System, Galápagos Islands, Palau Marine Sanctuary).

MPAs provide benefits to coral reefs through outcomes such as conservation of fish stocks (Spalding et al., 2001) and protection of reef ecosystems from mining. However, not all MPAs are effective in protecting coral reef ecosystems, as a number of chapters in this book highlight. Mora et al. (2006), for example, found that many existing marine reserves are largely ineffective and, overall, they remain insufficient for the protection of coral reef diversity. The underlying issue with MPAs is that many exist on paper only and are therefore referred to as "paper parks" (Depondt & Green, 2006). Estimations suggest that around 70–80% of MPAs are protected only in name without active management or protection (Gallagher-Freymuth, 2002). One issue is that "fencing off" certain sections for conservation purposes is more difficult in marine environments than in terrestrial environments (Weaver, 2008). However, little is known about the carrying capacity and the ecology of many marine environments, even though there is a rapid growth in activities such as recreational boating, snorkelling, scuba diving and whale watching.

The management of coral reef ecosystems is also made more complicated by the "invisibility" factor, in other words activities cannot be as easily monitored from above the surface, unless it is through diving and observational submarines (Weaver, 2008). Issues related to carrying capacity are emerging as a major concern. As the quality of coral reef systems decline in coming decades (unless current trends in global warming are reversed), current understandings of carrying capacity will need to be adjusted to ensure that reef ecosystems do not suffer increased damage through human use.

Protection by designation of coral reef systems as World Heritage Areas (WHA) under the United Nations Educational, Scientific and Cultural Organization's (UNESCO) World Heritage system theoretically provides the highest level of protection for reefs because there is a capacity to apply mild sanctions on countries that fail to comply with the listing conditions of the reef systems. All natural heritage World Heritage sites are listed on the basis of their universal values as described by UNESCO. Values applying to natural areas include "contains superlative natural phenomena or areas of exceptional natural beauty and aesthetic importance" (criteria 7) and "contains the most important and significant natural habitats for in-situ conservation of biological diversity, including those containing threatened species of outstanding universal value for the point of view of science or conservation" (criteria 10). Governments that do not comply with the undertakings they make to the World Heritage Committee to protect their WHAs may be asked to show why they should not have WHA status revoked. Australia's Great Barrier Reef, listed as a WHA in 1981, came under significant scrutiny when the World Heritage Committee expressed concerns in 2012 about large-scale coal port development and associated dredging adjacent to the Great Barrier Reef WHA.

The Great Barrier Reef (see Chapter 4 in this volume) provides one example of a well-protected MPA, but even the Great Barrier Reef faces problems in gaining the full benefits from protection through the MPA system. However, many other MPAs are poorly managed, suffer from a lack of funding, receive little or no support through enforcement and in many areas lack the scientific infrastructure necessary to develop effective management structures (Depondt & Green, 2006; Spalding et al., 2001). Several authors have highlighted that many MPAs focus mainly on controlling direct impacts from human activates on coral reefs such as fishing and tourism, with some stating that, overall, they are insufficient for the protection of coral reef diversity (Spalding et al., 2001; Mora et al., 2006; Edgar et al., 2014). Other threats such as sedimentation and pollution, coral disease and climate change are ignored. A further problem, highlighted by Brander et al. (2007, p. 209), is that many MPAs by their "open-access nature and public good characteristics" tend to be undervalued by decision-makers in relation to their use and conservation.

Threats to coral reefs

Anthropogenic threats

Climate change has been identified by many researchers as the leading cause for the current decline in coral reef systems globally. Chapter 4 illustrates how these threats are affecting the Great Barrier Reef. However, climate change is only the latest of a long list of human-caused factors that have adversely affected coral reefs. Daley et al. (2008) argue that coral reefs have been in decline since the nineteenth century, well before the cumulative effects of global warming began to be manifest in global-scale coral bleaching events, which are an outcome of climate change.

As Halpin et al. (2007) observed, there is an enormous range of anthropogenic threats to all marine ecosystems including coral reefs, from over-exploitation and continuing failure to address many of these concerns. In relation to coral reefs, Hughes et al. (2003) observed that these systems were under enormous threat from climate change. Apart from climate change, pollution also poses a significant threat.

One small example is a recent article in *The Telegraph* (dated 26 October 2017) with the title "Shocking photo shows Caribbean Sea being choked to death by human waste," which portrayed graphic images of garbage in the Caribbean Sea. There is also growing evidence that plastics are becoming a major problem with microplastics entering the food chain. Concerns have also been raised about indigenous hunting activities in protected areas that, in some locations, have moved from hunting for family consumption to semi-commercial harvesting. The failure of individual citizens to adopt proactive climate mitigation on an individual basis is another major, though largely unreported, danger.

While some argue that climate change on its own will probably not completely destroy coral reefs, particularly where some biodiversity remains (Scott, 2014), others agree that the accumulation of impacts such as increased development, over-exploitation, disturbances from major storms and coral bleaching events will weaken the resilience of global coral reef systems (GBRMPA, 2014). There is widespread agreement that threats should be tackled at their source to improve coral reef system's capacity to recover. This includes strategies at reef-wide, regional, national and international levels to deal with issues such as water quality and fishing. The massive threat posed by climate change is beyond the scope of any single nation to challenge and requires long-term globally agreed actions such as those agreed in the Paris Climate Change Accord 2015.

Natural threats

Coral reefs face a number of natural threats, including coral diseases, predation by COTS, natural weather variability and the impact of large windstorms. Discharge from river catchments may also pose a threat when onshore flooding events occur. The discharge of large quantities of silt can affect corals and, in some cases, smoother them. Anthropogenic factors such as climate change can amplify the impacts of natural effects such as seasonal changes in water temperature and sediment discharge following rainfall and flooding. For example, intensive farming or grazing can lead to increased erosion and sediment (McCulloch et al., 2003).

Demand and supply perspectives

As with any tourism resource, coral reefs compete in local, national and international tourist markets for customers. Figure 1.2 illustrates a range of relationships that affect the development and delivery of coral reef experiences. The model takes a demand and supply perspective and seeks to illustrate how a range of factors such as ownership, threats and resilience can impact the supply of coral reef experiences, which in turn impacts the demand for coral reef experiences.

Major factors that influence the ability of communities to offer coral reef experiences include ownership, protection status and the quality of management systems. From a demand-side perspective the appeal of coral reefs is often associated with images of spectacular coral colours, iconic marine species and recreation such as

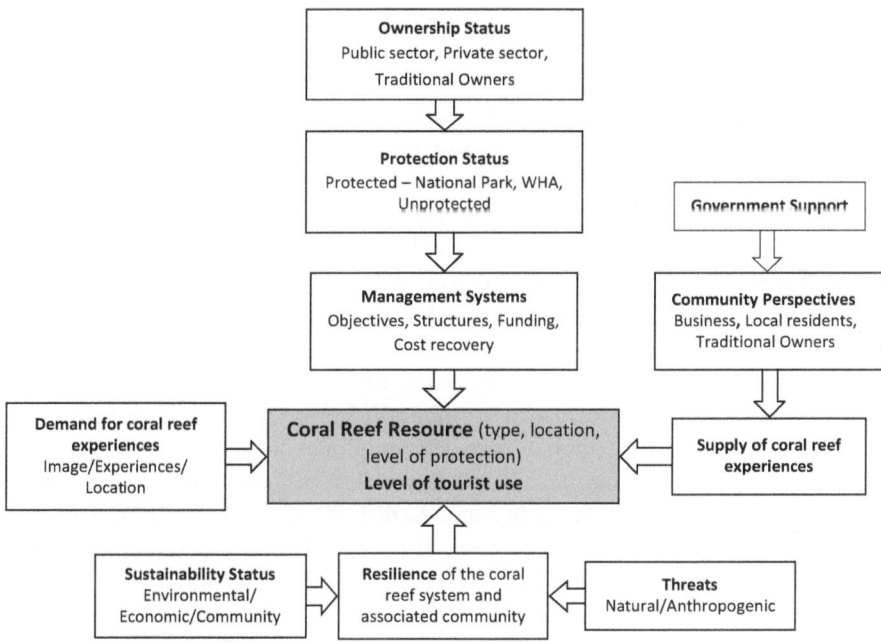

FIGURE 1.2 Factors that affect the development and delivery of coral reef tourism experiences

swimming, snorkelling and scuba diving. Images of a relaxed beach lifestyle often complement these images. Events such as destructive windstorms and coral bleaching can weaken these images with graphic pictures of destruction of resorts and dead coral. If the ability of coral reef communities to provide quality coral reef experiences is degraded, the most likely response by tourists will be a fall in demand.

From a supply-side perspective, a range of factors can affect the type of coral experiences offered. These include the role that local communities, including Indigenous Traditional Owners, may have in making decisions about how natural resources such as coral reefs are used. The role of investors is also important in developing the depth of experiences and infrastructure necessary to support the tourism industry. The public sector also has an important role in ensuring effective governance of issues that includes land zoning and also enforcement and supply of a range of supporting community infrastructure such as health, public safety, conservation and education.

Community resilience

Previous discussion in this chapter has considered a range of threats that are faced by coral reef systems and the value of these systems to the human populations they support. The scientific literature on the threats faced by coral reef systems is vast and growing. Sadly, rather less consideration has been given to how threats to coral reefs may affect the communities that rely on coral reefs as a source of food, recreation or tourism. Community resilience is an important issue that has received very little attention in the literature, particularly in relation to coral reef systems. The following discussion highlights a range of community resilience issues that require further research to enable both communities and the public sector to understand the impact that a decline in coral reef ecosystems may have on human welfare.

Human resilience is a measure of how humans respond to unexpected shocks and disturbances by triggering renewal and innovative thinking (Stockholm Resilience Centre, 2015). Resilience is also a measure of the capacity of a human, or natural system to withstand change and recover from a disturbance while continuing to retain the integrity, structure and functions that were characteristic of the system before the disturbance. From a tourism perspective, a resilient destination must have the capacity to maintain functional linkages between the social system (individuals, communities and industries) and supporting ecosystem, as well as the capability of returning to a high level of economic health following a disturbance or other form of major change (RAND Corporation, 2017). A resilient community will also demonstrate a capacity to adjust to economic, physical and social environment through learning from experiences and continually improving response mechanisms over time (Price-Robertson & Knight, 2012). Communities that have a low level of resilience have limited adaptive capacity to recover (Becken, 2013).

Measuring community resilience also requires an understanding of the magnitude of the shock or disturbance, the timing and length of the shock and the capacity of coping mechanisms within the community and from outside in assisting in how to deal with the shock or disturbance. In the case of coral reef dependent destinations, a rapid decline in coral health can be expected to have a significant impact on all another sectors of the local economy. At best, impacts will be short-lived and recovery will be speedy. At worst, if a key tourism resource is lost, the flow on effects to the economy may mean loss of jobs, business failure and out-migration of displaced workers.

Measures to enhance community resilience should commence with an assessment of the capacity of the community to respond to and recover from a shock or disturbance. Lew (2014) suggested that some destinations will require a change in the composition of tourism experiences to facilitate recovery. Factors that will determine the ability of a community to recover include the willingness of the destination to recognise the problem and respond; the speed of decline; support for innovation; the willingness of community members to participate in recovery; the strength of public and private sector leadership; and the resources available to assist with recovery. If recovery is able to be based on rehabilitating the affected resource, the community will not need to search for substitute resources. If adaption involving the developing of new resources is required, then a transition stage will be required. Depending on the time and extent of change required, the community may experience a range of hardships arising from growing unemployment and business failure. From this perspective, reducing vulnerability will require building a destination's adaptive capability. In the case of coral reef destination's this will require changing the destination's supply mix. In some Pacific Ocean destinations described in Chapter 2 (this volume), this will be difficult because of the limited nature of alternative resources.

Conclusion

It is difficult to be optimistic rather than pessimistic about the long-term future of coral reefs and the tourism industry they support given the range of anthropogenic and natural threats that global reef systems face. As the following chapters in this book highlight, many coral reefs have been over-exploited by a range of actors whose focus is directed towards short-term gain rather than long-term sustainability. As the evidence for the need to move from exploitation to protection mounts and the effects of the substantial economic losses that accrue through short-term exploitation become even more apparent, there has been a noticeable shift in direction towards protection by many governments. This shift is reflected at the national level by new laws to protect reef systems and at the international level through the Paris Climate Change Accord 2015.

The question that must be asked is if this shift in direction will be effective or is it too late for many coral reef systems? The answer to this question will become apparent in coming decades. From a tourism perspective, the role of coral reefs as a major pull factor at both destination and national levels is significant and highlights the need for the tourism industry to become part of the solution to the problems faced by coral reefs, not part of the problem. This shift must be manifest in both land-side as well as marine tourism development. The desire for a sea view at the expense of a mangrove forest must be resisted and, equally, marine operators need to ensure that their clients do not harm the very ecosystems they have paid to see through inappropriate use of sun creams, touching coral and marine animals, etc.

The aim of this chapter has been to introduce a range of issues related to coral reef tourism. Later chapters in this book explore many of these issues in greater detail in a variety of settings. The final chapter of this book looks to the future and discusses some positive initiatives that can assist in ensuring that the long-term outlook for coral reefs leans to the positive side of the sustainability ledger rather than the negative side.

References

Acemoglu, D., & Robinson, J. (2008). Persistence of power, elites and institutions. *The American Economic Review*, *98*(1), 267–293.

Becken, S. (2013). Developing a framework for assessing resilience of tourism sub-systems to climatic factors. *Annals of Tourism Research*, *43*, 506–528.

Brander, L.M., van Beukering, P., & Cesar, H. (2007). The recreational value of coral reefs: A meta-analysis. *Ecological Economics*, *63*(1), 209–218.

Brown, B.E., & Dunne, R.P. (1988). The environmental impact of coral mining on coral reefs in the Maldives. *Environmental Conservation*, *15*(2), 159–165.

Burke, L., Reytar, K., Spalding, M., & Perry, A. (2011). *Reefs at risk revisited*. Washington, DC: World Resources Institute.

Burke, L., Selig, L., & Spalding, M. (2002). *Reefs at risk in Southeast Asia*. Washington, DC: World Resources Institute.

Coral Reef Alliance. (2017). How coral reefs grow. Retrieved 15 September 2017 from https://coral.org/coral-reefs-101/coral-reef-ecology/how-coral-reefs-grow/

Daley, B., Griggs, P., & Marsh, H. (2008). Reconstructing reefs: Qualitative research and the environmental history of the Great Barrier Reef, Australia. *Qualitative Research, 8*(5), 584–615.

Deloitte Access Economics. (2017). *At what price? The economic, social and icon value of the Great Barrier Reef*. Report to the Great Barrier Reef Foundation. Retrieved from https://www2.deloitte.com/content/dam/Deloitte/au/Documents/Economics/deloitte-au-economics-great-barrier-reef-230617.pdf

Depondt, F., & Green, E. (2006). Diving user fees and the financial sustainability of marine protected areas: Opportunities and impediments. *Ocean & Coastal Management, 49*(3–4), 188–202.

Dinsdale, E.A., & Harriott, V.J. (2004). Assessing anchor damage on coral reefs: A case study in the selection of environmental indicators. *Environment Management, 33*(1), 126–139.

Dubinsky, Z., & Stambler, N. (1996). Marine pollution and coral reefs. *Global Change Biology, 2*(6), 511–526.

Edgar, G.J., Stuart-Smith, R.D., Willis, T.J., Kininmonth, S., Baker, S.C., Banks, S., . . ., & Thomson, R.J. (2014). Global conservation outcomes depend on marine protected areas with five key features. *Nature, 506*, 216–220.

Fabricius, K.E. (2005). Effects of terrestrial runoff on the ecology of corals and coral reefs: Review and synthesis. *Marine Pollution Bulletin, 50*(2), 125–146.

Gallagher-Freymuth, L. (2002). The Bonaire National Marine Park. Solutions case study 2001. Retrieved from http://www.solutions-site.org/cat1_sol117.htm

Goldberg, J., Marshall, N., Birtles, A., Case, P., Bohensky, E., Curnock, M., . . ., & Visperas, B. (2016). Climate change, the Great Barrier Reef and the response of Australians. *Palgrave Communications, 2*, 1–8.

Great Barrier Reef Marine Park Authority (GBRMPA). (2014). *Great Barrier Reef outlook report 2014*. Townsville: Great Barrier Reef Marine Park Authority.

Halpin, B., Selkoe, K., Micheli, F., & Kappel, C. (2007). Evaluating and ranking the vulnerability of global marine ecosystems to anthropogenic threats. *Conservation Biology, 21*(5), 1301–1315.

Hodgson, G. (1999). A global assessment of human effects on coral reefs. *Marine Pollution Bulletin, 38*(5), 345–355.

Hoegh-Guldberg, O. (2015). *Reviving the ocean economy: The case for action – 2015*. Gland: WWF International.

Hughes, T., Baird, A., Bellwood, D., Card, M., Connolly, S., & Folke, C. (2003). Climate change, human impacts, and the resilience of coral reefs. *Science*, 301.5635.

International Coral Reef Initiative (ICRI). (n.d.). International Year of the Reef (IYOR). Retrieved 15 January 2018 from International Coral Reef Initiative, https://www.icriforum.org/about-icri/iyor

Knez, I., & Thorsson, S. (2006). Influences of culture and environmental attitude on thermal, emotional and perceptual evaluations of a public square. *International Journal of Biometeorology, 50*, 258–268.

Lew, A.A. (2014). Scale, change and resilience in community tourism planning. *Tourism Geographies, 16*(1), 14–22.

McCook, L.J., Ayling, T., Cappo, M, Choat, J.H., Evans, R.D., De Freitas, D.M., . . ., & Williamson, D.H. (2010). Adaptive management of the Great Barrier Reef: A globally significant demonstration of the benefits of networks of marine reserves. *Proceedings of the National Academy of Sciences of the United States of America*, *107*(43), 18278–18285.

McCook, L.J., Schaffelke, B., Apte, S.C., Brinkman, R., Brodie, J., Erftemeijer, P., . . ., & Warne, M.S. (2015). *Synthesis of current knowledge of the biophysical impacts of dredging and disposal on the Great Barrier Reef: Report of an Independent Panel of Experts*. Townsville: Great Barrier Reef Marine Park Authority (181 pp).

McCulloch, M., Fallon, S., Wyndham, T., Hendy, E., Lough, J., & Barnes, D. (2003). Coral record of increased sediment flux to the inner Great Barrier Reef since European settlement. *Nature, 421*, 727–730.

McKercher, B., & Prideaux, B. (2014). Academic myths of tourism. *Annals of Tourism Research, 46*, 16–28.

McKercher, B., Prideaux, B., & Pang, S. (2013). Attitudes of tourism students to the environment and climate change. *Asia Pacific Journal of Tourism Research, 18*(1–2), 108–143.

Mora, C., Andréfouët, S., Costello, M.J., Kranenburg, C., Rollo, A., Veron, J., . . ., & Myers, R.A. (2006). Coral reefs and the global network of Marine Protected Areas. *Science, 312*, 1750–1751.

Moskwa, E.C. (2012). Exploring place attachment: An underwater perspective. *Tourism in Marine Environments, 8*(1/2), 33–46.

Norris, P. (2000). *A virtuous circle: Political communications in post-industrial societies*. Cambridge: Cambridge University Press.

Price-Robertson, R., & Knight, K. (2012). *Natural disasters and community resilience: A framework for support*. CFCA Paper No. 3, Australian Institute of Family Studies (13pp). Retrieved from https://aifs.gov.au/cfca/publications/natural-disasters-and-community-resilience-framework-support

Prideaux, B., Carmody, J., & Pabel, A. (2018). *Impacts of the 2016 and 2017 mass coral bleaching events on the Great Barrier Reef tourism industry and tourism-dependent coastal communities of Queensland*. Report to the Reef and Rainforest Research Centre Limited, Cairns (97pp).

RAND Corporation. (2017). Community resilience. Retrieved from http://www.rand.org/topics/community-resilience.html

Reopanichkul, P., Schlacher, T.A., Carter, R.W., & Woranchananant, S. (2009). Sewage impacts coral reefs at multiple levels of ecological organization. *Marine Pollution Bulletin, 58*(9), 1356–1362.

Scott, C.M. (2014). *The Koh Tao ecological monitoring program* (2nd ed.). Koh Tao, Thailand: Conservation Divers Ltd. Pt., 160pp.

Spalding, M.D., Ravilious, C., & Green, E.P. (2001). *World atlas of coral reefs*. Prepared at the UNEP World Conservation Monitoring Centre. Berkeley, CA: University of California Press.

Spalding, M., Burke, L., Wood, S.A., Ashpole, J., Hutichison, J., & Zu Ermgassen, P. (2017). Mapping the global value and distribution of coral reef tourism. *Marine Policy, 82*, 104–113.

Stockholm Resilience Centre. (2015). What is resilience? Retrieved 2 August 2017 from http://www.stockholmresilience.org/research/research-news/2015-02-19-what-is-resilience.html

Stoeckl, N., Farr, M., Jarvis, D., Larson, S., Esparon, M., Sakata, H., . . ., & Costanza, B. (2014). *The Great Barrier Reef World Heritage Area: Its 'value' to residents and tourists Project 10–2 Socioeconomic systems and reef resilience. Final Report to the National Environmental Research Program*. Reef and Rainforest Research Centre Limited, Cairns (68 pp).

Sutherland, K.P., & Shaban, S., Joyner, J.L., Porter, J.W., & Lipp, E.K. (2011). Human pathogen shown to cause disease in the threatened eklhorn coral acropora palmate. *PLoS ONE, 6*(8), e23468.

Suzuki, A., & Kawahata, H. (2004). Reef water CO_2 system and carbon production of coral reefs: Topographic control of system-level performance. *Global Environmental Change in the Ocean and on Land,* 229–248.

Weaver, D. (2008). *Ecotourism* (2nd ed.). Milton, Qld: John Wiley & Sons.

Zillmann, D., & Vorderer, P. (2008). *Media entertainment: The psychology of its appeal.* New York: Routledge.

PART I

The ecology and governance of coral reefs

2

CORAL REEFS

Impacts and sustainability in the South Pacific Islands

Semisi Taumoepeau and Anja Pabel

Introduction

Coral reefs and other marine ecosystems such as islands are an important tourism resource and bring various benefits to the local communities. However, these natural resources are under threat from climate change and its related impacts, for example coral bleaching and ocean acidification, as well as anthropogenic stresses from human activities such as overfishing, anchor damage, coastal development, agricultural run-offs, sedimentation and coral mining. Despite these threats, coral reefs in many countries continue to be used in an unsustainable manner because of ignorance or greed or because of weak governance systems. This chapter examines the extent of coral reef tourism in the South Pacific region, including sustainability issues and the establishment of marine protected areas to provide protection for coral reef ecosystems.

Although the region's tourism industry is relatively small by world standards, tourism is the primary source of export income for many of the region's economies. In 2016, tourism arrivals recorded by the member countries of the South Pacific Tourism Organization (SPTO) grew by 3.7% to reach 2.03 million. The 2016 growth rate was slightly below the global growth rate of 3.9% (SPTO, 2017). Based on source market, arrivals from Australia and New Zealand accounted for 52% of market share in 2016. Images of tropical beaches and water-based activities, including coral reef related activities, are a common theme in the promotional material produced by almost all SPTO members.

The Pacific Ocean is very important to all inhabitants of the islands of the Pacific. Coral reefs continue to play a vital role in supporting subsistence livelihoods, the economic development in island economies and many cultural and traditional ways of life in the region (White et al., 2000; Whittingham et al., 2003). However, coral reefs are now facing unprecedented threat and challenges. In recognising the South

Pacific Ocean's essential role to the people who are living in this region, several South Pacific governments and agencies are now working together, with the help of the UN and other international agencies, to work out practical strategies both at national and regional levels to mitigate the impacts of climate change that have affected the small Pacific islands more markedly and rapidly during the past 20 years. The Secretariat of the Pacific Regional Environment Program (SPREP, 2016), as the leading agency for environmental programmes in the South Pacific, is working with partners to help protect, manage, maintain and sustain the cultural and natural integrity of the Ocean.

The value of ecosystem services in the South Pacific

Recognising the value of ecosystem services is becoming increasingly important for coral reef management, not simply for conservation actions but also for improving the management of coral reef ecosystems and to better inform policy makers (Laurans et al., 2013; Pascal, 2016). However, in a widespread region such as the South Pacific this can be a challenging undertaking. Ecosystem services provided by coral reefs are of primary importance for many South Pacific countries and deliver ecological resources for economic growth and other benefits to their inhabitants (South et al., 2012). The economic challenges faced by Pacific island states include high rates of poverty, low human capital capacity, demographic pressures and risks from natural disasters (Leisher et al., 2007; Pollnac et al., 2000). Any factors that reduce the value of ecosystem services (such as damaging coral reefs) will increase the other economic challenges faced by these countries.

The key coral reef ecosystem services in the South Pacific are generated from biodiversity, coastal protection and coral reef fisheries (Laurans et al., 2013). There tends to be considerable variations in coral reef tourism across the region. The variations in value generated from coral reef tourism stem from the ecological attributes of the reef but also the socio-cultural and developmental contexts in which the tourism activities take place (Laurans et al., 2013). When the link between the value of ecosystem services and the health status of the ecosystem providing the services is recognised, stakeholders might be more inclined to protect the conditions of coral reefs because they see the benefits of coral reef conservation (David et al., 2010).

A review of South Pacific coral reefs

The South Pacific region (see Figure 2.1) has the largest region of coral reefs in the world's oceans (Chin et al., 2011; Vieux et al., 2008). According to Spalding et al. (2001) the region accounts for about 40% of the world's coral reefs and includes a vast array of coral atolls. With population densities at low levels, biodiversity levels tend to be high. Humans in this regions rely on coral reefs as an important food source and as protection from tropical storms and cyclones. Overall, the region faces localised threats from coastal development, sedimentation and pollution from agriculture and logging, tourism overuse, climate change and acidification of the oceans (Mumby & Steneck, 2008). Although these issues tend to be localised, they still reduce the potential of certain reefs as a renewable source of food. Furthermore, most parts of the world's oceans have seen temperatures increase (NOAA, 2016), leading to coral bleaching events in the South Pacific.

Mangroves, with their dense root systems, protect coral reefs by filtering out sediment. Mangrove areas in the islands of Fiji, New Caledonia and the Solomon Islands have been greatly cut back due to agricultural development and land clearing for construction. Mangroves in other South Pacific islands are similarly affected mainly through clearing for construction and building of ports and wharf areas like in Samoa and Tonga. Moreover, many mangrove areas are polluted from oil spillages, contamination from heavy metals and run-off (Scott, 1993). In 2000 Ellison reported that the total mangrove area in the Pacific islands was about 343,735 ha, with the largest areas to be found in Papua New Guinea, Solomon Islands, Fiji and New Caledonia. However, the International Union for Conservation of Nature's (IUCN, 2018) Pacific Mangroves Initiative states that 50–80% of mangroves have been cleared in the last 20 years, making mangroves one of the most critically threatened ecosystems.

Overall, coral reefs in the South Pacific appear healthier than in other regions, for example the Indian Ocean and the Caribbean (Bryant et al., 2011). However, the value assigned to coral reef ecosystem services in the South Pacific is lower compared to other coral reef destinations, for example the Caribbean or South East Asia (Laurans et al., 2013). This can be explained by different levels in population density, tourist visitation and scale of fisheries. Tourism remains an important economic driver for many South Pacific countries, particularly activities focusing on coastal areas, such as snorkelling and scuba diving. Tourist arrivals remain low for this region and tend to be limited to areas with developed infrastructure (Spalding et al., 2001). The following discussion briefly reviews the current situation in the region's three main geographic groupings.

Melanesia forms a core part of the Coral Triangle region and is estimated to contain 75% of known coral species (Wildlife Conservation Society, 2018). The islands of Melanesia, Papua New Guinea, the Solomon Islands, New Caledonia, Vanuatu, Fiji, Tonga and Samoa contain over 35,000 km^2 of reef area supporting a large variety of marine biodiversity (Drew & Amatangelo, 2017; Chin et al., 2011). Since this area is known as a biodiversity hotspot, dive tourism in the area has been growing, particularly the availability of liveaboard dive vessels. The major environmental threats to biodiversity on Melanesian coral reefs are coral bleaching, sedimentation associated with logging, mining, palm oil plantations, sewage discharge, industrial pollution and overfishing (Foale, 2008; Chin et al., 2011). Traditional fishing controls, which often include complex and relatively effective management regimes, still predominate in coastal waters (Spalding et al., 2001). However, Foale (2008, p. 30) raised concerns about increasing pressures from subsistence and artisanal fishing practices which are likely to lead to irreversible degradation of reefs in this region in the future, "unless more culturally enlightened approaches to marine resource management and economic development are embraced."

Micronesia covers an area of 6.7 million km^2 in the Pacific Ocean (The Nature Conservancy, 2016). High levels of coral and species diversity can be found in Palau, but biodiversity appears to decline towards the east (Spalding et al., 2001).

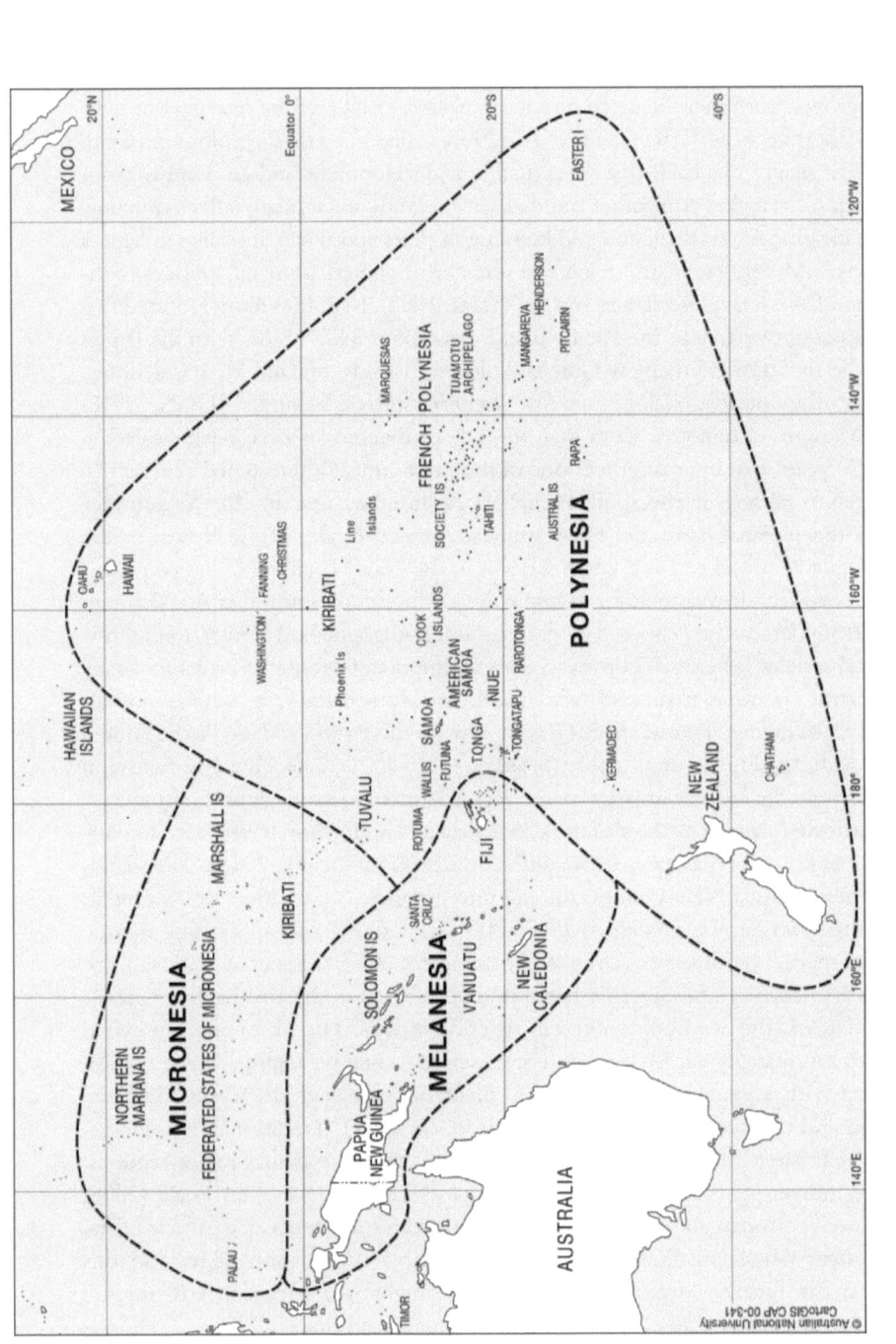

FIGURE 2.1 Micronesia, Melanesia and Polynesia Islands in the South Pacific

Source: CartoGIS, College of Asia and the Pacific, The Australian National University.

Reefs in this area are generally in good condition; however, poor reef development can be noted in areas affected by natural disturbances such as volcanoes, earthquakes and tropical cyclones. The long-term sustainability of Micronesian reefs is under threat from local and global stressors as well as insufficient resource and management capacity (The Nature Conservancy, 2016). In some areas, coral reefs are also stressed by resort and residential development, port construction, sedimentation, urban pollution, sewage/garbage disposal and overharvesting of certain species. The coral reefs around the populated areas of Guam tend to be more degraded with soil erosion from military activity, resort development, residential development and port construction. The atolls of the Marshall Islands were subjected to nuclear weapons testing by the USA in the 1940s and 1950s resulting in significant physical impacts (Spalding et al., 2001). Tourism is growing in the region, with visitors attracted by scuba diving and other marine reef tourism activities (Kostka & Gavitt, 2006). Visitors to this region are particularly attracted to destinations such as Guam, Saipan and Chuuk Atoll.

Polynesia contains 11,000 km^2 of coral reefs with diverse ecological reef communities including atolls, islands with fringing reefs and partially submerged volcanoes with barrier reefs (Spalding et al., 2001). Species diversity tends to be low and declines further towards the east. Countries in Polynesia are remote with small populations relying on fisheries. Tourism is an important economic driver for many countries in Polynesia particularly Tonga, French Polynesia and Hawai'i (Spalding, Ravilious & Green, 2001). Due to the widespread geographical distribution, low population density and relatively varied climates, Polynesia's coral reefs are considered to be in good overall ecological health. However, there are areas of concentrated development where coral reefs face the usual pressures from human activity including land clearing, overexploitation of resources, destructive fishing practices, pollution from waste water discharge and physical modification of the coastline (Ruppert et al., 2017; Vieux et al., 2008). Disturbances to coral reef growth is also caused by natural stressors such as cyclones and earthquakes. Furthermore, crown-of-thorns starfish infestations have reduced live coral cover.

Issues, challenges and management of coral reefs in Fiji, Solomon Islands and Tonga

For the South Pacific region, rising sea temperatures and coral bleaching could potentially become the main threat to tourism development in the islands as most tourism activities are based on coastal and marine recreational activities (Riegl et al., 2009; Vieux et al., 2008; Whippy-Morris, 2009). Sea-level rise and changes to ocean chemistry (e.g. acidification) lead to additional stress on coral reefs (Garrod & Gössling, 2008; Lück, 2008). Based on an assessment of current initiatives and coral reefs studies available in the South Pacific region, Table 2.1 outlines the major threats and mitigation measures in three countries, namely Fiji, Solomon

TABLE 2.1 Overview of status and management of coral reefs in Fiji, Solomon Islands and Tonga

Country	Major threats	Mitigation measures in place or recommended	Reports and regulations background
Fiji	Climate change and its related impacts such as coral bleaching and ocean acidification. Anthropogenic stresses from human activities including overfishing, anchor damage and coral mining. Land-based activities that threaten reefs in Fiji include mangrove clearance, poorly planned coastal development, run-offs from mines and agriculture and sedimentation (Spalding et al., 2001; Cumming et al., 2002). Fiji is a major exporter of live coral and live fish for the aquarium trade, and live fish for food trade into Asia.	Regulations, local education and training, climate change workshops and campaigns.	Endangered and Protected Species Act 2002 in addition to the Environmental Management Act 2005, which are administered by the Departments of Fisheries and Environment, respectively (Sykes & Morris, 2009). The aquarium trade must comply with the Endangered and Protected Species Act 2002 which lays down the requirements of permits for CITES-listed species (Lovell & Whippy-Morris, 2008; Lovell & McLardy, 2008).
Solomon Islands	Climate change and its related impacts such as coral bleaching and ocean acidification as well as anthropogenic stresses from human activities including overharvesting of certain key species, destructive fishing practices (i.e. poison fishing), coastal development, logging, agricultural run-offs, sedimentation and coral mining (Albert et al., 2012; Green et al., 2006; Masu & Vave-Karamui, 2012).	Regulations, local education and training, climate change workshops and campaigns.	The Wildlife Protection and Management Act 1998 implements the requirements of CITES; however, it does not appear to restrict the trade and export of corals (National Parliament of Solomon Islands, 1998). The Fisheries (Amendment) Regulations 1993 provide for the declaration of areas in which the collection of coral (dead or alive, or coral sand) is prohibited, and prohibits the use of machines for coral gravel or sand extraction. The Fisheries Act 1998 was reported to prohibit use of explosives or poison for fishing and the export of live corals without a licence (National Parliament of Solomon Islands, 1998).
Tonga	Climate change and its related impacts such as coral bleaching and ocean acidification as well as anthropogenic stresses from human activities including overfishing, destructive fishing practices, coastal development, pollution from untreated sewage and fertiliser run-off, sedimentation and sand/coral mining (Salvet, 2002; Lovell & McLardy, 2008).	Regulations, local education and training, climate change workshops and campaigns.	Fisheries Management Act 2002 (Tongan Ministry of Fisheries, 2002) and the Aquaculture Management Act 2003 (Tongan Ministry of Fisheries, 2003). The Aquaculture Management Regulations set out requirements for certification and licensing of operators and equipment (Tongan Ministry of Fisheries, 2008a). The Fisheries Management Regulations specify permitted fishing areas, quotas, equipment and methods (Tongan Ministry of Fisheries, 2008b). Tonga is not a signatory to CITES; however, it has designated competent authorities to issue comparable documentation and to undertake non-detrimental findings for exports of CITES-listed species.

Islands and Tonga. These countries reflect what is happening in the rest of the South Pacific region and offers some guidelines that may be worth considering for other islands to learn from in the future.

Regional initiatives for coral reef sustainability in the South Pacific

During the past ten years, a wide range of initiatives have been put in place for many South Pacific countries and were recently accepted globally at the 2015 Paris Climate Change Accord where the developed countries signed and declared their support and, more importantly, funding and sponsorship for the small countries of the South Pacific. This universal agreement's main aim was to keep global temperature rise below 2 degrees Celsius, as discussed by the UN Framework Convention on Climate Change (UNFCCC, 2015). The changing Pacific Ocean's health, acidity level and rising sea temperature affect marine life and account for the recent degradation of coral reefs in the South Pacific region. However, the Paris Convention was considered to be a win for the small developing countries of the South Pacific, especially the vulnerable small islands of Kiribati, Tuvalu, Nauru and Tonga.

The UN Green Climate Fund and the South Pacific

The Green Climate Fund (GCF) aims to advance the objective of reducing greenhouse gas emissions in developing countries and to help vulnerable small developing South Pacific countries (UNFCCC, 2015). The GCF is vital in guiding adaptation initiatives and efforts in the many South Pacific countries. The funding provides resources and technical capabilities to formulate and guide new projects for biodiversity conservation. The SPREP provides information and guidance on how to access these funds and helps with project formulations for the Pacific countries. Apart from the GCF, other governments and international funding agencies (including the World Bank and Asian Development Bank) are also funding renewable energy projects to enhance climate resilience in the South Pacific islands. Green Climate funding are either being accepted or planned for countries including the Cook Islands, Federated States of Micronesia, Marshall Islands, Papua New Guinea, Samoa and Tonga (Green Climate Fund, 2018).

The Phoenix Islands marine protected area in the Kiribati

A major conservation declaration made by the Kiribati Government in 2006 for about 408,250 km² in the Phoenix chain of islands (located almost halfway between Fiji and Hawaii) is an example of a move by a small government in the Pacific to protect and enhance its marine environment for the long-term benefit of all stakeholders. These chains of islands are uninhabited, except for Kanton Atoll which houses a small caretaker population and is located 1,750 km from Kiribati's capital Tarawa. The eight islands in the Phoenix group are still unspoiled and contain natural coral reefs with rich marine life. The Phoenix Islands Protected Area (PIPA) was officially adopted by Kiribati's government in 2009 and is widely supported by financial and conservation partners in the region and globally.

Sustainable tourism development

Tourism in the South Pacific is the main economic sector in these developing countries. Most tourism attractions and recreational activities are based on coastal and marine environments. Visitors are attracted to the region because of activities such as coral viewing, swimming, snorkelling, scuba diving, surfing, sailing, game fishing and dolphin and whale watching. However, the impacts of sea pollution and coral bleaching are affecting these tourism activities, potentially reducing the economic and social benefits which local communities could gain from these activities. Tourism operation is also affected by climate related events, including physical damage from cyclones and coral bleaching (Becken, 2004). The restoration and conservation of coral reefs and their associated biodiversity are therefore essential for the survival and growth of tourism in the Pacific islands. There are several regional agencies in the South Pacific region supporting initiatives for the better protection and management of the coral reefs, including the SPREP based in Apia, Samoa and the SPTO based in Suva, Fiji.

The SPREPs overall long-term plan is to ensure sustainable development by supporting renewable energy initiatives funded by agencies such as the UN GCF. The SPREP also advocates the education of local communities on the impacts of climate change and the adaptation of mitigation measures by their Pacific members. Importantly, the SPREP plays an important role in ensuring ongoing communication amongst stakeholders such as scientists, policy makers, tourism managers and other stakeholders.

The SPTO is a regional body consisting of government members as well as key tourism organisations and businesses in the region. The organisation recognises the significance of healthy coral reefs as a major part of the region's tourism product. They organise activities and seek sponsorships in tourism promotional programmes amongst its members to run projects to increase awareness on the importance of keeping the ocean clean and minimise pollution from tourism resorts. The division of sustainable tourism development is responsible for supporting the sustainable development of tourism in the South Pacific in terms of environmental, socio-cultural and economic aspects (SPTO, 2017). It offers guidance of tourism development that preserves the region's natural and cultural integrity by encouraging the use of renewable energy and sustainable practices by tourism businesses in the region.

Conclusion and future directions

This chapter assessed key studies, observations and stakeholders reports on the status and current effort to mitigate impacts of climate change and other natural and anthropogenic causes in the South Pacific region. The SPREP, the regional governments and overseas donor organisations mentioned in this chapter are continuously studying and reviewing effective mitigation measures and putting in place new initiatives and regulations to respond to the challenges and issues affecting the coral reefs of the islands. All the Pacific islands have been projected to experience, in varying degrees, coral bleaching, sea-level rises, ocean acidification, temperature increases, precipitation changes and cyclone intensity increases in the years to come. These will cause significant changes on tourism development and the overall economic and social well-being of the inhabitants of the region. With climate change issues becoming increasingly prominent, it is critical to include such considerations in environmental policies.

One important future direction is the need to develop more effective communications and cooperation amongst the various governments and stakeholders of the Pacific region. Further strategies may include improved information sharing, training for government planners on coastal resource management, community workshops and greater dialogues to enable local inhabitants to learn about new techniques and trends in marine parks conservation and overall biodiversity conservation. Countries of the South Pacific need to continue to embrace science and climate changes mitigation measures initiated under various national and regional environmental plans. As part of the adaptation measures needed to mitigate climate change, traditional lifestyles and reef fishing techniques, including harvesting of shell fish, coral products and sand mining, may have to be changed to accommodate current scientific understanding of the state of the Pacific Ocean coral reef ecosystems.

References

Albert, J., Trinidad, A., Cabral, R., & Boso, D. (2012). *Economic value of coral reefs in Solomon Islands: Case-study findings from coral trade and non-coral trade communities.* Solomon Islands: WorldFish Center, p. 24.

Becken, S. (2004). *Climate change and tourism in Fiji.* Suva, Fiji: University of the South Pacific.

Bryant, D., Burke, L., Mcmanus, J., & Spalding, M. (2011). *Reefs at risk: A map-based indicator of threats to the world's coral reefs.* World Resources Institute (WRI), International Centre for Living Aquatic Resources Management (ICLARM), World Conservation Monitoring Centre (WCMC). United Nations Environment Programme (UNEP).

Chin, A., Lison de Loma, T., Reytar, K., Planes, S., Gerhardt, K., Clua, E., Burke, L., & Wilkinson, C. (2011). *Status of coral reefs of the Pacific and outlook: 2011.* Washington, DC: Global Coral Reef Monitoring Network, p. 260.

Cumming, R., Aalbersberg, W.G., Lovell, E., Sykes, H., & Vuki, V. (2002). *Coral reefs of the Fiji Islands: Current issues.* IAS Technical Report 2002/11. Institute of Applied Sciences, University of the South Pacific.

David, G., Fontenelle, G., Dumas, P.S., Ferraris, J., Herrenschmidt, J.B., & Leopold, M. (2010). Integrated coastal zone management perspectives to ensure the sustainability of the coral reefs in New Caledonia. *Marine Pollution Bulletin,* 61(7e12), 323e334.

Drew, J.A., & Amatangelo, K.L. (2017). Community assembly of coral reef fishes along the Melanesian biodiversity gradient. *PLoS ONE,* 12(10), e0186123. https://doi.org/10.1371/journal.pone.0186123

Ellison, J.C. (2000). How South Pacific mangroves may respond to predicted climate change and sea-level rise. In A. Gillespie & W.C.G. Burns (eds) *Climate change in the South Pacific: Impacts and responses in Australia, New Zealand, and Small Island States*. Advances in Global Change Research, Vol. 2. Dordrecht: Springer.

Foale, S.J. (2008). Conserving Melanesia's coral reef heritage in the face of climate change. *Historic Environment*, 21(1), 30–36.

IUCN. (2018). Pacific Mangroves Initiative. Accessed 9 February 2018 from International Union for Conservation of Nature, https://www.iucn.org/regions/oceania/our-work/deploying-nature-based-solutions/water-and-wetlands/completed-projects/pacific-mangroves-initiative

Garrod, B., & Gössling, S. (eds) (2008). *New frontiers in marine tourism: Diving experiences, sustainability, management*. (1st ed.). Amsterdam: Elsevier.

Green, A., Lokani, P., Atu, W., Thomas, P., Almany, J., & Ramohia, P. (2006). *Solomon Islands marine assessment. Technical report of survey conducted May 13-June 17, 2004*. Brisbane, Australia: The Nature Conservancy, p. 519.

Green Climate Fund. (2016). Pacific Islands Renewable Energy Investment Program. Accessed 7 May 2017 from https://www.greenclimate.fund/-/pacific-islands-renewable-energy-investment-program

Holthus, P. (1996). Coral Reef Survey Vava'u, Kingdom of Tonga. (96). Apia, Samoa.

Kostka, W., & Gavitt, J.D. (2006). A threats- and needs-assessment of coastal marine areas in the states of Kosrae, Chuuk and Yap, Federated States of Micronesia. Available from https://data.nodc.noaa.gov/coris/library/NOAA/CRCP/project/1413/NA05NMF4631043_FinalReport_assessment_micronesia.pdf

Laurans, Y., Pascal, N., Binet, T., Brander, L., Clua, E., David, G., Rojat, D., & Seidl, A. (2013). Economic valuation of ecosystem services from coral reefs in the South Pacific: Taking stock of recent experience. *Journal of Environmental Management*, 116, 135–144.

Leisher, C., Beukering, P.v., Scherl, L.M., 2007. Nature's investment bank: How marine protected areas contribute to poverty reduction report. Arlington, VA: The Nature Conservancy.

Leisz, S.J. (2009). *Locations in Melanesia most vulnerable to climate change*. (May), Fort Collins, CO: Colorado State University, p. 16.

Lovell, E.R., & McLardy, C. (2008). *Annotated checklist of the CITES-listed corals of Fiji with reference to Vanuatu, Tonga, Samoa and American Samoa*. JNCC Report No. 415. Peterborough, UK: Joint Nature Conservation Committee.

Lovell, E.R., & Whippy-Morris, C. (2008). *Live coral fishery for aquaria in Fiji: Sustainability and management*. Suva, Fiji: University of the South Pacific.

Lück, M. (ed.). (2008). *The encyclopedia of tourism and recreation in marine environments*. Oxfordshire: CAB International.

Masu, R., & Vave-Karamui, A. (2012). *State of the Coral Triangle Report. Solomon Islands*. Coral Triangle Initiative, p. 4.

Mumby, P.J., & Steneck, R.S. (2008). Coral reef management and conservation in light of rapidly evolving ecological paradigms. *Trends in Ecology and Evolution*, 23(10), 555–563.

National Parliament of Solomon Islands. (1998). *Fisheries Act 1998*. National Parliament of Solomon Islands, p. 92.

NOAA. (2016). Extended reconstructed sea surface temperature (ERSST.v4). National Centres for Environmental Information. Accessed 16 June 2016 from National Oceanic and Atmospheric Administration, www.ncdc.noaa.gov/data-access/marineocean-data/extended-reconstructed-sea-surface-temperature-ersst

Pascal, N., Allenbach, M., Brathwaite, A., Burke, L., Le Port, G., & Clua, E. (2016). Economic valuation of coral reef ecosystem service of coastal protection: A pragmatic approach. *Ecosystem Services*, 21, 72–80.

Pollnac, R.B., McManus, J.W., del Rosario, A.E., Banzon, A.A., Vergara, S.G., & Gorospe, M.L.G. (2000). Unexpected relationships between coral reef health and socio-economic pressures in the Philippines: reefbase/RAMP applied. *Marine and Freshwater Research*, 51, 529e533.

Riegl, B., Bruckner, A., Coles, S.L., Renaud, P., & Dodge, R.E. (2009). Coral reefs – threats and conservation in an era of global change. In R.S. Ostfeld & W.H. Schlesinger (eds), *The year in ecology and conservation biology*. New York: New York Academy of Sciences, pp. 36–186.

Ruppert, J.L.W., Vigliola, L., Kulbicki, M., Labrosse, P., Fortin, M.J., & Meekan, M.G. (2017). Human activities as a driver of spatial variation in the trophic structure of fish communities on Pacific coral reefs. *Global Change Biology*, 24, e67–e79.

Salvet, B. (2002). Status of southeast and central Pacific Coral Reefs "Polynesa Mana Node": Cook Islands, French Polynesia, Kiribati, Niue, Tokelau, Tonga, Wallis and Fotuna. In C.R. Wilkinson (ed.), *Status of coral reefs of the world: 2002*. Townsville: Australian Institute of Marine Science, pp. 203–216.

Scott, D.A. (1993). *A directory of wetlands in Oceania*. Kuala Lumpur: Asian Wetland Bureau.

Secretariat for the Pacific Regional Environmental Programme (SPREP). (2016). *Annual Report 2016*. Apia, Samoa.

South, G.R., Veitayaki, J., Limalevu, L., Morris, C., & Bala, S. (2012). *Global change and coral reef management capacity in the Pacific: Engaging scientists and policy makers in Fiji, Samoa, Tuvalu and Tonga: Final report*. IMR Technical Report 01/2012. Suva, Fiji: Institute of Marine Resources, School of Marine Studies, FSTE, USP & Pacific Centre of Environment & Sustainable Development, p. 64.

Spalding, M.D., Ravilious, C., & Green, E.P. (2001). *World atlas of coral reefs*. Prepared at the UNEP World Conservation Monitoring Centre. Berkeley, CA: University of California Press.

SPTO. (2017). Sustainable tourism development. Accessed 15 December 2017 from South Pacific Tourism Organisation, https://corporate.southpacificislands.travel/sustainable-tourism-development/

SPTO. (2017). Annual review of visitor arrivals in Pacific island countries 2016. Suva, Fiji: SPTO.

Sykes, H., & Morris, C. (2009). Status of coral reefs in the Fiji Islands, 2007. In C. Whippy-Morris (ed.), *South-west Pacific status of coral reefs report 2007*, Noumea, New Caledonia: CRISP, pp. 1–52.

The Nature Conservancy. (2016). Micronesia. Accessed 17 January 2018 from https://oceanwealth.org/project-areas/micronesia/

Tongan Ministry of Fisheries. (2002). *Fisheries Management Act 2002*.

Tongan Ministry of Fisheries. (2003). *Aquaculture Management Act 2003*.

Tongan Ministry of Fisheries. (2008a). *Aquaculture Management Regulations*.

Tongan Ministry of Fisheries. (2008b). *Fisheries Management Regulations*.

UNFCCC. (2015). Historic Paris agreement on Climate Change: 195 Nations set path to keep temperature rise well below 2 degrees Celsius. Accessed 23 March 2018 from https://unfccc.int/news/finale-cop21

Vieux, C., Salvat, B., Chancerelle, Y., Kirata, T., Rongo, T., & Cameron, E. (2008). Status of coral reefs in Polynesia Mana Node countries: Cook Islands, French Polynesia, Niue,

Kiribati, Tonga, Tokelau and Wallis and Futuna. In C. Wilkinson (ed.), *Status of Coral Reefs of the World 2008*. Townsville: Global Coral Reef Monitoring Network, p. 304.

Whippy-Morris, C. (2009). *South-west Pacific status of coral reefs report 2007*. Noumea, New Caledonia: CRISP, p. 208.

White, A.T., Ross, M., & Flores, M., (2000). Benefits and costs of coral reef and wetland management, Olango Island, Philippines. In H. Cesar (ed.), *Collected essays on the economics of coral reefs*. Kalmar, Sweden: CORDIO, Kalmar University, pp. 215e227.

Whittingham, E., Campbell, J., Townsley, P., 2003. Poverty and reefs. *DFIDeIMMe IOC/UNESCO*, pp. 260.

Wildlife Conservation Society. (2018). Melanesia. Accessed 18 January 2018 from Wildlife Conservation Society, https://www.wcs.org/our-work/regions/melanesia

Worldfish. (2013). *Community-based marine resource management in Solomon Islands: A facilitator's guide. Based on lessons from implementation of CBRM with rural coastal communities in Solomon Islands (2005–2013)*. Penang, Malaysia, p. 52.

3

CORAL-BASED TOURISM IN EGYPT'S RED SEA

Nathalie Hilmi, Alain Safa, Stéphanie Reynaud and Denis Allemand

Introduction

The aim of this chapter is to examine a range of issues associated with tourism related resource management of Egypt's coral reef resources. With its unique geographic location midway between Africa and Asia, Egypt is home to a wide variety of ecosystems and terrestrial and aquatic life. Egypt has two coastlines, the Red Sea and the Mediterranean Sea. The Red Sea coasts have white soft sands and beautiful crystal waters with extremely colourful underwater life, making the Red Sea reefs one of the world's best diving spots. Egypt's coastline possesses a significant proportion of the coral reefs found in the Red Sea, with about 3,800 km² of reef area (Spalding et al., 2001) and 1,800 km of coastline (PERSGA, 2010). The Red Sea is home to approximately 300 species of hard coral and 125 species of soft coral. Among the 300 hard coral species found in the Red Sea, two-thirds are found in the Egyptian reefs, including a number of endemic species (Kotb et al., 2008). The species richness is higher than those recorded for the Caribbean and is equal to the Indian Ocean. Egyptian coral reefs are particularly well developed with large sizeable offshore reef complexes containing many islands, fringing reefs and other coral reef habitats. The reefs extend in the north to the Gulfs of Suez and Aqaba to Ras Hedarba and in the south to the border of Sudan. They are, however, not continuous because periodic flooding from valleys (wadies) has created gaps within the reef system. Because the northern part of the Red Sea has the highest coral diversity and number of islands (Shaalan, 2005), the government of Egypt established several marine protected areas (MPAs), including the Ras Mohammed National Park in 1983.

Egypt's reefs exhibit significant biodiversity with over 5,000 species. These include 800 species of seaweeds and seagrasses, 209 species of reef-building corals, more than 800 species of molluscs, 600 species of crustaceans, 350 species of

Echinodermata, 17 species of marine mammals, 4 species of marine turtles, more than 20 species of sharks, and bivalves (clams) (Nature Conservation Sector 2007, 2009, in Samy et al., 2011).

Economic importance of coral reef-based tourism in the Egyptian Red Sea

The importance of tourism in Egypt

The Egyptian tourism sector remains an important economic and social sector despite the problems that arose following the "Arab Spring" (Hilmi et al., 2012), with a significant percentage of tourism activity centred on the Red Sea. The direct contribution of Egypt's travel and tourism sector was EGP (Egyptian Pounds) 96.8 billion (5.6% of total gross domestic product [GDP]) in 2013. The contribution to GDP declined after 2013 to 4.9% of GDP in 2014 as a result of political and security factors that affected the Middle East region. The contribution to GDP rose by 0.7% in 2016 and is forecast to rise by 5.3% per annum to reach EGP 177.7 billion (5.3% of total GDP) in 2026.

The total contribution of the travel and tourism sector to GDP (including wider effects from investment, the supply chain and induced impacts) was EGP 259.7 billion (11.4% of GDP) in 2013 and rose by 1% in 2016 (11.2% of GDP). It is then expected to rise by 4.3% per annum to reach EGP 401.1 billion (12% of total GDP) in 2026 (Figure 3.1)

Travel and tourism generated 1,251,000 direct jobs (5.1% of total employment) in 2013, before falling to 1,110,500 jobs (4.4% of total employment) in 2014

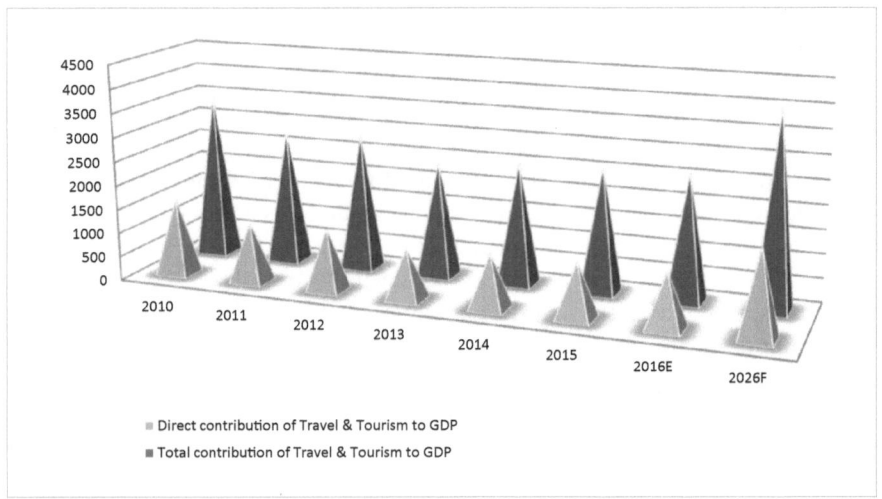

FIGURE 3.1 Direct and total contribution of travel and tourism to GDP, constant EGP billion 2015

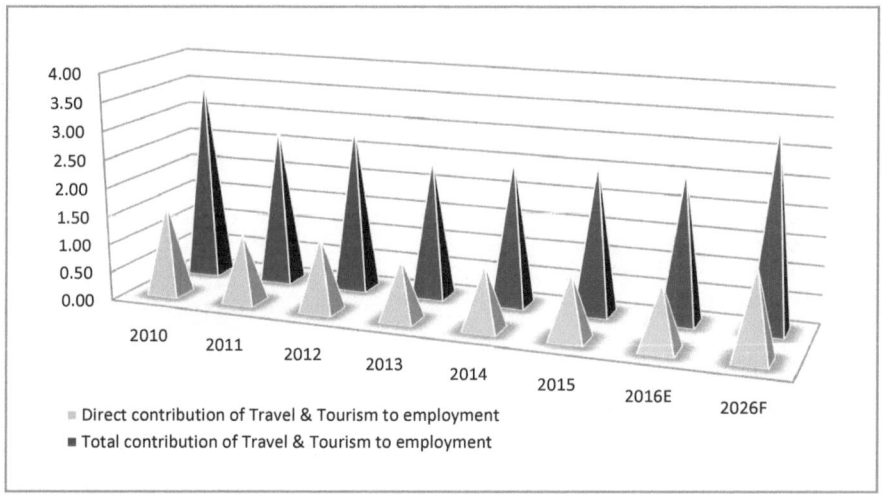

FIGURE 3.2 Direct and total contribution of travel and tourism to employment, millions of jobs

although it is expected to rise by 3.3% per annum during the next ten years to reach 1,522,000 jobs (5% of total employment) in 2026 (World Travel and Tourism Council [WTTC], 2016). The total contribution to employment (including effects from investment, supply chain and induced income impacts) was 2,620,000 jobs in 2015 (10.5% of total employment) and forecast to rise to 3,471,000 jobs (11.2% of employment) by 2026 (see Figure 3.2).

Money spent by foreign visitors (visitor exports) is a key component of the direct contribution of the travel and tourism sector. In 2015, Egypt generated EGP 57.5 billion in visitor exports (20.7% of total exports). By 2026, Egypt is forecast to attract 15,738,000 foreign tourists who will generate an estimated expenditure of EGP 103.7 billion, an increase of 6.5% per annum (Figure 3.3).

Coastal tourism related to coral biodiversity in the Egyptian Red Sea

The Red Sea governorate represents about 33% of the nation's total hotel capacity, followed by the south governorate (North Sinaï) at about 32%, the Greater Cairo governorate at about 13%, the governorates of Luxor and Aswan at about 4% and floating hotels at about 8% of the total hotel capacity. More than 90% of Egypt's tourism investment is concentrated in coastal resorts or in the southern Sinaï, with a large concentration on dive tourism and beach resorts around the Red Sea Gulf of Aqaba. In "normal" times, European countries are the main source of incoming tourism. Russia is the primary source followed by Germany, England, Italy and France (Hilmi et al., 2012).

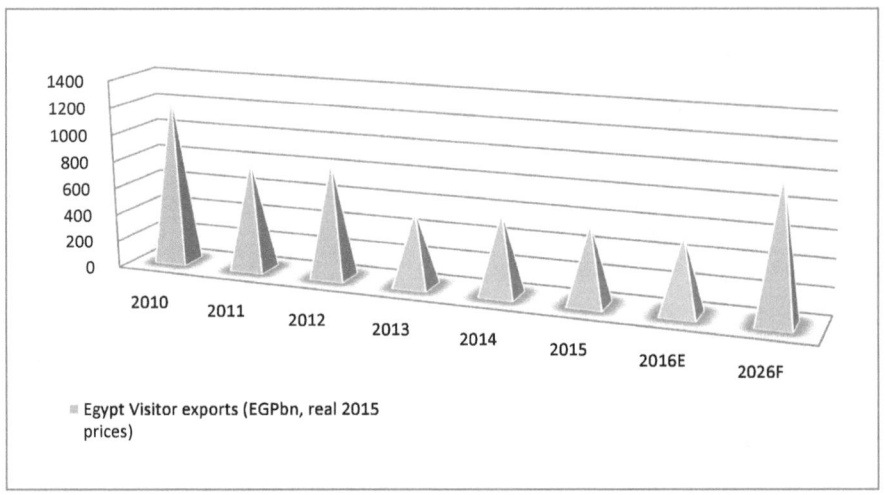

FIGURE 3.3 Visitor exports contribution of travel and tourism to GDP, constant EGP billions 2015

Hurghada was the first tourist resort to be established on the Egyptian Red Sea. It extends about 62 km along the Red Sea and mainly attracts tourists interested in water-based sports and activities. Safaga (50 km to the south) has also become a popular destination. Both cities attract scuba divers because of the abundance of coral reefs, white sandy beaches, exotic fish, clear water and year-round warm climate and it boasts some of the best diving sites in the world (Hilmi et al., 2012).

The Red Sea has unique marine habitats, including coral reefs, mangroves and seagrass beds. They provide key resources for coastal populations such as food, shoreline protection and stabilisation, as well as economic benefits from tourism. For these reasons the nation's coral reefs are considered a natural capital. The total economic value (TEV) of this natural capital can be derived from the value of all goods and services provided by marine ecosystems including tourism, fisheries, research, biodiversity and bioprospecting. In the case of the Red Sea, tourism is the most important generator of goods and services (85%) followed by fisheries (11%) (MVE Unit, 2003).

Biodiversity is essential for healthy ecosystems, human health, prosperity, security and well-being. Identifying the economic values of biodiversity in terms of the goods and services that it provides is relatively new and slowly gaining recognition in Egypt. Since ancient times, Egypt has relied on the wealth of its natural resources to sustain its civilisation. The contribution of biodiversity to the national economy is substantial. Much of the country's economy is built on its natural resource base. Biodiversity is also crucial for fisheries activities and supports the production of food, which contributes to the overall well-being of the Egyptian people. Moreover, biodiversity supports the development of nature-based tourism and recreational activities, which provide high economic returns on national, regional and local scales.

The aquatic resource base in Egypt is extensive and consists of marine, fresh water and brackish water. Marine resources are located in the Mediterranean Sea, the Red Sea, the Suez Canal and the Great Bitter Lakes. Fisheries provide a major source of food, revenues and employment (Hilmi et al., 2014). Local seafood is also an important attraction of the region and is popular in the nation's restaurants.

Threats to coral reefs

Current health

The status of coral reefs along the whole Egyptian coast was reviewed by Kotb et al. (2008) and PERSGA (2010). In addition, Mohamed et al. (2012), Kotb et al. (2015) and Attalla et al. (2015) provided regional studies for the northern and southern Egyptian coast, respectively. Although there has been significant tourism development south of Marsa Alam over the last ten years, the reef systems are still healthy (Kotb et al., 2015). Total coral cover ranged between 52 and 84% (branching corals were dominant) and coral diversity (number of species) ranged between 4 and 13 per transect (125 m^2). They also observed that impacted sites showed more physically damaged colonies than others sites (74 vs. 4 colonies/125 m^2) and that coral disease was recorded in less than half of the studied sites (between 1 and 36 colonies/125 cm^2). This level of coral disease is much lower than those reported from other reefs in the Indo-Pacific and Great Barrier Reef (GBR) (Willis et al., 2004), India (Thinesh et al., 2009) and in the Caribbean (Nugues, 2002). This finding confirms previous observation about the state of coral disease in the northern Red Sea (Mohamed et al., 2012). Live coral (hard plus soft) cover on Egyptian reefs averages 48% (Kotb et al., 2008). By comparing the survey data between 2002 and 2008 provided by PERSGA (2010), it can be seen that the health state of Egyptian reefs is stable (in term of fish abundance, including butterflyfish, grouper, sweetlips, parrotfish and hard coral cover). Outbreaks of the starfish *Acanthaster planci* (crown-of-thorns) were recorded in 2000, 2002 and 2004 but due to remedial actions taken by the Egyptian government and some non-governmental organisations (NGOs), no new outbreaks have been observed since 2008. Populations of dugong (*Dugong dugon*) are present in a number of areas, including the Tiran Islands, Nabq and Abu Galum Marine Park in the north, and El Quseir in the south. The presence of marine turtles (hawksbill, green turtle, leatherback and loggerhead) and sharks supports the finding that the ecosystem is healthy. Globally, it appears that there has been no major change in reef health since 2008 (Kotb, pers. comm).

A more recent study undertaken in the northern part of the Egyptian coast, including the Gulf of Aqaba, confirms this area is also in relatively good health. By measuring different indicators of coral diseases (black-band disease, white syndrome, pink line syndrome, ulcerative white spots, skeletal eroding band, as well as coral bleached), Mohamed et al. (2012) concluded that there was low coral disease prevalence of 0.63%. This level is lower than other major reef systems: 7–10% in the GBR of Australia; 8.9% in south-eastern India; and 11% in St Lucia in the

Caribbean. Attalla et al. (2015) found hard coral cover in the southern reefs around Marsa in the range of 56.7–78.9%, a figure that is much higher than in the GBR (De'ath et al., 2012).

Global and local threats

Both global and local threats can affect coral reefs. Among global threats are global warming and ocean acidification. Reefs in the northern Red Sea and the Arabian Gulf are especially vulnerable to degradation due to limited water circulation and temperature extremes. Studies have shown that Egypt's climate has changed greatly over the last 10,000 years (Bubenzer et al., 2007), changing gradually from a wet climate (rainfall was more than 300 mm/year) to a more arid climate (less than 50 mm/year). According to a study of seasonal temperatures distribution in Egypt (Hegazy et al., 2008), it is anticipated that temperatures will rise in all four seasons, moving from the southern to the northern parts of Egypt, in the coming 100 years.

Global warming induces increase of sea surface temperature (SST), which leads to dysfunctional reef-building corals and particularly of the symbiosis between the coral host and its microalgae Dinoflagellate symbionts (zooxanthellae). This leads to a loss of the symbionts, causing the coral to turn completely white (coral bleaching) and ultimately to coral death. Coral bleaching is probably the most significant threat to the world's coral reefs. The 2014–16 El Niño event is considered to be the third global bleaching event and probably the most destructive. In the Red Sea, SST normally varies between 24 and 30°C, with the highest values in the central Red Sea (Sofianos et al., 2002). However, very few bleaching events were reported for Red Sea coral reefs (Kotb et al., 2008; Kotb, pers. comm.) possibly because of their reported greater tolerance of high sea temperatures (Coles & Riegl, 2013). This high bleaching threshold suggests that the northern part of the Red Sea could be a reef refugium for tolerant corals species and could be used for reseeding of lost reef areas (Fine et al., 2013). Nevertheless, if global warming continues at the present rate, it is quite possible that bleaching events will occur on a more frequent basis in the near future.

The second global threat to coral reefs is ocean acidification. This process results from the dissolution of excess anthropogenic CO_2 in the ocean waters, leading to the decrease of oceanic pH (Doney et al., 2009). While it is apparent that ocean acidification is likely to be a major threat in the future, it is difficult to determine its current impact.

Local threats are also increasing. Almost two-thirds of the Gulf reefs face some level of risk because over 30% of the world's oil tankers each year traverse the area. Increasing human pressure, particularly from an expansion of tourism activities, also poses a major threat for Egyptian coral reefs. On the Israeli coastline of the Gulf of Aqaba, Rinkevich (2005) identified the impacts of tourism as the major cause of the decline of coral reefs. A study by Jameson et al. (2007), which compares four coral sites exposed to extensive tourism with a site that is fairly unexposed (all located near Hurghada), found that all four of the tourism-exposed sites suffered from physical damage, reflected in consistently having a lower frequency of hard coral (especially *Acropora* coral), a higher percentage of soft coral and a higher percentage of algae. Coral reefs within protected areas have shown to be relatively healthier than their counterparts in non-protected areas. Since 2001, the status of Egypt's coral reefs has been monitored at more than 120 sites in the Red Sea and the Gulf of Aqaba using environmental indicators (living vs. non-living coral reefs, number of species) and other indicators such as fish and vertebrates. Studies conducted in 17 dive sites (8 around Hurghada, 2 in Safaga, 4 in Marsa Alam and 3 in the Wadi El Gemal area) from 2001 to 2011 revealed that sites farther away from human activities witnessed a 15% increase in coral reef cover compared to corals exposed to human activities (5–7%) where soft corals have increased at the cost of hard ones (Attalla et al., 2015).

Tourism may impact on coral reefs in two ways. Direct impacts include irresponsible snorkellers and divers causing coral breakage (by walking on them or touching them, even stirring up the sand in reef areas can cause damage to corals), fish capture (for aquariophilly) and shark (or fish) feeding. There is also a series of indirect impacts caused by the development of tourism facilities, including landfill, dredging for artificial beaches, boat anchors, excavations for creating artificial swimming pools, sewage run-off, pollution and increased sedimentation (Frihy et al., 2006). Riegl and Velimirov (1991) found that coral breakage was the most common form of damage, especially on highly frequented reefs. Coral breakage was most frequent within the first 10 metres of depth, suggesting that inexperienced divers and snorkellers pose a major threat to coral reefs rather than experienced divers who practise a more eco-friendly tourism (Jobbins, 2006). In addition to coral breakage, scuba diving is known to induce coral diseases (Lamb & Willis, 2011; Lamb et al., 2014). A survey of the impact of diving tourism near Dahab (northern Red Sea) showed that intensive scuba diving induced a net decrease of hermatypic coral cover, together with an increase of damaged corals (Hasler & Ott, 2008). Branching corals are particularly sensitive. These results agree with Kotb et al. (2015).

Increased commercial fishing is also an important factor in relation to the health of Egyptian reefs of the Gulf of Aqaba (Kotb et al., 2008). Shark fishing (for shark fins) emerged in the late 1990s but was banned in 2004 by the Egyptian government because of its adverse impact on tourism. Kotb et al. (2008) noted that the annual commercial value of an individual shark at the diving site of Brother Island exceeds US$300,000 because these sharks represent the main attraction for divers.

Modification of the coastline by the construction of marina jetties or land extension offshore affects the marine environment in three ways: (i) by changing seawater currents, (ii) by increasing the water turbidity through sediment suspension, and (iii) by accelerating coastal erosion due to the destruction of coral reefs. For example, on the downdrift side of a protruding structure in a resort village near Hurghada, shoreline recession of about 10 m was observed over a 2-year period (Frihy et al., 1996). Increasing sediment suspension increases turbidity, then decreases the supply of light to corals and other phototroph ecosystems.

Moreover, corals are very sensitive to sediment, which induces tissue necrosis (Nugues & Roberts, 2003). Both a decrease in light and an increase in sediment events lead to coral death and decrease of coral cover. Because coral reefs provide natural coastal protection (Frihy et al., 2004), their destruction may induce net coastal erosion. Holden (2000) estimated that 73% of the coral along the Egyptian coast has been damaged as a result of construction.

In addition to impacts linked to tourism, pollution by heavy metals (cadmium, lead, copper, nickel, etc.) is important in some areas due to natural outflows from wadis (Qola'an Reefs, Kalawye Reefs) or anthropogenic activities in Safaga and Hurghada harbours and Ras El-Behar (Ali et al., 2011; Madkour 2013). Pollution from nutrients (nitrogen and phosphorous) and other sources is also a continuing and growing threat to biodiversity in coastal ecosystems. Pollutants, such as fertiliser and pesticides from agriculture and also wastewater treatment systems and industry, including mining and oil or gas extraction, harm biodiversity directly through mortality and reduced reproductive success. Pollution also has indirect effects through habitat degradation. Many coastal marine habitats face a major threat from waterborne pollutants.

Although some reefs are highly impacted, particularly those in the north, others remain relatively remote and inaccessible and, until now, have not been impacted by human activities. However, the demand of "virgin" spots for tourism may accelerate urban and coastal development.

Invasive species continue to be a major threat to all Egyptian ecosystems and species (Shaltout, 2008). Information on invasive species in Egypt is still insufficient or is not readily available. Efforts to control and eradicate existing invasive species and to prevent the introduction of new invasive species remains limited in spite of the threat they present to Egypt's marine ecosystems and the economy and human health.

Sustainability and governance

Reef management

Three main types of management can be used to reduce human pressure on reefs and increase the sustainability of tourism through restrictions and protection measures, environmental planning as well as educational programmes. Since tourism is a crucial component of Egypt's economy and because reefs are particularly important tourist attractions, protection of natural resources is a priority issue.

Restriction / protection measures

The first measure required is to reduce the number of tourists / divers per site on a year-by-year basis (Hasler & Ott, 2008). Such a measure can be determined by science-based approaches of the type designed by Leujak and Ormond (2008) for the Ras Um Sidd reef at Sharm El Sheikh and Ras Mohammed National Park.

TABLE 3.1 List of Marine Protected Areas in the Egyptian waters of the Red Sea

MPA name	Date of creation	Total surface of the marine portion (km²)	Number and surface of closed areas / no-take-zones (km²)
Ras Mohammed National Park	1983	597	3 / 2.99
Elba Protected Areas	1986	2,000	no data
Abu Galum Resources Protected Area	1992	121	4 / 52.72
Nabq Managed Resources Protected Area	1992	121.9	5 / 97.27
Wadi El Gemal-Hamata Protected Area	2003	1,600	3 / 305.57

Source: Adapted from Samy et al. (2011).

By using a carrying capacity concept, these authors determined that the limit should be set at approximately 50 tramplers/m²/year. This type of measure can be also integrated within MPAs.

Egypt's MPAs are quite well developed and form a good network of reefs (Samy et al., 2011). To date, five MPAs have been declared in the Egyptian Red Sea (see Table 3.1) and the Gulf of Aqaba zones with a further MPA (El Sallum PA) located in the Mediterranean Sea. These MPAs cover an area of approximately 50,000 km². Fishing is prohibited or restricted to local populations and for some recreational activities. The result of this protection is generally positive, with a higher abundance of fish (Roberts & Polunin, 1992) and a decreased prevalence of coral disease (Mohamed et al., 2012). These MPAs generate income from entrance fees, penalties and sanctions applied for violations (Samy et al., 2011); however, some parks are victims of their own success and, for example, some diving sites within Ras Mohammed National Park regularly exceed the recommended level of divers per year by over ten times (Samy et al., 2011).

Environmental planning

As outlined previously, many of the recreation facilities on the Egyptian coast have been implemented without considering environmental parameters, and have thus created disturbances within coastal ecosystems. However, in some places environmentally friendly solutions have been used. Frihy et al. (2006) reviewed different practical examples of eco-designed facilities selected from resorts at Sokhna, Sharm El Sheikh and Hurghada. These examples included using natural lagoons as swimming pools instead of excavating them, creating access to the reefs without environmental damage, creating sandy beaches and building marinas without coastal damage. To avoid anchor damage, the Hurghada Environmental Protection and Conservation Association implemented a mooring system at all key dive sites in Egypt (PERSGA, 2010). In 1992 approximately 100 mooring buoys were installed. This local mooring project has evolved into the world's largest mooring system with over 1,000 moorings installed and maintained throughout Hurghada, Safaga and the southern Red Sea. To ensure correct use of the buoys, this project has also provided ongoing training to boat captains and crews on the proper use of these buoys. Since 1998 more than 800 boat skippers have received best practice training in their use and additional environmental awareness. The presence and ongoing maintenance of the mooring system has proved to be effective and the number of coral colonies damaged by anchoring has been significantly reduced. An effective patrolling system has also helped reduce damage to corals caused by swimming tourists and aided in the increase of coral cover.

To restore degraded reefs or reduce the number of divers on natural reefs, artificial reefs may be used (Kotb, 2013). Artificial reefs are used worldwide to promote marine life (Seaman & Sprague, 1991). They can be made with concrete blocks, rocks, car tyres and so on. The Reef Ball Foundation (www.reefball.org), a non-profit organisation, has developed specific protocols for deploying, anchoring and transplanting corals to Reef Ball products. Biorock is a patented method that uses electrolytic deposition of calcium carbonate to build artificial structures. To enhance reef restoration, *in situ* or *ex situ*, coral farming has proved to be efficient (Kotb, 2016; Rinkevich, 2005). In all cases, coral recruits may be transplanted on artificial reefs or on various natural or artificial structures. The cost of coral reef restoration projects varies depending on the methods used and can range from US$6 to US$15,000/m^2 (2012 unit cost; Ferrario et al., 2014; see Table 3.2).

TABLE 3.2 The effectiveness of coral reefs for coastal hazard risk reduction and adaptation

Restoration technique	Location	Year	Original cost (US$/m^2)	2012 Unit cost (US$/m^2)	2012 Linear unit cost (US$/m^1)
Paving slabs + chain-link fencing	Maldives	1994	40	62	620
Armorflex	Maldives	1994	103	159	1,590
Armorflex + coral transplantation	Maldives	1994	151	233	2,330
Concrete blocks	Maldives	1994	328	508	5,080
Concrete structures + coral transplantation	Florida	1991	550	927	927
Concrete structures + coral transplantation	Florida	1994	10,000	15,500	155,000
Rock stabilisation	Indonesia	2005	5	6	60
Reef ball	Various	2005	40	47	470
EcoReef	Various	2005	70	82	820
Biorock	Various	2005	1.6–110	2–129	20–1,290

Source: Adapted from Ferrario et al. (2014).

Kotb (2013) studied a site close to Hurghada where fibreglass and car tyres were used to build artificial reef units. Hard and soft coral colonies were transplanted on these units. He observed that both coral diversity and fish abundance increased over a number of years and concluded that such artificial reefs may be used as an effective rehabilitation and conservation method to reduce the impact of recreational activities on Egypt's natural reefs.

Educational programmes through ecotourism

The International Ecotourism Society (TIES) subscribes to the following definition of ecotourism: "responsible travel to natural areas that conserves the environment and improves the well-being of local people." To develop sustainable tourism, it is necessary to develop environmental educational programmes for both the tourism industry and for tourists. These programmes:

- Enhance awareness by the tourism industry (hoteliers, restaurateurs, shopkeepers, diving clubs, dive guides and staff members) of the importance of healthy coral reefs. Implementation of coral conservation policies, coral conservation and coral reef-building, as well as educational projects, can also be undertaken with the assistance of NGOs.
- Provide tourists with full and responsible information to enhance their respect for the natural, social and cultural environments and helps enhance customer satisfaction (Ragheb, 2015).

Sustainable tourism

Considering Egypt's commitment to achieving the targets of the Millennium Development Goals (MDG), several national committees were established (sustainable development, integrated management of coastal zones, climate change, wetlands and conservation of biodiversity) to achieve harmonisation between policies, strategies and national action plans of development.

National responses to the continuing loss of biodiversity are varied and threats to biodiversity are addressed through a number of activities. Some of the most significant results achieved are:

- The Convention on Biological Diversity (CBD), which came into force at the end of 1993. The Convention requires the development of a National Biodiversity Strategy and Action Plan (NBSAP) with the aim of stimulating conservation action at the national level. The NBSAP (1997–2017) was developed using a wide participatory approach and was included in the plan for biodiversity 2011–2020.
- Joining international and regional agreements and strategies for cooperation, such as the United Nations Convention on the Law of the Sea, the CBD and its biosafety and access- and benefit-sharing protocols, and the United Nations Framework Convention on Climate Change.

- Protected area based conservation, which has been one of the primary responses for maintaining biodiversity in Egypt. These have expanded over the past 30 years in number and area. By 2013, 30 protected areas were established, extending over 14.6% of the total land and marine areas of the country.
- Managing the impacts of climate change on biodiversity through mitigation and adaptation to climate change, which requires the adjustment of natural or human systems in response to actual or expected climate stimuli. The Ministry of State for Environmental Affairs established the "Climate Change Central Department," which includes General Directorates for Risks and Adaptation, Mitigation and Clean Development Mechanism, Research and Technology of Climate Change, and Climate Change Information. The government is pursuing a low-carbon development strategy for sustainable development in the context of national plans of different sectors through pilot and operational adaptation and mitigation projects with the support of the United Nations Development Programme (UNDP) and other agencies.

Conclusion

Egypt has a unique coastal biodiversity that contributes to the economy and supports human well-being. A significant portion of Egypt's GDP is directly related to the use of biological resources and the role of biodiversity in the supply of ecosystem goods and services is gaining recognition. Loss of biodiversity will affect the ability of ecosystems to deliver their valuable goods and services and will have serious social, economic, cultural and ecological implications. Tourism activity in the Red Sea is related to the good health of the reefs, but its growth jeopardises the ecosystem that attracts the tourists.

Globally, climate change has huge consequences for the world's environment, placing a strain on resources and, as a consequence, directly threatening the well-being of local populations. It has been described by the UN Secretary General at the Climate Change Summit in New York in 2009 as "the pre-eminent geopolitical and economic issue of the twenty-first century. It rewrites the global equation for development, peace and prosperity." One of the outcomes of the focus on climate change impacts on tourism activities has been that a number of countries, including Egypt, have adopted sustainable tourism policies and strategies for the short and long term. In 2007, the Davos Declaration called for "a clear commitment for action to respond to the climate change challenge, including the urgent adoption of a range of sustainable tourism policies" (UNWTO, 2007). The tourism sector needs to reduce its CO_2 emissions and to develop the use of renewable energy. Climate change acts synergistically and, along with other threats, it poses a serious threat to biodiversity. Egypt needs to fully apply the goal 14 of the UN Sustainable Development Goals to conserve and sustainably use oceans, seas and marine resources for its economic development.

References

Ali, A.A.M., Hamed, M.A., & Abd El-Azim, H. (2011). Heavy metals distribution in the coral reef ecosystems of the Northern Red Sea. *Helgoland Marine Research*, 65(1), 67–80.

Attalla, T., Kotb, M.M.A., Hanafy, M.H., & Mohammed, S.Z. (2015). Status of the fringing coral reefs in the southern Egyptian coast of the red sea. *Egyptian Journal of Aquatic Biology and Fisheries*, 19(4), 51–67.

Bubenzer, O., Bolten, A., & Darius, F. (2007). *Atlas of Cultural and Environmental Change in Arid Africa*. Köln, Germany: Heinrich-Barth-Institut, 239 pp.

Coles, S.L., & Riegl, B.M. (2013). Thermal tolerances of reef corals in the Gulf: A review of the potential for increasing coral survival and adaptation to climate change through assisted translocation. *Marine Pollution Bulletin*, 72, 323–332.

De'ath, G., Fabricius, K.E., Sweatman, H., & Puotinen, M. (2012). The 27-year decline of coral cover on the Great Barrier Reef and its causes. *Proceedings of the National Academy of Sciences of the USA*, 109(44), 17995–17999.

Doney, S.C., Fabry, V.J., Feely, R.A., & Kleypas, J.A. (2009). Ocean acidification: The other CO_2 problem. *Annual Review of Marine Science*, 1, 169–192.

Ferrario, F., Beck, M.W., Storlazzi, C.D., Micheli, F., Shepard, C.C., & Airoldi, L. (2014). The effectiveness of coral reefs for coastal hazard risk reduction and adaptation. *Nature Communications*, 5, 3794.

Fine, M., Gildor, H., & Genin, A. (2013). A coral reef refuge in the Red Sea. *Global Change Biology*, 19(12), 3640–3647.

Frihy, O.E., El Ganaini, M.A., El Sayed, W.R., & Iskander, M.M. (2004). The role of fringing coral reef in beach protection of Hurghada, Gulf of Suez, Red Sea of Egypt. *Ecological Engineering*, 22(1), 17–25.

Frihy, O.E., Fanos, A.M., Khafagy, A.A., & Abu Aesha, K.A. (1996). Human impacts on the coastal zone of Hurghada, northern Red Sea, Egypt. *Geo-Marine Letters*, 16(4), 324–329.

Frihy, O.E., Hassan, A.N., El Sayed, W.R., Iskander, M.M., & Sherif, M.Y. (2006). A review of methods for constructing coastal recreational facilities in Egypt (Red Sea). *Ecological Engineering*, 27(1), 1–12.

Hasler, H., & Ott, J.A. (2008). Diving down the reefs? Intensive diving tourism threatens the reefs of the northern Red Sea. *Marine Pollution Bulletin*, 56(10), 1788–1794.

Hegazy, A.K., Medany, M.A., Kabiel, H.F., & Maez, M.M. (2008). Spatial and temporal projected distribution of four crop plants in Egypt. *Natural Resources Forum*, 32, 316–326.

Hilmi, N., Allemand, D., Cinar, M., Cooley, S., Hall-Spencer, J.M., Haraldsson, G., Hattam, C., Jeffree, R.A., Orr, J.C., Rehdanz, K., Reynaud, S., Safa, A., & Dupont, S. (2014). Exposure of Mediterranean countries to ocean acidification. *Water*, 6(6), 1719–1744.

Hilmi, N., Safa, A., Reynaud, S., & Allemand, D. (2012). Coral reefs and tourism in Egypt's Red Sea. *Topics in Middle Eastern and African Economies*. Retrieved from http://www.luc.edu/orgs/meea/volume14/meea14.html

Holden, A. (2000). *Environment and Tourism*. London: Routledge.

Jameson, S.C., Ammar, M.S.A., Saadalla, E., Mostafa, H.M., & Riegl, B. (2007). A quantitative ecological assessment of diving sites in the Egyptian Red Sea during a period of severe anchor damage: A baseline for restoration and sustainable tourism management. *Journal of Sustainable Tourism*, 15(3), 309–323.

Jobbins, G. (2006). Tourism and coral-reef-based conservation: Can they coexist? In: I.M. Côté & J.D. Reynolds (Eds.), *Coral Reef Conservation*. Conservation Biology 13. Cambridge: Cambridge University Press, pp. 237–263.

Kotb, M.M.A. (2013). Coral colonization and fish assemblage on an artificial reef off Hurghada, Red Sea, Egypt. *Egyptian Journal of Aquatic Biology and Fisheries*, 17(4), 59–70.

Kotb, M.M.A. (2016). Coral translocation and farming as mitigation and conservation measures for coastal development in the Red Sea: Aqaba case study, Jordan. *Environmental Earth Sciences*, 75(5), 1–8.

Kotb, M.M.A., Attalla, T.M., Hanafy, M.H., & Mohamed, S.Z. (2015). Resilience of coral reefs at the Southern Egyptian coast of the red sea. *Egyptian Journal of Aquatic Biology and Fisheries*, 19(4), 77–89.

Kotb, M.M.A., Hanafy M.H., Rirache H., Matsumura S., Al-Sofyani A.A., Ahmed A.G., Bawazir, G., & Al-Horani, F. (2008). Status of coral reefs in the Red Sea and Gulf of Aden Region. In: C.E. Wilkinson (Ed.), *Status of Coral Reefs of the World: 2008*. Townsville, Australia: Global Coral Reef Monitoring Network and Reef and Rainforest Research Centre, p. 67–78.

Lamb, J.B., True, J.D., Piromvaragorn, S., & Willis, B.L. (2014). Scuba diving damage and intensity of tourist activities increases coral disease prevalence. *Biological Conservation*, 178, 88–96.

Lamb, J.B., & Willis, B.L. (2011). Using coral disease prevalence to assess the effects of concentrating tourism activities on offshore reefs in a tropical marine park. *Conservation Biology*, 25(5), 1044–1052.

Leujak, W., & Ormond, R.F.G. (2008). Quantifying acceptable levels of visitor use on Red Sea reef flats. *Aquatic Conservation: Marine and Freshwater Ecosystems*, 18(6), 930–944.

Madkour, H.A. (2013). Impacts of human activities and natural inputs on heavy metal contents of many coral reef environments along the Egyptian Red Sea coast. *Arabian Journal of Geosciences*, 6(6), 1739–1752.

Mohamed, A.R., Abdel-Hamid, A.M.A., & Abdel-Salam, H.A. (2012). Status of coral reef health in the northern Red Sea, Egypt. *Proceedings of the 12th International Coral Reef Symposium*, Cairns, Australia.

Monitoring, Verification and Evaluation (MVE) Unit. (2003). *Economic Evaluation of the Egyptian Red Sea Coral Reef*. Policy brief for the Egyptian Environmental Policy Program.

Nugues, M.M. (2002). Impact of a coral disease outbreak on coral communities in St. Lucia: What and how much has been lost? *Marine Ecology Progress Series*, 229, 61–71.

Nugues, M.M., & Roberts, C.M. (2003). Partial mortality in massive reef corals as an indicator of sediment stress on coral reefs. *Marine Pollution Bulletin*, 46, 314–323.

PERSGA. (2010). The status of coral reefs in the Red Sea and Gulf of Aden: 2009. *PERSGA Technical Series Number 16*, PERSGA, Jeddah, 105 pp.

Ragheb, R.A. (2015). Sustainable tourism development: Assessment of Egyptian sustainable resorts. *International Journal of Social, Behavioral, Educational, Economic, Business and Industrial Engineering*, 9(12), 4151–4162.

Riegl, B., & Velimirov, B. (1991). How many damaged corals in Red Sea reef systems? A quantitative survey. *Hydrobiologia*, 216/217, 249–256.

Rinkevich, B. (2005). Conservation of coral reefs through active restoration measures: Recent approaches and last decade progress. *Environmental Science & Technology*, 39, 4333–4342.

Roberts, C.M., & Polunin, N.V.C. (1992). Effects of marine reserve protection on Northern Red Sea Fish populations. *Proceedings of the 7th International Coral Reef Symposium*, 2, 969–977.

Samy, M., Sanchez Lizaso, J.L., & Forcada, A. (2011). Status of marine protected areas in Egypt. *Animal Biodiversity and Conservation*, 34(1), 165–177.

Seaman, W., & Sprague, L.M. (1991). *Artificial Habitats for Marine and Freshwater Fisheries*. San Diego, CA: Academic Press.

Shaalan, I.M. (2005). Sustainable tourism development in the Red Sea of Egypt: threats and opportunities. *Journal of Cleaner Production*, 13, 83–87.

Shaltout, S. (2008). Ecological study on the alien species in the Egyptian flora. Thesis submitted to the Faculty of Science, Tanta University, in partial fulfilment of MSc Degree in Botany (Ecology).

Sofianos, S.S., Johns, W.E., & Murray, S.P. (2002). Heat and freshwater budgets in the Red Sea from direct observations at Bab el Mandeb. *Deep-Sea Research*, 49(7–8), 1323–1340.

Spalding, M.D., Ravilious, C., & Green, E.P. (2001). *World Atlas of Coral Reefs*. Berkeley, CA: University of California Press. 424 p.

Thinesh, T., Mathews, G., & Patterson, E.J.K. (2009). Coral disease prevalence in Mandapam group of islands, Gulf of Mannar, Southeastern India. *Indian Journal of Geo-Marine Science*, 38(4), 444–450.

UNWTO. (2007). Davos Declaration: Climate change and tourism – Responding to global challenges. Davos, Switzerland: 2nd International Conference on Climate Change and Tourism.

Willis, B., Page, C., & Dinsdale, E. (2004). Coral disease on the Great Barrier Reef. In: E. Rosenberg & Y. Loya (Eds.), *Coral Health and Disease*. Berlin: Springer-Verlag, pp. 69–104.

World Travel & Tourism Council (WTTC). (2016). Travel and tourism economic impact 2016 world. Retrieved from https://www.wttc.org/-/media/files/reports/economic%20impact%20research/regions%202016/world2016.pdf

4

AUSTRALIA'S GREAT BARRIER REEF – PROTECTION, THREATS, VALUE AND TOURISM USE

Bruce Prideaux and Anja Pabel

Introduction

The Great Barrier Reef (GBR) has a long history of attracting travellers but a much shorter history as a mass tourism destination. The following passage written by Dr A Rattray Royal Navy in 1869 provides an insight into how early travellers to the GBR described what they experienced:

> When opportunity offers, a trip by boat to the outer edge of the reef well repays the trouble. Mollusca, crustaceans, and especially fishes, team, whose splendid colouring, intensified by the tropical sun and beautiful transparent waters, rivals that of the ledges and grottoes of many-hued coral amidst which they sport, forming a fairy-like scene, surpassing all that the imagination can conceive.

A more contemporary view drawn from Trip Advisors' Viator tour booking site for the GBR (https://www.viator.com/Cairns-and-the-Tropical-North-attractions/ Great-Barrier-Reef/d754-a1443) paints a somewhat grimmer picture:

> Encompassing roughly 3,000 individual reefs and dotted with almost 900 islands and coral cays (small sandy isles), Australia's Great Barrier Reef is one of the world's most unforgettable natural treasures. Snorkelers and certified divers often place the reef at the top of their bucket lists due to the unparalleled array of marine life in its underwater world, ranging from thousands of different varieties of fish, birds, and clams to hundreds of types of birds, seaweed, and turtles. *And with experts expecting much of this diversity to dwindle in the next decade, there's never been a better reason to plan a visit to this natural wonder* (italics inserted by author).

Reports about global scale coral bleaching over the period 2014–2017 highlight the vulnerability of many coral reef destinations to climate change in both the short and long term. In Australia, the gravity of the situation is summed up in a report (Great Barrier Reef Marine Park Authority [GBRMPA], 2017a) on the impact of the 2016 coral bleaching event on the GBR which stated that "bleaching events are expected to increase in frequency and severity as a result of climate change" (p. 25). This observation raises concern about the long-term sustainability of coral reefs as a major destination pull factor in regions where coral reefs constitute an important part of destination image. The preceding warning from Trip Advisor's Viator booking site about the future of the reef should generate considerable concern in destinations that rely on the GBR as one of their main destination pull factors. This chapter focuses on Australia's GBR and examines how the coral reef system is protected, the threats it faces and the importance of this system to the tourism industry it supports.

The GBR is widely regarded as one of Australia's best known tourism icons and is almost always included in international marketing campaigns undertaken by government funded tourism promotion bodies as well as the private sector. As a result of this marketing effort, about 10% of international visitors find their way to the GBR, making it one of the most important non-metropolitan tourism attractions in the nation. In 2016, 2.4 million domestic and international tourists visited the GBR (GBRMPA, 2017b). From a tourism dispersal perspective, the GBR plays a major role in attracting visitors to regional destinations. However, the GBR faces a number of problems that may affect its ongoing capacity to attract tourists to regional Queensland. These problems include climate change (see Chapter 6 in this volume for a discussion of the impact of coral bleaching on tourism destinations), declining water quality, coral diseases, coastal development, crown-of-thorns starfish (COTS) infestations and natural events such as cyclones. This chapter includes a brief discussion on key biophysical characteristics of the GBR, how it is protected and the threats that it faces from both natural and anthropogenic causes.

Figure 4.1 illustrates the matrix of boundaries that are used to describe the Great Barrier Reef Marine Park (GBRMP) and its marine subregions, and the mainland Natural Resource Management (NRM) regions that are discussed in this chapter. The GBRMP has four management areas, as illustrated in Figure 4.1. The adjacent coast region is divided into six NRM catchment regions (Burdekin, Burnett-Mary, Cape York, Fitzroy, Mackay-Whitsunday and Wet Tropics). The major coral reef destinations in the GBR and catchment region are Port Douglas, Cairns and the Whitsundays. The remaining towns and cities have limited involvement in coral reef tourism but many offer a large range of non-coral reef tourism related activities.

Biophysical characteristics

The GBR is the world's largest coral reef system, extending over 2,300 kilometres from north to south and includes a small area that extends beyond Australian territorial waters into Papua New Guinea. Most of the GBR is located within the

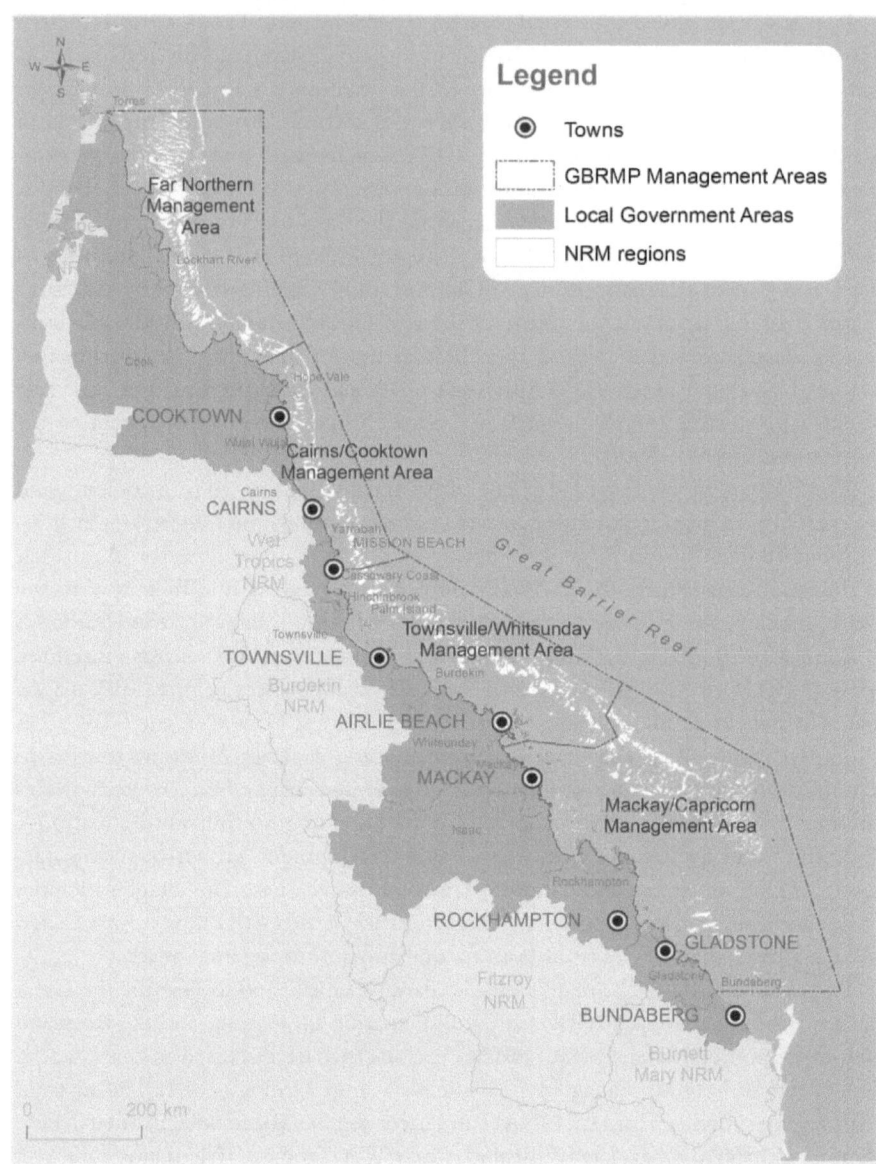

FIGURE 4.1 Great Barrier Reef Marine Park sections and NRM regions

Source: Marshall et al. (2014).

boundaries of the GBRMP, which is managed by the GBRMPA. Coral comprises about 7% of the GBRMP's total area, with the remainder comprised of the continental slope, seagrass beds, mangrove forests, inter-reefal areas and the GBR lagoon. The GBR includes over 2,900 individual reefs, 600 continental islands, 300 coral cays and about 150 inshore mangrove islands covering an area of approximately 344,400 square kilometres (GBRMPA, 2018a).

The geological history of the eastern Australian seaboard indicates that after the separation of Australia from Gondwanaland, eastern Australia experienced a period of tectonic uplift and associated volcanic activity that culminated in the creation of the Coral Sea Basin (Queensland Museum, 2013). As the Australian continent drifted north driven by plate tectonics, the area where the GBR is now located drifted north into the warmer oceans of the tropics creating the conditions that led to the formation of the region's first coral reef systems. The extent of the reef system waxed and waned as the earth's climate has oscillated between glacial periods, rising and falling sea levels, the impact of erosion of the Great Dividing Range and Australia's continued drift northwards. Over the last 500,000 years there have been 17 cycles of sea level oscillations and sea levels have been closest to current levels five times during this period, the last being about 120,000 years ago (Queensland Museum, 2013).

Between 20,000 and 6,000 years ago the sea level rose by 120 metres, submerging extensive areas of the adjacent coastal plain and creating a large number of continental islands, the best examples of which are Hinchinbrook Island and the islands of the Whitsundays. Other submerged areas provided platforms for cays and individual reefs. The age of the reef in its current form is estimated to be about 6,000 to 8,000 years old, paralleling the end of post-glacial ocean level rises (Queensland Museum, 2013). In the most recent period of reef development, coral reefs built on limestone platforms that were part of earlier coral reef systems (Hopley et al., 2007) and created the GBR lagoon that lies between the inner and outer reef systems. The outer reef lies on the edge of the continental shelf while the inner lagoon has much shallower waters that support high levels of marine biodiversity.

The northern section of the GBR features numerous deltaic and ribbon reefs that are generally located quite close to the coast. There are no atolls in this part of the reef system. The southern part of the GBR has a large number of continental islands such as those found in the Whitsundays with many of the islands surrounded by fringing reefs. Other types of reef systems found on the GBR include crescentic reefs, which are quite common in the northern end of the GBR marine park, and plana reefs found in the northern section near Cape York Peninsula, Cairns and Princess Charlotte Bay.

The GBR supports a wide diversity of terrestrial and marine habitats containing many endemic, vulnerable or endangered species (Queensland Museum, 2013). While corals and marine animals are the most commonly recognised GBR species, the reef system also supports a wide range of terrestrial flora and fauna, including mangrove forests, seagrass meadows and rainforests as well as a range of small marsupials, birds and crocodiles. Ongoing research continues to locate and name new species of corals, fish and even a new species of dolphin (Australian snubfin dolphin first described in 2005) hence the following discussion often uses terms such as "estimated" or "about" to describe the number of species.

The system has the world's richest diversity of coral species, including an estimated 600 species of soft and hard corals (GBRMPA, 2018a). Most corals in the inner reef spawn during the week after the first full moon in October triggered by rising sea temperatures as the region moves into summer. Corals in the outer reef have a different pattern of spawning, which usually occurs in late November or early December. The GBR also has an estimated 3,000 varieties of molluscs ranging from giant clams to nudibranchs (a soft-bodied marine gastropod that often exhibits an extraordinary variety of colours and shapes), 500 species of worms, 100 species of jelly fish (GBRMPA, 2018a) and 330 species of ascidians (sea squirts, a sac like invertebrate filter feeder). Other marine creatures that demonstrate the diversity of the GBR include about 500 species of bryozoans (an invertebrate filter feeder typically up to 0.5 millimetres in length) and 630 species of echinoderms (star fish and sea urchins) (GBRMPA, 2018b).

Marine animals are also abundant in the GBR and include six of the world's seven known species of sea turtles, 30 species of whales, dugongs, dolphins and porpoises, 133 varieties of sharks and rays and 14 breeding species of sea snakes (GBRMPA, 2018b). From a tourism perspective, whales and dolphins, sharks and rays, marine turtles and large fish are the most popular of the GBRs marine animals (Farr et al., 2014). Whales, including the humpback and dwarf minke, migrate to the warmer winter waters of the GBR to give birth before returning to the Antarctic during the summer. The reef is an important site for turtle hatcheries, with a number of turtle hatcheries such as Mon Repos being popular with tourists. Turtle species include the green sea turtle, hawksbill turtle, the Olive Ridley turtle, loggerhead turtle, leatherback sea turtle and the flatback turtle. Although protected in the GBRMP, many of these turtle species travel widely throughout the Pacific and Indian Oceans and many are in decline because of hunting, ocean pollution and commercial harvesting. The Olive Ridley turtle is an example of a turtle species in decline because of disturbances to its traditional hatchery areas and commercial harvesting. The Olive Ridley is classified as threatened by the International Union for Conservation of Nature [IUCN] (2012).

Fish life is particularly abundant with over 1,624 recorded species, including 1,400 reef species found in the GBR system (GBRMPA, 2018b). Several of these species, including clown fish and sharks, have attained iconic status as a result of featuring in Hollywood movies such as *Finding Nemo*. Other marine animals present in the GBR region include about 125 species of stingrays, sharks and skates (Queensland Museum, 2013).

The region also features large seagrass meadows, fringing mangrove forests and salt marshes. These areas are important hatcheries for many species of fish. The region's seagrass meadows sustain large but possibly declining populations of dugongs. Coastal development, particularly in areas not protected by the GBR marine park or terrestrial parks, has reduced the size of the region's mangrove forests while run-off from rivers and in some areas dredging for shipping channels has adversely affected many seagrass meadows. The region's mangrove forests also provide habits for saltwater crocodiles.

A wide variety of birds, both endemic and migrating, have been observed in the GBR region, including 215 species that visit to nest or roost on the islands (GBRMPA, 2018b). The GBRMP also supports a wide range of plant life on the region's islands. Of the region's approximately 2,000 known species of vascular plants, the most diverse are found in the tropical forests of the Whitsunday Islands (Turner & Batianoff, 2007).

Protection

The ecological significance of the GBR is recognised by the Australian Federal Government through the Great Barrier Reef Marine Park Act 1975. The Act also established the GBRMPA to manage the GBRMP. In the four decades since the gazetting of the Great Barrier Reef Marine Park Act 1975, enormous resources have been allocated by both the Federal and Queensland state governments for the management and protection of the GBR. In 1981, the GBR was proclaimed as a World Heritage Site, an action that firmly established the GBR as a world scale tourism attraction. The proclamation of World Heritage Area status also alerted government to the need to invest in the science needed to understand the complexities of the GBR ecosystem.

The role of the GBRMPA is to manage the park. The Authority has developed a range of policy instruments and agreements to manage the park, including zoning plans, plans of management for specific areas of the GBRMP, agreements with Traditional Owners, partnerships, education, and issue of use permits for commercial tourism operators, compliance and monitoring. Funding for the GBRMPA is provided by the Australian Federal government and through a management charge levied on tourists who visit the GBR on commercial tourism vessels.

Every five years the GBRMPA publishes an outlook report which discusses a range of GBR related issues, including threats. The *2014 Outlook Report* (GBRMPA, 2014) nominated climate change, land-based run-off, coastal land-use change and residual impacts of commercial fishing as the most significant threats that will affect the future vitality of the GBR. Concerns identified in the *2014 Outlook Report* were subsequently addressed by the *Great Barrier Reef Water Quality Protection Plan (Reef 2050 Plan)* (Commonwealth of Australia, 2015), which also responded to concerns raised by the World Heritage committee about the long-term sustainability of the GBRs Outstanding Universal Values.

The *Great Barrier Reef Water Quality Protection Plan* (Commonwealth of Australia, 2015) generally referred to as the Reef 2050 plan was a major policy initiative designed to halt and then reverse the decline in the quality of the water entering the reef. Actions outlined include reducing nitrogen, pesticide and sediment loads. Other actions include building resilience and funding additional research. In the first decade of the plan, total spending is estimated to be AU$2 billion. The plan was supported by a commitment from the Queensland state government and the Australian Federal government to invest AU$375 million over five years to improve water quality. The overall aim of the plan was to ensure "that by 2020 the quality of the water entering the reef from adjacent catchments has no detrimental impacts on the health and resilience of the Great Barrier Reef" (Commonwealth of Australia, 2015). The need for plans of this type are important to halt the continued decline of the GBR, which the *Great Barrier Reef Water Quality Protection Plan* acknowledged has resulted in a 50% reduction in coral cover over the 30 years to 2013.

Examples of other policies adopted by the GBRMPA include the *Biodiversity Conservation Strategy 2013* (GBRMPA, 2013) and the *Great Barrier Reef Climate Change and Adaption Strategy and Action Plan 2012–2017* (GBRMPA, 2012). The *Biodiversity Conservation Strategy 2013* was developed in response to threats posed to the GBR and concerns about declining biodiversity. The aim of the strategy is to provide a framework for biodiversity protection, conservation and management within the GBR. A key part of the plan is vulnerability assessments of at risk habitats including mangroves, seagrass meadows and the reef lagoon floor.

The *Great Barrier Reef Climate Change and Adaption Strategy and Action Plan 2012–2017* (GBRMPA, 2012) builds on earlier climate change action plans and outlines a series of initiatives to improve the resilience of the GBR. Actions supported as part of this plan include supporting GBR reliant industries to implement adaption strategies, funding research aimed at supporting effective management decisions in relation to climate change and, more generally, supporting the need for climate change adaptation.

The marine park is divided into a range of use zones, each of which have a specific set of rules governing permitted and prohibited activities. The zoning plan was introduced in 2004 and was based on four management areas (see Figure 4.1). The zoning plan also specified permitted activates in each zone. A key aim of the zoning plan was to protect the GBRs biodiversity. The zones are largely self-explanatory: general use, habitat protection, conservation park, marine national park, preservation, scientific research, buffer and Commonwealth Islands (Queensland Museum, 2013). An important aspect of the zoning plan is that it assists with regulation of activities such as fishing and shipping and prevents conflicting use. The zoning scheme has enabled the GBRs 30 protected reef bioregions and 40 protected non-reef bioregions to be protected under the Representative Areas Program that ensures that at least 20% (and often more) of each habitat is represented within a highly protected zone.

Human use

Traditional indigenous use of the GBR included a range of activities such as fishing, hunting, quarrying of stone, cultural activities and religious ceremonies. The Queensland Museum (2013) reported that Aboriginal people used watercrafts to voyage around their "sea country" at least 9,000 years ago. In the north, archaeological evidence (Queensland Museum, 2013) indicates that Torres Strait Islanders were also making extensive use of watercrafts to travel throughout the Torres Straits region. Chapter 16 provides a detailed discussion of indigenous use of coral reefs.

In the nineteenth century, European settlers moved into the region displacing many of the Aboriginal people living in the area at that time. The first wave of European settlers were mainly interested in developing natural resource-based export industries and included farmers, pastoralists, miners and timber cutters. These settlers were supported by coastal settlements that acted as ports for the exports generated by agriculture, timber and mining. Extractive industries continue to be an important part of the economies of these coastal settlements. During the latter decades of the twentieth century tourism emerged as a major industry in many of the GBR NRM catchment regions, aided by advances in transport technology including rail, roads and commercial aviation. Easy access to the GBR from Cairns, Port Douglas and the Whitsundays encouraged the tourism industry in those communities to develop a very reef focused tourism offer. In the southern section of the GBR NRM catchment region, the tourism industry has primarily focused on terrestrial tourism experiences. Other industries in the GBR hinterland that can impact on the reef include extensive coastal farming and grazing as well as large scale mining. Exports of farm output, as well as coal, occur through a large number of ports located in the GBR lagoon.

Threats

Although regarded as one of the world's most pristine reef systems, there is evidence that the GBR has been in decline for many decades (Daley et al., 2008). As Daley et al. (2008) note, there was little systematic scientific study of the reef prior to 1970 when systematic scientific monitoring began. Prior to that, the GBR had been affected by a number of activities, although the extent of impacts is difficult to gauge as available evidence is in qualitative rather than quantitative form. Factors leading to the post-European settlement decline of the GBR were identified by Daley et al. (2008) and others as coral mining, coral and shell collection, guano and phosphate mining, the introduction of exotic species and commercial harvesting of fish, turtles, bêche-de-mer and dugongs, coastal urban development, recreational fishing, sewerage discharge, run-off from farming, dredging and reclamation of wetlands and mangrove forests. Daley et al. (2008) make the point that reconstruction of ecological changes prior to the commencement of systematic scientific monitoring indicates that the GBR was already in decline as early as 1900. Pandolfi et al. (2003) noted that on a global scale most reefs were substantially degraded before 1900 through over-fishing and land-derived pollution. Pandolfi et al. (2003, p. 957) also observed that "maintenance of the status quo within partially protected areas such as the Great Barrier Reef is at best a weak goal for management, which should strive instead for restoring reefs that are clearly far from pristine." Considering the current level of decline and the future impact of even more extensive bleaching, Pandolfi et al. (2003) call seems a target that is becoming more difficult to achieve as each year passes.

As the following discussion highlights, many of the factors that have contributed to post-1970s coral decline continue to pose problems despite the best efforts of the

GBRMPA and the state and national governments. Climate change is of particular concern because of the impacts that will be generated by rising water temperature, sea level rise, coral bleaching and ocean acidification. Collectively, these impacts will continue to reduce biodiversity and cause significant change in the structure of reef ecosystems.

Threats to the GBR can be classed as anthropogenic and natural. Anthropogenic threats can be classed as localised or global, while natural threats can be grouped into biological and weather related. The creation of the GBRMPA has played a major role in reducing a wide range of non-nature based threats to the GBR, such as coral mining, over-fishing, oil drilling and coastal development including wetland conversion and more recently agricultural run-off. Dealing with the impacts of climate change is more problematic because of the global nature of the threat and the need for mitigation on a global scale. In its response to climate change, the GBRMPA has implemented a number of strategies that have focused on adaptation. More recently, a number of non-governmental organisations (NGOs) and research organisations have begun looking at the potential for coral regeneration. Compared to other coral reef systems, including the Maldives and Thailand (see Chapter 10 in this volume), programmes supporting reef regeneration are in an early stage of development but may provide a useful method of regenerating damaged coral reefs in areas of high tourism use.

Natural threats to the GBR include a range of climatic factors such as wind storms and discharge from flooded coastal river systems. Wind storms such Tropical Cyclone Debbie, which struck the Whitsundays region in March 2017, may cause extensive damage to coral reef systems by increasing water turbidity and sediment load and through the pounding action of the large waves generated by cyclonic systems. Many of the Whitsunday's inner coral reefs were covered with a fine layer of silt which led to significant coral death. Coral disease can also have a significant impact on the health of reefs as can the presence of COTS. Although described as part of the normal cycle of life in coral reef ecosystems, the magnitude of many of these threats is increased by climate change. Wind storms, for example, are predicted to increase in intensity in coming decades (IPCC, 2014) as are the number and intensity of El Niño events.

Coral bleaching

The first major bleaching event in the GBR occurred in 1998, followed by a second event in 2002. More recently, the GBR suffered two consultative years of bleaching in 2016 and 2017. These bleaching events were part of a much larger global scale bleaching event that occurred over the period 2014–2017 (Eakin et al., 2016. During the 2016 event, average ocean temperatures rose 1.35°C above the average of ocean temperatures between 1951 and 1980 (Hoegh-Guldberg & Ridgway, 2016). As Figure 4.2 illustrates, the most severe bleaching during 2016 occurred in reefs north of Cairns. During the 2017, the area of bleaching extended to the vicinity of Townsville but did not affect reefs further south.

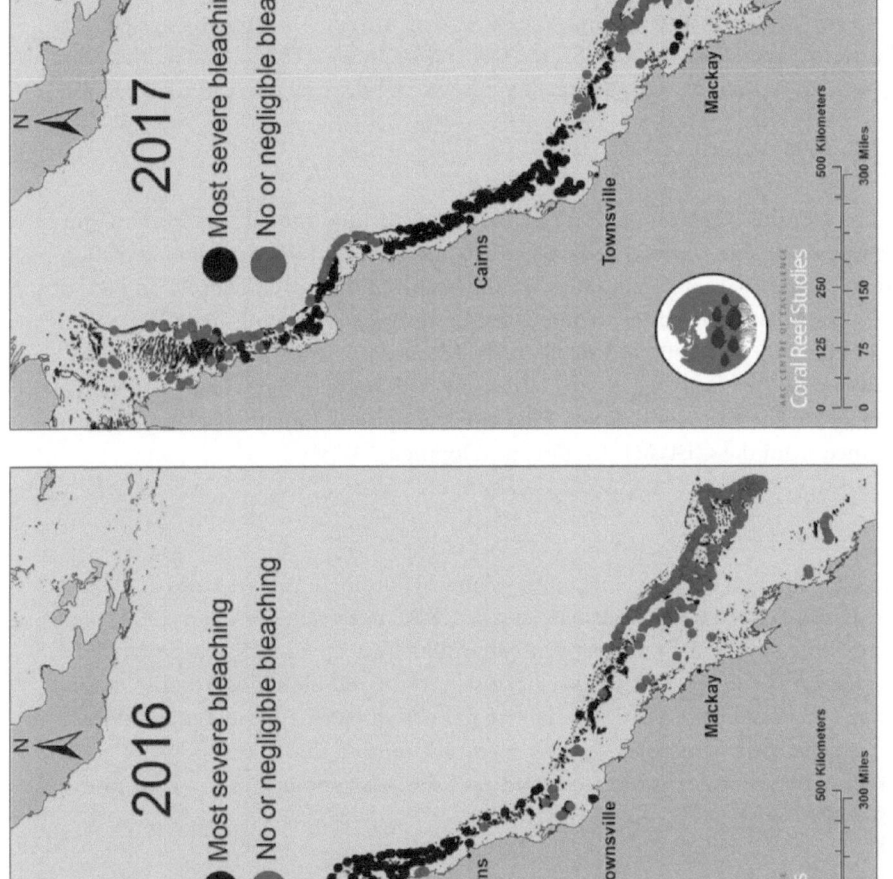

FIGURE 4.2 Coral bleaching of the Great Barrier Reef in 2016 and 2017

Source: ARC Centre of Excellence for Coral Reef Studies (2017).

Declining water quality

The GBRMPA has identified land-based run-off as one of the most significant threats to the long-term health and resilience of the GBR. Thirty-five major river catchments drain into the GBR and in periods of high rainfall these river systems may deposit significant sediment load into the reef lagoon. Intensive farming in these regions also adds to the volume of sediment, nutrients and pesticides carried by these systems. To address the problems the *Great Barrier Reef Water Quality Protection Plan* (GBRMPA, 2013) was implemented as a long-term strategy to improve water quality through measures that include encouraging agricultural industries to reduce their use of fertilisers and pesticides. Considerable effort has also been given to improving the quality of sewerage disposal into the GBR lagoon.

Coastal development

Land uses in the GBR catchment region, which include agriculture, mining, port construction and associated dredging, urban and industrial development and aquaculture, can have a significant effect on the health of the GBR. More than 80% of the GBR catchment supports some form of agriculture ranging from sugar cane production to pastoralism and small crops. Mining is also an important economic activity in the area from Mackay to Gladstone with some of the world's largest coal mines located in this region. Since 1981 about 8 square kilometres of land has been reclaimed from the GBRMP for port development.

Coral disease

A number of coral diseases, including White Syndromes, Brown Band and Black Band Disease, have been recorded on the GBR. Between 1995 and 2009, coral diseases are estimated to have been responsible for up to 6.5% of coral death recorded on the GBR. Corals that have survived coral bleaching remain stressed for some time and have a lower resilience to coral diseases. Although coral diseases are a normal part of reef ecosystems there is evidence that anthropogenic activity such as poor water quality can increase the susceptibility of coral to disease (GBRMPA, 2018c).

Crown-of-thorns starfish

COTS are found in reefs throughout the Indo-Pacific region and in normal circumstances assist slower growing corals to form colonies by feeding on faster growing staghorn and plate corals. However, outbreaks of COTS can cause significant damage to coral reefs. Surveys by the Australian Institute of Marine Science (De'ath et al., 2012) found that in the 30 year period to mid-2010s coral cover has declined by about 50%, with COTS responsible for half of this loss. Outbreaks of COTS occurs about every 17 years, with the last outbreak commencing in 2010 and was still underway when this chapter was written in 2018. One method to contain COTS, and operated by the Association of Marine Park Tourism Operators (AMPTO), uses teams of divers to kill COTS. Given the volume of coral cover that has been lost through COTS in the last 30 years it is difficult to understand why substantially more resources have not been allocated to COTS control programmes. As De'ath et al. (2012, p. 17995) state, "reducing COTS populations, by improving water quality and developing alternative control measures, could prevent further coral decline and improve the outlook for the Great Barrier Reef".

Tourism

Tourism has emerged as a major industry in the GBR region and the GBR is widely promoted as a unique attraction by Tourism Australia and Tourism and Events Queensland, as well as by regional tourism organisations such as Tropical Tourism North Queensland based in Cairns and Tourism Whitsunday located in the Whitsundays. Most coral reef focused activity is located in Port Douglas, Cairns and the Whitsundays. In 2015–2016, 17.8 million people visited the GBR and its catchment region staying for 43.2 million visitor nights (Australian Bureau of Statistics [ABS], 2017). About 2.4 million visits were made to the GBR on commercial tourism vessels during this period (GBRMPA, 2017b). The number of tourists visiting the GBR region compared to the number of tourists visiting the GBR is relatively small. This poses an interesting question for destinations such as Port Douglas, Cairns and the Whitsundays – are tourists visiting the GBR and taking the opportunity to engage in other tourism activates after their visit to the GBR, or are tourists visiting these destinations and while there taking the opportunity to visit the GBR?

Tourism activity in the GBR includes both land and water-based activities. The type of water-based activities varies between destinations but usually includes day and overnight tours that offer opportunities for swimming, snorkelling, scuba diving and coral reef viewing via glass-bottom boats and semi-submersible coral viewing vessels. A number of tourism operators use moored platforms located on popular reefs as the focus of their reef experience (see Figure 4.3). Glass-bottom boats and semi-submersibles operate from these platforms which also serve as a base for swimming, snorkelling and scuba diving. Other operators moor their boats adjacent to coral reefs, with customers able to participate in snorkelling and other activities from the boat. The other main form of water-based activity is sailing, which includes day and overnight options. Sailing includes both crewed expedition boats and bare boat charters where tourist hire and crew sailing boats. The majority of operators are located in the Whitsundays.

The strong supervision exercised by the GBRMPA over all groups that use the GBR, including tourists, has ensured that damage from commercial tourism operations has been minimised. However, snorkelling and diving can cause damage through touching coral. The use of sun cream also continues to be a problem.

FIGURE 4.3 Reef pontoon on the Great Barrier Reef

Source: Photo courtesy of Bruce Prideaux.

In recent years, ocean cruising has emerged as a popular and fast growing form of tourism in the GBR region and is supported by state and local tourism authorities. However, further expansion of cruising will require significant dredging, with associated risks in the areas where dredging occurs. This appears to be at odds with efforts to improve the quality of the GBR. Land-based activities are mainly focused on beach activities including sunbathing, swimming and snorkelling and visits to islands that feature fringing reefs.

A recent report by the ABS (2017) estimated the total economic contribution of tourism in the GBR and its catchment to the Australian economy in 2015–2016 was AU$7.976 billion and that tourism accounted for 8.2% of total employment in the GBR and catchment region. Direct tourism employment was estimated to be 46,000 persons in 2015–2016. In the Cairns region, tourism expenditure was estimated to be AU$3.2 billion (ABS, 2017).

Apart from economic value, other values that can be placed on natural systems such as the GBR include social and iconic values. A report by Deloitte Access Economics ([DAE], 2017) calculated that GBR's asset value was AU$56 billion based on its social, economic and iconic values, including the value of the reef to tourists who had visited the GBR (AU$29 billion), Australians who had yet to visit the GBR (AU$24 billion) and recreational users (AU$3 billion).

Given that tourism accounts for 8.2% of total employment in the GBR and its catchments, it is apparent that any threats to the quality of the reef experience are likely to have an adverse impact on destinations such as Port Douglas, Cairns and the Whitsundays. Much of the current research into the impact of climate change on the GBR has focused on biological impacts and has given relatively little attention to investigating how a decline in the quality of the GBR experience will affect coastal communities. From a tourism perspective, this is a significant gap which has only just begun to be addressed. Research into community resilience by Prideaux et al. (2018) found Cairns residents were well informed about the causes of coral bleaching, concerned about the future impacts of coral bleaching and supported efforts to protect the GBR. Two-thirds of respondents thought that coral bleaching will affect them on a personal level. The research also found that most respondents did not believe that their employment will be affected by future decline of the GBR.

Extensive media coverage of the 2016 and 2017 coral bleaching events alerted both domestic and international tourists, and the travel trade in general, to the bleaching event and raised fears that the GBR was either in terminal decline or had suffered extensive damage. Media coverage began with a report posted by the ARC Centre of Excellence Coral Reef Studies that stated that 93% of the GBR had suffered bleaching based on an aerial survey of 911 of the GBRs 2,700 reefs (ARC, 2017). The release was quickly reworked into media stories such as that posted by Smail (2016) which stated that "Only 7% of the Great Barrier Reef has avoided coral bleaching." The event was given extensive coverage by the print and electronic media as well as social media. The impression that the entire GBR rather than one-third of the reef had suffered bleaching was never corrected by the media nor did affected destinations mount an effective crisis management strategy to counter that view. The result of what can only be described as "sloppy" and factually incorrect media that was not effectively challenged led to the type of report found on the Viator website and reported earlier in this chapter. Surprisingly, Viator's websites for the Florida Keys, Maldives and Koh Tao (Thailand) do not contain similar warnings, despite these destinations having low coral cover (see Chapters 8 and 10 in this volume). There are a number of clear messages emulating for this event. Any communications from research institutions should be factual and based on peer reviewed research, not findings that are open to interpretation and lacking rigour. Destinations also need to be prepared for media sensationalism and be able to rapidly deploy crisis recovery strategies to deal with the crisis in an honest, timely and factual manner.

Discussion and conclusion

The GBR exhibits the greatest biodiversity of any coral reef system and is claimed to be the most well managed coral reef system in the world. Although both state and national governments have allocated substantial financial resources to the protection and management of the GBR, it continues to show signs of ongoing decline through problems with water quality, the effects of climate change and COTS. While significant funds have been allocated to monitoring, science-based research and management, very little funding has been allocated to active control measures such as the AMPTO-managed COTS eradication programme. Regeneration projects which have enjoyed varying degrees of success elsewhere are just beginning to be considered by the GBRMPA. Only a very small percentage of research funds have been allocated to investigate how threats to the GBR may impact the tourism industry and coral reef dependent tourism destinations. The ABS (2017) and DAE (2017) reports are only the latest of a large number of reports that have confirmed the importance of the GBR in terms of jobs and income for the regional economy.

The overwhelming majority of research into the GBR has taken a reef centric science view and largely neglected the impact on the GBR on human populations. This situation has arisen because of the science capture of research programmes and funding schemes and the seeming inability of governments to recognise the needs of human communities in the GBR catchment. Funding further scientific study of the GBR's decline, but also neglecting to study how the decline is caused by and will impact on human communities, seems counterproductive at best and verging on stupidity at worst.

It is apparent from the discussion in this chapter that tourism is a significant industry in the GBR region and that a range of threats to the GBR may have serious spill over effects on local communities, particularly if the quality of coral cover and the tourism experiences associated with the ability to view live and healthy coral is diminished. Chapter 6 discusses in greater detail how coral bleaching poses a significant threat to the major coral reef destinations of Port Douglas, Cairns and the Whitsundays. The potential for further coral bleaching is concerning and suggests that these destinations need to commence studies into alternative tourism experiences.

This chapter has also touched on the problems caused by COTS and the first experiments in coral reef regeneration. Combined, these two initiatives have the potential to increase the resilience of the GBR and enhance the long-term economic sustainability to coral reef tourism in the GBR region. The GBRMPA has been slow to react in both areas and, as a consequence, has lost valuable time discharging its responsibilities to protect the GBR. Ultimately, however, the fate of the GBR and all other coral reef systems rests on the ability of the global community to come together and implement meaningful programmes to reduce global warning and hold global temperature increase to under 2°C. The withdrawal of the US from the 2015 Paris Climate Change Accord is a stark reminder of how difficult this will be to achieve in a time frame that will prevent significant damage to coral reef ecosystems.

References

ARC Centre of Excellence Coral Reef Studies. (2017). *Coral bleaching and the Great Barrier Reef.* Retrieved 6 April 2017 from https://www.coralcoe.org.au/for-managers/coral-bleaching-and-the-great-barrier-reef

Australian Bureau of Statistics (ABS). (2017). *4680.0 – Experimental environmental-economic accounts for the Great Barrier Reef, 2017.* Australian Bureau of Statistics, Canberra.

Commonwealth of Australia. (2015). *Reef 2050 Long-Term Sustainability Plan*, Commonwealth of Australia, Canberra.

Daley, B., Griggs, P., & Marsh, H. (2008). Reconstructing reefs: qualitative research and the environmental history of the Great Barrier Reef, Australia. *Qualitative Research*, 8(5), 584–615.

De'ath, G., Fabricius, K., Sweatman, H., & Puotinen, M. (2012). The 27-year decline of coral cover on the Great Barrier Reef and its causes. *PNAS*, 109(44), 17995–17999.

Deloitte Access Economics (DAE). (2017). *At what price? The economic, social and icon value of the Great Barrier Reef.* Report to the Great Barrier Reef Foundation. Retrieved from https://www2.deloitte.com/content/dam/Deloitte/au/Documents/Economics/deloitte-au-economics-great-barrier-reef-230617.pdf

Eakin, C.M., Lui, G., Gomez, A., de la Cour, J., & Heron, S. (2016). Global coral bleaching 2014–2017. *Reef Currents*, 31(1), 20–26.

Farr, M., Stoeckl, N., & Beg, R. (2014). The non-consumptive (tourism) 'value' of marine species in the northern section of the Great Barrier Reef. *Marine Policy*, 43, 89–103.

Great Barrier Reef Marine Park Authority (GBRMPA). (2012). *Great Barrier Reef climate change and adaption strategy and action plan 2012–2017.* Great Barrier Reef Marine Park Authority, Townsville.

Great Barrier Reef Marine Park Authority (GBRMPA). (2013). *Great Barrier Reef water quality protection plan.* Great Barrier Reef Marine Park Authority, Townsville.

Great Barrier Reef Marine Park Authority (GBRMPA). (2014). *Outlook report 2014.* Great Barrier Reef Marine Park Authority, Townsville.

Great Barrier Reef Marine Park Authority (GBRMPA). (2017a). *Final report 2016 coral bleaching event on the Great Barrier Reef.* Great Barrier Reef Marine Park Authority, Townsville.

Great Barrier Reef Marine Park Authority (GBRMPA). (2017b). Great Barrier Reef tourist numbers. Retrieved 30 May 2017 from http://www.gbrmpa.gov.au/visit-the-reef/visitor-contributions/gbr_visitation/numbers

Great Barrier Reef Marine Park Authority (GBRMPA). (2018a). Facts about the Great Barrier Reef. Retrieved 11 November 2017 from http://www.gbrmpa.gov.au/about-the-reef/facts-about-the-great-barrier-reef

Great Barrier Reef Marine Park Authority (GBRMPA). (2018b). Animals. Retrieved 11 November 2017 from http://www.gbrmpa.gov.au/about-the-reef/animals

Great Barrier Reef Marine Park Authority (GBRMPA). (2018c). Coral disease on the Great Barrier Reef. Retrieved 11 November 2017 from http://www.gbrmpa.gov.au/managing-the-reef/threats-to-the-reef/climate-change/what-does-this-mean-for-species/corals/coral-disease/coral-disease-on-the-great-barrier-reef

Hoegh-Guldberg, O., & Ridgway, T. (2016). Coral bleaching comes to the Great Barrier Reef as record-breaking global temperatures continue. Retrieved on 11 March 2017 from *The Conversation*, https://theconversation.com/coral-bleaching-comes-to-the-great-barrier-reef-as-record-breaking-global-temperatures-continue-56570

Hopley, D., Smithers, S.G., & Parnell, K.E. (2007). *The geomorphology of the Great Barrier Reef: development, diversity, and change.* Cambridge University Press, Cambridge.

Intergovernmental Panel on Climate Change (IPCC). (2014). *Climate change 2014 synthesis paper.* Intergovernmental Panel on Climate Change, Geneva.

International Union for Conservation of Nature (IUCN). (2012). *Lepidochelys olivacea.* Retrieved 17 November 2017 from http://dx.doi.org/10.2305/IUCN.UK.2008.RLTS.T11534A3292503.en

Marshall, N.A., Bohensky, E., Curnock, M., Goldberg, J., Gooch, M., Nicotra, B., Pert, P.L., Scherl, L., Stone-Jovicich, S., & Tobin, R.C. (2014). *The social and economic long term monitoring program for the Great Barrier Reef. Final report.* Report to the National Environmental Research Program. Reef and Rainforest Research Centre Limited, Cairns, Australia.

Pandolfi, J., Bradbury, R., & Jackson, J. (2003). Global trajectories of the long-term decline of coral reef ecosystems. *Science*, 301(5635), 955–958.

Prideaux, B., Carmody, J., & Pabel, A. (2018). *Impacts of the 2016 and 2017 mass coral bleaching events on the Great Barrier Reef tourism industry and tourism-dependent coastal communities of Queensland.* Report to the Reef and Rainforest Research Centre Limited, Cairns, Australia.

Queensland Museum. (2013). *The Great Barrier Reef.* Queensland Museum, Brisbane.

Rattray, A. (1869). From Sydney to Singapore. In: H. Bates (Ed.), *Illustrated travels: A record of discovery, geography, and adventure.* Cassell, Petter, and Galpin, New York.

Smail, S. (2016, 20 April). Great Barrier Reef: Only 7 per cent not bleached survey finds. *ABC News* [online]. Retrieved 7 March 2017 from http://www.abc.net.au/news/2016–04-20/great-barrier-reef-bleaching/7340342

Turner, M., & Batianoff, G. (2007). Vulnerability of island flora and fauna in the Great Barrier Reef to climate change. In: J. Johnson & P. Marshall (Eds.), *Climate change and the Great Barrier Reef, a vulnerability assessment.* Great Barrier Reef Marine Park Authority, Townsville.

5

STRATEGIC CARIBBEAN CORAL REEF TOURISM MANAGEMENT

Dennis J. Gayle and Bernadette E. Warner

A strategic management perspective

This chapter will consider the impact of coral reef tourism on the sustainability of coral reefs in six Caribbean countries using a strategic management perspective. Strategic management includes the five basic stages of goal setting, analysis, strategy formation, strategy implementation, outcome evaluation and control, which may lead to process adjustments (Rothaermel, 2014). Goal setting begins by defining short, medium and long-term objectives in relation to the environment as well as all relevant resources at hand. The results may be articulated as a mission statement for stakeholders. In turn, analysis should lead to the clearest possible understanding of issues that might impact on implementation. Methods of undertaking this form of analysis include strengths, weaknesses, opportunities and threats (SWOT) analysis and the balanced scorecard approach.

A more refined assessment of the available resources relative to the defined goals and objectives follows and includes identifying gaps that need to be addressed and the range of potential options, given the status of the environment that is being evaluated and the projected timeline. Strategic plan implementation tests the extent to which the goal setting, analysis and strategy formulation stages have been effectively executed. Finally, outcome evaluation and control includes performance measurement and assessment, culminating in a comparison of expected versus actual results and the implementation of corrective action which might lead to goal adjustment (Rothaermel, 2014).

Coral reefs

Live coral reefs are populated by polyps that grow in the photic zone of sea floors in tropical waters between 30 degrees north and 30 degrees south of the equator.

Coral reefs consist of calcium carbonate secreted by these polyps. Each cell includes symbiotic algae that photosynthesise the carbon dioxide produced, providing food and maintaining a nearly constant pH balance. Coral reefs may be defined as major geomorphic features generated by live coral and coralline algal growths that constitute actively growing, wave-resistant structures.

Coral reefs constitute less than 0.01% of the marine environment but provide a habitat for some 25% of marine species and generate up to 30% of the export earnings of some of the economies that promote reef-related tourism (Cohen, 2014). Coral reef habitats may be thousands of years old and expand over several square kilometres. While a single coral head may take 20 years to cover 24 square metres, it may be irretrievably damaged in minutes (Manfrino, 2008). A coral reef may be classified as either a patch, bank, barrier, fringing, shelf-edge or atoll reef. Others may be large linear offshore structures that are not shallow water barriers; be separated from the shoreline by a deep lagoon or channel; border a shoreline; be located at the edge of a continental shelf or on a ridge in the open sea that forms a complete or open ring around a lagoon with a fore reef that drops off into deep water.

Coral reefs have been severely stressed by a litany of adverse factors: shoreline development; industrial pollution; urban run-off; agricultural pesticides; dredging or land-clearing sediments and sewage; direct damage from scuba diving, snorkelling and swimming; overfishing; marine ornamental fish collections for aquaria; commercial shipping and oil spills; disease and thermal stress; and cruise and land-based mass tourism. Apart from anthropogenic factors, natural hazards include hurricanes and tropical storms as well as coral diseases (Aronson & Precht, 2001; Webster et al., 2005). As a result of cumulative stresses, and despite the increasingly widespread development of marine protected areas and sanctuaries, coral reefs worldwide continue to be threatened.

In the case of coral reefs and their marine habitats, a strategic management approach allows for goal setting, including the systematic reduction of threats posed by related tourism. Analyses as to the specific nature of the threats and practicable mitigation measures may follow with formulated strategies that have gained support among the target governments, non-governmental organisations (NGOs) and private sectors. Special attention to tourists, coastal zone developers, cruise operators and fishermen will be required in the strategy implementation stage. The extent to which prior agreement is actually achieved, and the means adopted for action, constitute primary independent variables that affect the extent to which success is probable. In the outcome evaluation and control stage, the viability and vitality of national coral reefs will determine the result achieved, and the gaps that remain to be filled.

Caribbean coral reefs

The Caribbean Sea hosts a variety of coral reefs along the coastal zones of its 28 regional countries. This region is home to over 40 species of stony corals, although coral taxonomy is subject to change. For instance, *Montastraea annularis*, often considered the most abundant and wide-ranging Caribbean coral, may actually consist of three sibling species, two with significantly different growth rates and one with an unusual colouration that may be confused with bleaching (Knowlton et al., 1992). Although regional coastal zone ecology often includes mangroves and sea-grasses as well as coral reefs, the analysis discussed in this chapter considers only coral reefs and their immediate marine environment.

Coral reefs are critical to Caribbean regional economies, generating over US$3 billion annually from tourism and fisheries alone, multiplied many times by related goods and services in other economic sectors and providing employment for an estimated 43 million people. Indeed, the Caribbean is more dependent upon tour-ism than any other region of the world. However, many expert estimates indicate that the Caribbean may have lost more than 50% of its coral reefs since 1970 (Jackson et al., 2014). The drivers of this change include coastal land use for homes, hotels and other tourism purposes, commercial activities, shipping, pollution and overfishing, as well as invasive species such as lionfish and, to a lesser extent, ocean acidification and global warming.

At the same time, coral reef grazers such as sea urchins (which have suffered waves of mass mortality) and parrotfish (which have been overfished throughout the twentieth century) are disappearing. As a consequence, grazers are less able to consume the algae that otherwise smother coral. Turtles that also ingest algae are declining in numbers, while sponges remain inadequately protected (Burke et al., 2011). During 1970–2012, Caribbean locations with more than the median value of 1,500 visitors per square kilometre per year suffered continuing losses of coral cover to a median residual level of 14% of the 1970 coverage, except for the islands of Bermuda (39%) and Grand Cayman (31%), reflecting relatively progres-sive environmental regulations and the infrastructure for implementation in those countries (Jackson et al., 2014).

The Caribbean Coastal Marine Productivity Program (CARICOMP) is a regional network of 29 marine laboratories, parks and reserves, as well as data management centres, across 13 islands and 9 mainland countries and was established in 1982 to provide scientific data relevant to sustainable resource man-agement. Network participants study interactions between coastal zones and the contiguous sea, including coral reefs. They examine the structure and functions of coral reefs, seagrasses and mangroves, in order to differentiate between nega-tive anthropogenic changes and those resulting from long-term natural causes. This programme exemplifies an integrated, multidisciplinary approach to resource management that incorporates the perspectives of governments, private sectors, NGOs and concerned individuals (Creel, 2003). CARICOMP was supplemented in 1998 by Atlantic and Gulf Rapid Reef Assessment (AGRRA) with the goals of conducting improved regional surveys of coral reef health, providing easy access to the resulting data base and promoting a supportive learning network to inform reef policies, legislation, management and conservation (http://www.agrra.org).

This chapter reviews comparative strategic Caribbean coral reef tourism man-agement in the CARICOMP countries of Belize, the Cayman Islands, Antigua and Barbuda, Jamaica, Trinidad and Tobago, and Barbados. It draws on a one-page customised survey sent to expert practitioners within each of the case study countries discussed. Participants were selected on the basis of their knowledge of coral reef tourism and ecology in each of the case study countries. The survey instrument was developed following literature review and was administered by email. The chapter concludes with a general discussion on the extent to which a strategic management perspective has informed the continuing development of coral reef tourism in each country. The analysis includes the effectiveness of the marine conservation areas in each country.

Belize

Coral reefs comprise diverse reef habitats, including patch, semi-emergent dry, shallow and deep slopes that gives way to open ocean, deep seafloor, mudflats, sandy beaches, seagrass meadows, mangrove channels or bogues, and inter-tidal red mangrove roots. Belize still has some 18% of its original coral reef cover within its territorial waters (Potts, 2016).

Forty-four species of reef-building corals are found in the Belize Barrier Reef providing a home environment to more than 700 species of vertebrate fish, as well as macro-invertebrates, including the Queen Conch, the Spiny Lobster and the Donkey Dung Sea Cucumber. This ecosystem shelters over 700 West Indian manatees (the largest such group in Central America and the insular Caribbean) as well as frequently observed bottlenose dolphins.

In 2007 coral reef tourism employed 20% of the national labour force and represented up to 15% of the country's gross domestic product (GDP). The fisheries sector employed 4,000 fishers and processors, providing another 3% of total employment (UNDP, 2008). By 2016 reef-related tourism generated US$250 million, directly employing some 20% of the country's labour force. Tourism directly and indirectly impacted upon more than half of Belize's population (Potts, 2016).

Tourists are attracted by the availability of flora and fauna with intriguing names such as the Azure Vase Sponge, the Christmas Tree Worm, the Caribbean Reef Octopus, the Queen Angelfish and the Blue Chromis. Reefs also offer subsistence fishing and land reclamation material to most coastal communities. Fish remains a staple in local ethnic community diets, such as those with Mayan, Spanish, African and European heritage, as well as the Garifunas who combine African with Amer-Indian ancestry.

The Belize Barrier Reef provides a buffer against storm surges and ocean wave erosion and also acts as a "carbon sink" by removing greenhouse gases from the atmosphere. This ecosystem is threatened by sedimentation, nutrient enrichment or eutrophication, thermal stress, as well as direct damage and destruction. For instance, land-based run-off and seafloor dredging significantly increased during 2000–2010 because of tourism resort development.

Traditional agricultural practices continue to combine with land-clearing, deforestation and urbanisation to generate additional sedimentation that impacts on the Belize Barrier Reef. Over 20% of the country's territorial sea is now included in a protected marine park and Belize has a coastal zone integrated management plan, including a coral reef restoration plan implemented by the Fisheries Department as a mitigation measure (Potts, 2016). Chapter 7 in this book discusses issues faced by the Belize Barrier Reef System in greater detail.

The Cayman Islands

By 1986 a system of marine parks had been established to protect approximately one-third of all coastal waters. One park on Grand Cayman was also designated as an Environmental Zone. Measuring 4,169 acres in size, this park is dominated by mangrove and seagrass. Water-based recreational activities, except for the passage of boats at 5 knots or less, are prohibited (Husemann & Bent-Hamilton, 2011). Nevertheless, a significant loss of live coral cover was observed between 1999 and 2006, especially on the shallow reefs surrounding Little Cayman (Manfrino & Whiteman, 2008).

Within the three Cayman Islands (Grand Cayman, Cayman Brac and Little Cayman) of the Western Caribbean, tourism and financial services underpin the economy. During the period 2007–2013 tourists spent on average US$524 million per annum, amounting to almost US$17,000 per resident (The Tourism Company, 2014). This expenditure supported a wide range of businesses, generating employment opportunities for Caymanians and expatriates.

The total number of visitors to the Cayman Islands has grown each year since Hurricane Ivan in 2004; however, most of this growth has been in cruise ship passengers. The number of higher spending stay-over visitors has declined with economic and visitor management implications. Rapid growth in cruise ship arrivals began in 1990 and significantly expanded after the 11 September 2001 terrorist attack on the World Trade Centre in New York. This event encouraged the redeployment of cruise ships to Caribbean waters, often perceived as safer, from 2002 onwards. This change in visitor arrival patterns was supported by government policy that was directed at attracting cruise ship visitors to compensate for the decline in stay-over visitors. The Cayman Islands became the second most visited Caribbean country by cruise passengers after the Bahamas – 1.9 million such visitors by 2009.

Expenditure by cruise ship passengers is much lower than stay-over visitors. In 2009, for example, 18% of the stay-over passengers had incomes of US$100,000 or more, 69% were university graduates and 42% were professionals. Cruise visitors were more likely to consist of families, with traditional "sun, sea and sand" vacation motivations. About 75% of cruise boat visitors focused on shopping, and more than 50% also swam and/or snorkelled (The Tourism Company, 2014). Within the Cayman Islands, residents continued to articulate keen interest in the implementation of a national conservation law that included expanded marine parks and require rigorous environmental impact assessments prior to any development (Husemann & Bent-Hamilton, 2011). However, the tension between marine environment protection and growing dependence upon cruise tourism development has continued as the quality of the coral reefs has declined (Manfrino, 2008).

Antigua and Barbuda

The islands of Antigua and Barbuda are located in the Eastern Caribbean, 17 degrees north of the equator and consist of low-lying coral and limestone formations. Large bank, patch and fringing reefs extend over 180 square kilometres. There are at least 32 species of spiny corals in the territorial sea, and beds of stag horn coral are common on both shallow and deep bank reefs. Six marine parks have been designated to protect 13% of the local marine habitat.

Coral reefs and their associated resources are essential to the economic development, growth and sustainability and growth of Antigua and Barbuda, which have one of the highest levels of reef dependence in the Caribbean. Tourism accounted for approximately 63% of GDP by 2013 and generated US$528 million for the economy (World Travel and Tourism Council, 2014). Boating tourism is especially popular and the availability of many sheltered bays and inlets makes it ideally suited to fishing, snorkelling and scuba diving.

One example of shipwreck diving is *The Andes*, a three-masted merchant ship that sank in 1905 and now rests in Deep Bay in less than 10 metres of water. Dive facilities and operations in Antigua are more established than in Barbuda. Across the twin-island nation, the consumption of seafood such as conch, parrotfish, lobster, marine turtles and finfish is high. Commercial fishing accounted for almost 2% of GDP by 2003 when this sector employed 2% of the population. However, under post-2013 fisheries regulations, fishing for protected species, such as lobsters, turtles, parrotfish and conch, requires special permits for spear guns and beach seine nets.

Antigua has a deeply indented coastline, entirely surrounded by reefs, except on parts of the west and south coasts, and tourism promotional advertisements regularly mention the availability of an attractive beach for each day of the year. On the much smaller island of Barbuda, an extensive algal ridge runs along the east, and reefs are found along most of the coast. Much of the reef structure consists of large, interlocking branches of dead elk horn coral, suggesting prior luxuriant growth. In Antigua, although only 1% of coral colonies showed signs of disease by 2006, more that 20% of shallow coral colonies now display yellow blotch disease. A major coral bleaching event during 2005 eventually led to a reduction in average coral cover from 16% to 7% in 2007. However, some new elk horn coral reef growth in both shallow and deep reefs is now evident in many marine habitats, representing reef regeneration.

The main environmental threats include Caribbean hurricanes and coral bleaching events. For instance, Hurricanes Hugo (1989) Luis and Marilyn (1995) caused extensive damage to reefs in the south and south-east, especially in areas with shallow branching corals. All reefs remain under continuing anthropogenic threat, including overfishing and recreational diving. Up to 70% of the nation's reefs are threatened by coastal development and 30% by marine-based pollution and sedimentation that results in turbid coastal waters and elevated algal cover.

Six marine parks managed by the Fisheries Department, a division of the Ministry of Aquaculture, Lands and Fisheries, have been established in Antigua and Barbuda under the Marine Parks Act 1972. The Antigua National Trust is responsible for marine resource conservation within these marine parks, which include 13% of the total reef area. The main protected coral reefs are the Diamond Reef Marine Park, the Palaster Reef Marine Park (both 1973); the Cades Bay Marine Park (1999), the Codrington Lagoon and the North Sound (2005). Luxuriant mangrove, seagrass and coral growths can still be found within the Nelsons Dockyard National Park where a Snorkelling Reef Trail has been established to help raise awareness as to the types of coral reefs to be observed. However, there is little active management of these marine park conservation areas.

Jamaica

Jamaica is 230 kilometres in length and 80 kilometres in width with a coral reef area of 1,240 square kilometres. The north and east coasts feature well-developed fringing reefs, accompanied by patchy fringing reefs on the broader south coast shelf. Coral reefs are also located on neighbouring banks and shoals within the country's Exclusive Economic Zone and include the Pedro Cays to the south, the Morant Cays to the south-west and the Formigas Banks to the north-east (Woodley et al., 2000).

There has been significant degrading of Jamaica's coral reefs since the 1960s as a result of eutrophication, increased sedimentation, poor coastal development planning, overfishing, hurricane damage, coral bleaching episodes and diseases. As a result, nutrient-indicating algae dominated the island's reefs and coral cover declined from a high of 50% of the originally observed extent in the 1970 to less than 5% in the early 1990s, before rebounding somewhat to an average of 15% (The Nature Conservancy, 2006).

Jamaican reef resilience has also been impacted by continuing decline in the fish population because of poor fishing practices. In response, the Jamaican government has implemented a stringent permit and licensing system for coral reef-impacting activities. The number of coral reef sites visited by the Jamaican Coral Reef Monitoring Network has been increased and a public education campaign that emphasises the importance of coastal ecosystems as an economic asset has been funded. By June 2007 monitoring capacity had been significantly improved by the installation of a Coral Reef Early Warning Station at West Fore Reef in Discovery Bay on the island's north coast (Government of Jamaica, 2009).

Jamaica's coastal ecosystems have also been impacted by natural forces, including hurricanes. For instance, between 1980 and 2007 four major hurricanes made landfall in Jamaica: Hurricane Allen (1980); Hurricane Gilbert (1988); Hurricane Ivan (2004); and Hurricane Dean (2007). Hurricanes Ivan and Dean were powerful Category 4 weather systems that severely damaged the Port Royal Cays, the Portland Bight Cays and the Negril Reefs. Many fragmented and dead branching corals resulted, especially in shallow waters. Soft corals were battered and uprooted by stormy waves and boulder corals were tossed around leading to scars, lesions and overgrowth by algae (Centre for Marine Sciences [CMS], 2006). If the Pedro Cays constitute one of Jamaica's last remaining healthy marine ecosystems, and a haven for nesting seabirds and sea turtles, this is partly because the waters and reefs around Southwest Cay were declared a Special Fishery Conservation Area in 2011 and a result of its distance from the mainland (United Nations Environment Programme

[UNEP], 2013). In summary, it is apparent that an integrated approach to environmental (not only marine) management is not fully supported by the relevant agencies and the related financial resources are inadequate (Ross, 2016).

Trinidad and Tobago

The islands of Trinidad and Tobago lie south of the Caribbean hurricane belt, in the vicinity of Venezuela, and only occasionally experience hurricanes – for instance, in Tobago during September 1963. The best example of contiguous reef, seagrass and mangrove wetland consists of the Buccoo Reef which covers an area of 7 square kilometres, and the Bon Accord Lagoon system at the south-western end of Tobago. The Buccoo Reef is a fringing reef with a patchy distribution of coral communities in a shallow sandy lagoon, accompanied by the mangrove-fringed Bon Accord Lagoon, which includes an extended seagrass community.

Buccoo Reef is exposed to moderate to high wind and waves, especially during the dry season when winds are stronger, and when strong ocean swells are common, leading to turbidity in the area (Kenny, 1976). Water salinity and turbidity is strongly influenced by the Orinoco River flowing from Venezuela and, in part, Colombia, during the November–December wet season. During the dry season, the sea is clearer with typically constant temperatures.

The Buccoo Reef–Bon Accord Lagoon area is unique to the southern Caribbean due to its size, attractiveness and easy accessibility. This led to its development as a major tourist attraction, with guided tours to the reef commencing in the 1930s. The primary tourist activities include glass-bottom boat tours to the Outer Reef flat, the Coral Gardens and the Nylon Pool, along with reef-walking and snorkelling on the shallow back reef areas, as observed by the co-authors themselves in 1999. Sport-diving usually occurs at other reef locations in Tobago.

The promotion of Buccoo Reef area as a major tourist attraction, as well as hotel and residential development in adjacent coastal areas, has had a negative impact on the ecosystem. Impacts include broken or crushed coral reefs as a result of trampling, falling anchors and occasional boat groundings, as well as the indirect effects of untreated sewerage discharge (which is a hazard for swimmers) and increasing surface run-off (Goreau, 1967; Kenny, 1976; Laydoo & Heileman, 1987). This is unsurprising, given the proximity of the villages of Buccoo and Bon Accord, together with many hotels and guest houses along the coast from Plymouth to Crown Point. Nutrient enrichment of the seawater and increased algal growth threatens reef viability, while reef-walking reduces the potential for coral regeneration in damaged areas.

In 1973 the Buccoo Reef system was designated as the country's only marine protected area under the Marine Areas (Preservation and Enhancement) Act of 1970. The Institute of Marine Affairs subsequently collaborated with the Tobago House of Assembly to develop a management plan for the Buccoo Reef Marine Park. A draft, yet to be finalised, national policy framework for Integrated Coastal Zone Management was prepared in 2014 (Juman, 2016).

Barbados

Barbados is an island in the south-eastern Caribbean with an area of 430 square kilometres and a population of 258,000, making it one of the most densely populated in the world. This relatively flat island has a maximum elevation of 340 metres above sea level and 88% of it is covered by a Pleistocene coral cap with shallow, fringing coral reefs along most of the west coast of Barbados generally located about 300 metres from shore. The main reef consists of the Bellairs fringing reef, with live coral cover directly in front of the Folkestone Park Beach where major current streams from the north and south of the island meet. This reef is within the Barbados Marine Reserve, established in 1981, and is regularly monitored by the Bellairs Research Institute, which has observed that the back reef area of the fringing reef mostly consists of dead coral rubble covered by algae filaments and mobile sand (Roberts et al., 1975).

Within the Barbados Marine Reserve, fishing is prohibited and boat use limited to authorised vessels. However, Folkestone Park beach has remained very popular for purposes such as recreational swimming, snorkelling and diving. The Bellairs fringing reef is also regularly exposed to significant sediment loads that originate from surface water run-off from the nearby Holetown Hole (a narrow brackish water lagoon to the south) and has been impacted by a continuing deterioration in water quality near the shore. The largest proportion of living coral is found on the reef spurs, but the coral species composition has changed significantly after the damage caused by Hurricane Allen in 1980. In summation, contaminated surface water run-off from this densely populated island, ground water seepage and wastewater effluent pipes along the coast have resulted in elevated nutrient levels, and signs of eutrophication have been obvious in the reef community (Tomascik & Sander, 1985; Johnson, 1994). These issues have yet to be effectively addressed, especially beyond the area encompassed in the Marine Reserve.

Comparative results

Over 29 marine laboratories, parks and reserves, as well as data management centres, have been established under the CARICOMP network, including the reviewed examples of Belize, the Cayman Islands, Antigua and Barbuda, Jamaica, Trinidad and Tobago, and Barbados. In many of the CARICOMP countries a strategic management approach has been adopted but with varying degrees of success. The strategic management approach is considered essential because of the interactive relationships between the benthic organisms such as clams, sponges, lobsters and crabs, together with related fish communities, and the tourism and fishing activity that occurs in these reefs (Alcolado et al., 2001). The member countries have engaged in goal setting, analysis, strategy formation and attempted strategy implementation, together with outcome evaluation and control. Despite the complementary contributions of AGRRA, the extent to which actual rather than planned process adjustments have been implemented remains limited.

A key intermediate variable consists of coral reef fish, which constitute the most diverse vertebrate communities on Earth. Their distribution is influenced by biological and physical factors, including larval supply, competition, wave exposure, depth and habitat complexity. The declining availability of reef fish relative to corals and algae, together with other herbivores and predators, can decrease coral cover and increase algal abundance. Monitoring and management programmes therefore focus upon reef fish as a key independent variable in evaluating the condition of coral reef communities (Alemu, 2014).

The Global Coral Reef Monitoring Network has noted that coral reef health requires an ecological balance between corals and algae, and that reef fish, especially parrotfish, constitute a critical component of coral reef marine environments. This has been especially so since the decline of Diadema sea urchins during the early 1980s. Since the main causes of parrotfish mortality consist of fishing techniques such as spearfishing, and particularly fish trap use, national and local government action to reverse overfishing, as exemplified in the cases of Bermuda, Belize and Bonaire, as well as Sea Park in the Bahamas, was strongly recommended by the Network (Jackson et al., 2014).

Some Caribbean coral reef ecosystems are relatively intact compared with median regional conditions. Many reefs in the Cayman Islands as well as the Netherlands Antilles still possess 30% or more of their original live coral reefs, with little macroalgae, and at least a moderate availability of reef fish. In contrast, Jamaican coral reefs, as well as those in the US Virgin Islands, cover less than 10% of their original area, with abundant macroalgae and few fish larger than a few centimetres. Such differences are associated with land-based pollution, more fisheries regulation and enforcement, moderate economic prosperity and a lower frequency of hurricanes, as well as coral bleaching and disease. However, coral taxa are not identical in resilience and significant variability in coral responses to stresses have persisted across the Caribbean region (Jackson et al., 2012).

Strategic goal setting and analysis presupposes the existence of universally accepted standards for monitoring the ecological status of coral reefs. Further work is required to assure comparability and inclusiveness – for instance, by recording observations of particular kinds of coral such as Diadema and Acropora (each containing subspecies) within specific types of reef, such as patch, bank or barrier reefs, at specified depths. Stakeholder participation in Caribbean Marine Park planning and management has typically involved relatively narrow groups of resource users and activists, rather than significant cohorts of community members (Dalton et al., 2012). In addition, even where best management practices are reflected in the regulations that govern marine parks and reserves, problems of implementation persist, as a consequence of internal stakeholder management problems that are not uniform and vary between the cases discussed in this chapter, as demonstrated in the previous country by country discussion.

Conclusion

Coral reefs within marine protected areas that are protected by legislation and regulations against negative anthropogenic activities, such as shoreline development and overfishing, have a better chance of survival than areas that do not enjoy legislative protection. Even so, hurricanes and ocean warming, as well as acidification, will continue to damage Caribbean coral reefs. Bleached coral reefs are not necessarily dead and may be able to regrow if the initial cause of bleaching is removed. Since coral varieties vary in response to such changes in the marine environment, based on their cohabiting algae and bacteria, experiments such as those at the Hawai'i Institute of Marine Biology with coral ecotypes represent an interesting approach that might eventually aid in the rehabilitation of damaged reefs. However, with the exception of some coral reefs in Belize and the Cayman Islands, attempts at strategic management of coral reef tourism have failed to include many stakeholders in the strategy formulation stages. Strategy implementation has often remained ineffective, with clear implications for the long-term sustainability of coral reef tourism.

References

Alcolado, P.M., Alleng, G., Bonair, K., Bone, D., Buchan, K., Bush, P.G., . . ., & Zieman, J.C. (2001). The Caribbean coastal marine productivity program (CARICOMP). *Bulletin of Marine Science, 69*(2), 819–829.

Alemu, J. (2014). Fish assemblages on fringing reefs in the Southern Caribbean: Biodiversity, biomass, and feeding types. *International Journal of Tropical Biology, 62*, 169–181.

Aronson, R.B., & Precht, W.F. (2001). White-band disease and the changing face of the Caribbean coral reefs. *Hydrobiologia, 460*, 25–38.

Burke, L., Reytar, K., Spalding, M., & Perry, A. (2011). *Reefs at Risk Revisited*. Washington, DC: World Resources Institute.

Centre for Marine Sciences (CMS). (2006). *Monitoring of Transplanted Corals at Rackham's Cay, Kingston Harbor, Jamaica*. Centre for Marine Sciences, University of the West Indies.

Cohen, B.R. (2014). *Cayman Islands Coral Reefs Experiencing Dramatic Comeback*. Washington, DC: National Center for Public Policy Research.

Creel, L. (2003, September). The population reference bureau measure. Retrieved 12 May 2017 from www.measurecommunication.org

Dalton, T., Forrester, G., & Pollnac, R. (2012). Participation, process quality, and performance of marine protected areas in the wider Caribbean. *Environmental Management, 49*, 1224–1237.

Goreau, T.F. (1967). *Buccoo Reef and Bon Accord Lagoon, Tobago: Observations and Recommendations Concerning the Preservation of the Reef and Its Lagoon in Relation to Urbanisation of the Neighboring Coastal Islands*. Memorandum to Permanent Secretary (Agriculture), Economics Planning Unit, Prime Minister's Office, Government of Trinidad and Tobago.

Government of Jamaica. (2009). *Vision 2030: Natural Resources and Environmental Management*. Kingston: Pear Tree Press.

Husemann, A.J., & Bent-Hamilton, M. (2011). *Economic Value of Coral Reefs and Mangroves*. Cayman Islands: International College of the Cayman Islands.

Jackson, J.B.C., Cramer, K.L., Donovan, M., Friedlander, A., Hooten, A., & Lam V.V. (2012). Tropical Americas Coral Reef Resilience Workshop. April 29–May 5. International Union for the Conservation of Nature, Panama City, Republic of Panama.

Jackson, J.B.C., Donovan, M.K., Cramer, K.L., & Lam, V.V. (Eds.) (2014). *Status and Trends of Caribbean Coral Reefs: 1970–2012*. Gland, Switzerland: Global Coral Reef Monitoring Network, IUCN.

Johnson, A. (1994). *Compilation, Critical Review and Summary of Water Quality Studies on the West and Southwest Coasts of Barbados from 1968 to 1993*. Report for Delcan International CCPU, Government of Barbados.

Juman, R. (2016). Trinidad and Tobago Institute of Marine Sciences. Response to Authors' Questionnaire: Coral Reef Management in Trinidad & Tobago.

Kenny, J.S. (1976). *A Preliminary Study of the Buccoo Reef/Bon Accord Complex, with Special Reference to Development and Management.* Department of Biological Sciences, University of the West Indies, Trinidad.

Knowlton, N., Weil, E., Weigt, L., & Guzman, H. (1992). Sibling species in Montastraea annularis, coral bleaching, and the coral climate record. *Science, 255,* 300–333.

Laydoo, R.S., & Heileman, L. (1987). *Environmental Impacts of the Buccoo and Bon Accord Sewerage Treatment Plants, Southwestern Tobago. A Preliminary Report.* Institute of Marine Affairs and Crusoe Reef Society.

Manfrino, C. (2008). *Green Guide to the Cayman Islands: The Marine Environment.* Central Caribbean Marine Institute (CCMI Special Publication), Grand Cayman, Cayman Islands.

Manfrino, C., & Whiteman, L. (2008). *Report on Live Coral Cover in Little Cayman.* Proceedings of the International Coral Reef Symposium, Fort Lauderdale, FL.

Potts, R. (2016). Belize Coordinator, Healthy Reefs Initiative. Response to Authors' Questionnaire: Coral Reef Management in Belize.

Roberts, H.H., Murray, S.P., & Suhayda, J.N. (1975). Physical processes in a fringing reef system. *Journal of Marine Research, 33,* 233–260.

Ross, A. (2016). Seascape Caribbean Founder and Principal. Response to Authors' Questionnaire: Coral Reef Management in Jamaica.

Rothaermel, F.T. (2014). *Strategic Management: Concepts.* New York: McGraw Hill Education.

The Nature Conservancy. (2006). *Pedro Bank Coral Reefs – Status of Corals Reefs and Reef Fishes 2005.* The Ocean Research and Education Foundation.

The Tourism Company. (2014). A New Focus for Cayman Islands Tourism: A Revised National Tourism Management Plan, 2009–2013. Retrieved on 6 August 2015 from www.thetourismcompany.com

Tomascik, T., & Sander, F. (1985). Effects of eutrophication on reef-building corals. 1. Growth rate of the reef-building coral Montastraea annularis. *Marine Biology, 87,* 143–155.

United Nations Development Programme (UNDP). (2008). *Belize Country Report on the Protection of Coral Reefs.* The UN Secretary-General's Report. New York: UNDP.

United Nations Environment Programme (UNEP). (2013). *Caribbean Large Marine Ecosystem Pilot Reef Fisheries and Biodiversity Project: Best Practices and Lessons Learnt.* UNEP, Caribbean Regional Coordinating Unit.

Webster, P.J., Holland, G.J., Curry, J.A., & Chang, H.R. (2005). Changes in tropical cyclone number, duration, and intensity in a warming environment. *Science, 309*(5742), 1844–1846.

Woodley, J.P., Alcolado, T., Austin, J., Barnes, R., Clara-Madruga, G., Ebanks-Petrie, R., . . ., & Wiener, J. (2000). Status of coral reefs in the Northern Caribbean and Western Atlantic. In: C. Wilkinson (Ed.), *Status of Coral Reefs of the World: 2000.* Townsville, Australia: Australian Institute of Marine Sciences, pp. 261–285.

World Travel and Tourism Council. (2014). Travel and Tourism: Economic Impact 2014 Anguilla, Antigua and Barbuda, and British Virgin Islands Reports. Retrieved 16 June 2016 from http://www.wttc.org

PART II

Threats and sustainability issues facing coral reef tourism

6

THE POTENTIAL FOR CORAL BLEACHING TO AFFECT LONG-TERM DESTINATION SUSTAINABILITY

Bruce Prideaux, Anja Pabel, Michelle Thompson and Leonie Cassidy

Introduction

This chapter discusses a range of issues related to the long-term sustainability of coral reef destinations that have been, or are likely to be, affected by coral bleaching. The Brundtland Report (1987, p. 18) stated that "humanity has the ability to make development sustainable to ensure that it meets the needs of the present without compromising the ability of future generations to meet their own needs." Achieving this aspirational level of sustainability is becoming increasingly difficult in an era where global temperatures are rising rapidly and where global actions to reduce warming continue to be subject to political agendas that place domestic economic growth and other considerations ahead of global climate change mitigation. One impact of climate change is increased warming of oceans, which in turn has created ideal conditions for coral bleaching.

Over the period 2014–2017, coral reef systems in many countries experienced coral bleaching, with a number of reef systems, including Australia's Great Barrier Reef (GBR), experiencing multiple bleaching events. This multiyear bleaching event was described by the United States National Oceanic and Atmospheric Administration (NOAA, 2017) as the longest, most widespread and most damaging coral bleaching event on record. In a report on the impact of the 2016 coral bleaching event on the GBR, the Great Barrier Reef Marine Park Authority (GBRMPA, 2017, p. 25) warned that ". . . bleaching events are expected to increase in frequency and severity as a result of climate change." This grim news should raise concern across all destinations where coral reefs are promoted as a key destination attraction and hasten strategies to enhance coral resilience in the short-term, and in the longer-term encourage the development of alternative attractions to supplement coral reef experiences.

The science-based view taken in this chapter is that coral bleaching will become more frequent in coming decades and that it is just one of a number of climate-related events that will affect coral reef systems (GBRMPA, 2017). Other climate-related events that will affect coral reef destinations include rising sea levels, more intense wind storms, migration of species seeking cooler water and ocean acidification. Based on this assessment, it can be expected that coral reef reliant communities affected by coral bleaching will experience a range of economic, social and resource-related problems as tourists look for other experiences that offer a better value proposition. The aim of this chapter is to examine how tourists respond to coral bleaching events and to consider the implications of long-term coral reef decline on coral reef dependent tourism destinations.

Cairns, Australia was selected for this study because of the importance of the GBR to the destination's domestic and international tourism sectors. Lessons learnt from the Cairns experience can be applied to other coral reef destinations. This book also discusses aspects of the impact of coral bleaching on other coral reef destinations, including the Maldives and Koh Toa, Thailand (see Chapters 10 and 11). This chapter is organised into five sections commencing with a literature review, followed by a brief discussion on the value of the destination's tourism industry to the local economy, methodology, results and conclusion.

Literature review

The following discussion highlights a number of theories and models that can be used to understand and deal with the issues that confront coral reef dependent destinations facing future climate change related crisis events such as coral bleaching. The review discusses how location, path dependence, post-crisis recovery and comparative and competitive advantage may be used as theoretical tools to evaluate destination vulnerability and responses.

While modern mobilities, including jet aircraft, private cars and high speed trains, have reduced travel times and increased passenger comfort, distance between generating regions and destinations continues to be a major issue for destinations that are located on the edge of national and international peripheries. The concept of periphery has been used by a number of researchers (Harrison, 2015; Wall, 2015; Weaver, 2015) to explain how location, relative to tourism origin regions, can shape as well as inhibit destination development.

Periphery refers to the distance between tourism generating regions and destinations and is measured by physical distance and travel time. The degree of peripherality can be measured on a scale that ranges from near, middle and far (Prideaux, 2002). From this perspective, Cairns may be described as being located on a near periphery in the domestic tourism market, a middle periphery to west Asian markets and a far periphery to European and North American tourists. There is also an element of how individuals perceive distance and how this is influenced by the uniqueness or novelty of a destination's key attractions (Prideaux, 2013).

In general, tourists will only invest extra time and money in visiting distant locations if they are offered attractions or experiences that are unique or novel. Loss of unique attractions may lead to a decline in interest if consumers believe that the benefits of visiting a peripheral location are outweighed by the additional time and cost of travel (Prideaux, 2013). Coral reefs are able to offer a high level of uniqueness and have the capacity to attract long-haul tourists in search of experiences of this type, provided that the coral reef is in good condition. The degrading or loss of coral cover may negate the appeal of uniqueness and reduce the perceived value of the destination in the minds of consumers.

Path dependence

Destinations that rely on one or a small number of key attractions face the danger of "lock-in" (Hassink, 2010; Martin, 2010; Martin & Sunley, 2006), described as the situation where a dominant attraction defines the path that the destination takes in developing its suite of destination pull factors. From this perspective, a destination may become so closely tied to a particular attraction, or attractions, that it becomes difficult to develop new ideas and attractions. As Martin and Sunley (2006, p. 416) observe, a destination may become "stuck in established practices, ideas, and networks of embeddedness that no longer yield increasing returns and may even induce negative externalities". The tourism system thus becomes over-attached to a particular industry structure, resulting in ideas, practices, networks and business structures becoming so rigid that there is little space for new or innovative paths to develop. Grabher (1993) observed that lock-in may occur in a number of interconnected ways, including functional lock-in (where existing ideas and networks hinder alternative connections), cognitive lock-in (where entrenched visions and accepted norms prevent the acknowledgement of alternatives) and political lock-in (where vested political interests are best served by retaining the status quo).

The difficulty for destinations experiencing path dependency where key attractions have become "locked in" is that it is both difficult and expensive to create new pathways using alternative experiences and resources. Scott, Dawson and Jones (2008) observed that a number of winter resorts have suffered a loss of competitiveness because of a decline in snow cover. Breaking out of snow-determined path dependency has been possible where destinations are able to refocus their experiences away from snow to other forms of tourism activity, including adventure tourism (Hendrikx et al., 2013).

The process of developing new pathways may include layering, where new features are added to the experience facing decline, or the development of new experiences (Martin, 2010). Other approaches to amending current pathways include delayering (where some elements of the existing experience suite are removed) and conversion (where existing experiences are changed to broaden their appeal). A final approach to amending current pathways is through recombination, where old and new elements are combined to create a new and hopefully more appealing version of existing experiences. Adoption of strategies of this type are able to assist existing pathways to evolve, leading to new pathways that are more flexible and able to respond to changes in consumer demand as well as changes in the nature of the attractions.

As Carson et al. (2014) observe, ideas for changing the trajectory of path dependency resonate with the emerging literature on "learning regions" and "regional innovation systems". The key idea underling the notion of "learning regions" is that organisations, institutions and key actors learn from each other by exchanging information and adapting their behaviour to respond to changes in circumstances (Doloreux & Parto, 2005; Tödtling & Trippl, 2005). These processes, if successful, can lead to the emergence of new markets, products and networks that bring together the private and public sectors, as well as the community, and offer a more balanced understanding of the opportunities available to escape declining attraction structures (Carson et al., 2014). However, previous investment of time, energy and capital into existing pathways may generate institutional, business, community and political inertia, creating a range of obstacles that are difficult to overcome.

Recent research (Carson et al., 2014; Prats et al., 2008) has explored path dependence and economic evolution from a tourism perspective particularly in peripheral regions. The need to extend current thinking about tourism innovation by adopting a tourism innovations system approach has been advocated by Carson et al. (2014) who explored the potential of the tourism innovation systems approach in remote areas of South Australia. Carson et al. (2014) caution that in circumstances where path dependence was created by a history of dependence on a particular resource, the introduction of new experiences may be met by institutional and industry inertia. In some circumstances, political inertia may also inhibit institutional change as political careers and institutional world views reflect a desired rather than a rational perspective of reality.

Comparative and competitive advantage

In their seminal work on the concept of comparative and competitive advantage, Ritchie and Crouch (2003) argued that destinations have two broad classes of resources described as comparative and competitive. Competitive resources are those that confer a competitive advantage to a destination vis-à-vis other competing destinations. Competitiveness assists destinations to differentiate their products by creating a sense of uniqueness or conferring of iconic status (Prideaux, 2013). The degree of competitiveness is often associated with the level of uniqueness of key attractions. Beijing's Great Wall or New York's Statue of Liberty are examples of uniqueness that add to the destination's overall competitive position. The degree of competitiveness may also be an outcome of a destination's ability to use marketing strategies to convince tourists that the destination offers a unique experience. As Ritchie and Crouch (2003) observed, competitiveness can be lost if destinations fail to refresh their suite of experiences to meet the changing needs of visitors or new competitors succeed in developing attractions that are more appealing to tourists.

Ritchie and Crouch (2003) describe comparative resources as those that are currently underused or latent but if developed have the potential to attract tourists.

Resources of this type might include a pleasant climate, attractive landscapes, natural and built heritage and local lifestyles. Converting potential resources into tourism experiences or products enables a destination to broaden its suite of experiences and increase its competitive advantage in local, national and international tourism markets, which acts as a hedge against existing attractions losing their appeal. Over time, tourism experiences become unfashionable and new forms of experiences emerge to attract the attention of tourists. One example of changing tastes are amusement piers that were once a central feature of coastal resorts. Most now lie abandoned or have been demolished and replaced by other forms of amusement, including theme parks (Wismer, 2001). Similarly, as travellers have become more experienced, their consumption of experiences has changed. For example, moving from organised packaged group tours to independent travellers who enjoy the freedom of destinations and seeking out experiences on their own.

Post-crisis recovery

In recent decades the global tourism industry has suffered a number of crisis events on a scale that extends from local to national and global. The Global Financial Crisis of 2008–2009 is an example of a global event that affected the tourism sector in most countries. Coral bleaching can be described as a local event that has occurred on a global scale. Responding to crisis events is an important activity that should occur at all levels of the tourism industry; however, as Ritchie (2009) notes, crisis planning and crisis recovery is often neglected.

In one of the first papers to advocate the need for a specific tourism crisis management framework, Faulkner (2001) suggested a six stage model based on sequenced tourism disaster planning, management and recovery. Faulkner (2001) advised that planning should commence in the pre-crisis stage followed by disaster management when a crisis occurs and, importantly, emphasised the need for post-crisis communications in the recovery stage. While many modifications have been suggested to the original model (Scott et al., 2008), it still stands as an exemplar for crisis management including the post-crisis stage. Surprisingly, many of the elements of crisis planning and recovery advocated by Faulkner (2001) were absent in the response to the 2016 and 2017 coral bleaching events in Cairns. In particular, crisis recovery communication was ad hoc and relatively uncoordinated.

Considering the statement by the GBRMPA on coral bleaching outlined earlier in this chapter, it is apparent that further coral bleaching will occur. Destinations that promote coral reef tourism as a key destination pull factor will need to reconsider how coral reef experiences are delivered and if alternative experiences need to be developed. This will become particularly important for coral reef destinations that are located in peripheral areas. Many of the more remote coral reef destinations discussed in Chapter 2 can be described as being located on the periphery of international tourism and are not well serviced by airlines. A decline in the quality of their existing coral reefs and lack of substitute experiences will make it difficult for them to retain existing tourist numbers.

If tourists become concerned that the vivid images of coral, which have been central to the image of coral reef destinations, are no longer valid they may look for destinations that can deliver on the image promise made in their marketing campaigns. This is not to say that coral reef dependent destinations should turn away from promoting coral reef activities. There are many elements that constitute a coral reef experience, including viewing marine animals and a range of activities such as snorkelling and scuba diving. The layering of new activities (Martin, 2010) onto existing activities is one strategy that can be adopted to reduce the threat of lock-in.

The value of Great Barrier Reef tourism

In a 2017 report the Australian Bureau of Statistics estimated the contribution of tourism to the Wet Tropics Natural Resource Management Region (where Cairns is the major tourism centre) as AUD$3.2 billion in the 2015/16 financial year. Economic value is not the only indicator of the "value" of resources such as the GBR. Value can be interpreted in a broad sense to include its scientific value as a unique ecosystem and its value to local, national and global communities. A recent report by Deloitte Access Economics (2017) estimated the asset value of the GBR, based on its economic, social and iconic values, as AUD$56 billion. This figure was based on the value of the GBR to people who had visited the GBR (AUD$29 billion), Australians who had yet to visit the GBR (AUD$24 billion) and recreational users (AUD$3 billion). The report also estimated that the GBR contributed AUD$6.4 billion in value-added activity and had created over 64,000 direct and indirect jobs.

This more encompassing concept of the value of the GBR has underpinned the support that the community and the public sector have given to understanding the reef as an ecological system, and its economic value as a tourism experience. Based on a community-wide consensus that the GBR is an important national icon and ecological system, the need to protect the GBR has received considerable support from the scientific community, the tourism industry, the local community and from local, state and federal politicians.

The GBR is managed by the GBRMPA, with day-to-day operations in onshore locations, including islands, undertaken by the Queensland Parks and Wildlife Service (QPWS). Considerable financial resources have been allocated to institutions that undertake research in the GBR (including the Australian Institute for Marine Science, the Coral Reef Centre of Excellence and many research groups located in Australian universities). Economic arguments based on job creation and the value of the tourism industry in the GBR region underpin much of the discussion that has taken place on the need for further protection of the GBR (see Chapter 4 in this volume for a more detailed discussion of the role of the GBRMPA and the threats that face the GBR).

The Cairns tourism industry

Europeans began settling the Cairns region in the 1870s, attracted by the potential for logging the region's extensive rainforests. Other industries soon followed and a port was built to service the growing range of extractive industries that included minerals, logging and agriculture. These industries continued to be the mainstays of the local economy for many decades. The town was officially gazetted as a settlement in 1876. For the next eight decades the town's peripheral location and length of time taken by shipping to reach the area did little to encourage tourism development.

Tourism remained very much a sideshow to the region's main economic sectors of agriculture and mining until well after the opening of the rail line to the south and the connection of the city to the national highway system by the early 1930s. During the first seven decades of the twentieth century, the city's population grew steadily based on the prosperity of the region's agricultural and mining sectors, reaching 30,000 by 1967 (Schofield, 2009). Growth accelerated in the next few decades rising from 48,557 in 1991 to 161,932 in 2016 (Australian Bureau of Statistics [ABS], 2016). A substantial part of the post-1970 growth can be attributed to the growth in tourism and the various sectors that support the tourism industry, including construction, education, health and retail.

By the early 1980s, the city's growing reputation as a tropical tourism destination led to a decision to build an international airport. The opening of the Cairns International Airport in 1984, and substantial investment by Japanese entrepreneurs, provided the trigger for the rapid growth of international tourism in the Tropical North Queensland (TNQ) region, including nearby Port Douglas. Between 1965 and 2017 passenger movements through the Cairns International Airport increased from 85,000 to more than 5.2 million (Cairns Airport Corporation, 2017). During this period, the destination leveraged its location between the GBR and the Wet Tropics Rainforests (WTR) as its key selling proposition. Brochures of the period focused on colourful images of coral, rainforest animals and rainforest landscapes. The declaration of the GBR and the WTR as World Heritage Areas (WHAs) added to the appeal of the destination's key attractions. Promotion of the region, which extends from Mission Beach in the south to the top of Cape York in the north, is undertaken by Tropical Tourism North Queensland (TTNQ) and its predecessors. While the physical size of the TNQ tourism region is enormous, encompassing approximately 20% of the state (340,645 km^2) (TTNQ, 2014), most tourism activity occurs in the coastal strip between Cairns and Port Douglas.

In the decade following the opening of the Cairns International Airport, domestic and international tourism grew rapidly but hit several road bumps after 2000, as illustrated in Figure 6.1. Between 1999 and 2014 the destination recorded very modest growth of 12% in international visitor numbers compared to the doubling of the global number of international arrivals from 664 million to 1.2 billion

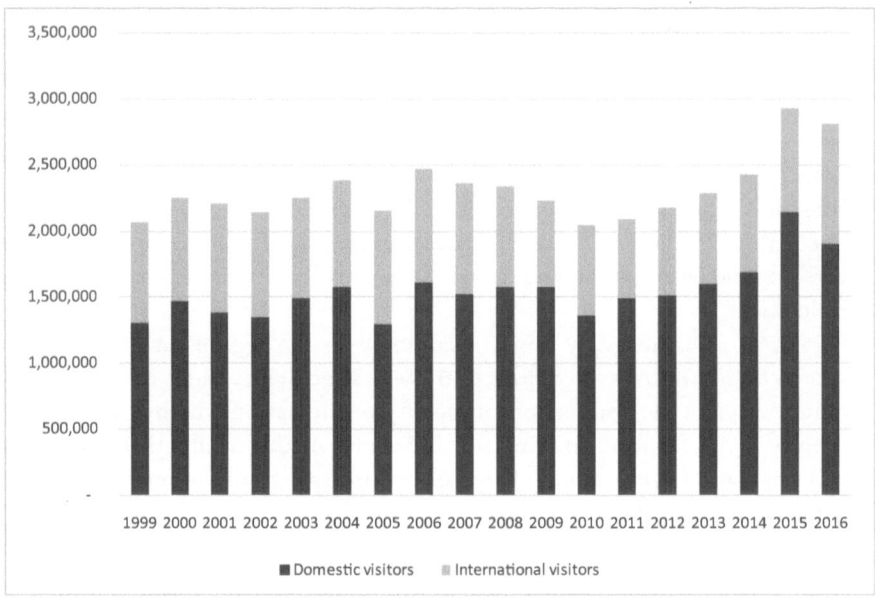

FIGURE 6.1 Domestic and international visitors to Tropical North Queensland between 1999 and 2016

Source: Data from Tourism Research Australia and Tourism and Events Queensland.

(United Nations World Tourism Organization [UNWTO], 2017). Over the same time period, Australia's international arrivals increased by 93% from 4.4 million to 8.5 million (Tourism Research Australia [TRA], 2017). These data highlight the extent of the destination's failure to keep pace with increases in national and international arrivals over the period 1999–2014. It also calls into question the strength of the "pull" power of its key tourism attractions over this period and illustrates the path dependency of the destination based on its key coral reef and rainforest attractions. Over this entire period the destination continued to promote its natural attractions as its key pull factors.

A further road bump was encountered by the destination in the year following the 2016 coral bleaching events. In the 12-month period to June 2017 the number of interstate visitors declined by 20%, while expenditure by this segment fell by AUD$211 million (to AUD$1.9 billion) compared to the same period in 2016 (TRA, 2017). In comparison, the international market grew by 4.5% over the 12 month period to June 2017 (TRA, 2017). During this period, overall visitor numbers to Queensland grew by 6.2% while bed nights increased by 6.3%. These figures point to the impact that coral bleaching had on the domestic sector of the region's tourism industry during the latter part of 2016 and the first part of 2017. It is not clear if this decline is temporary or of a more long-term nature.

The 2016 and 2017 coral bleaching events

The 2016 coral bleaching event was preceded by a significant increase in ocean temperature during a very strong El Niño event. By February 2016, the average sea temperature in the northern section of the GBR had risen 1.35°C above the average ocean temperatures between 1951 and 1980 (Hoegh-Guldberg & Ridgway, 2016). The extent of coral bleaching was first made public in a press release by the ARC Centre of Excellent for Coral Reef Studies in early March 2016 and received extensive international coverage (Smail, 2016). Ocean temperatures also increased in February 2017 leading to a second coral bleaching event that affected reefs north of Townsville. Figure 4.2 (see Chapter 4) highlights the extent of the 2016 and 2017 events. Conflicting views emerged about the extent of coral bleaching when research commissioned by the Reef and Rainforest Research Centre and the Association of Marine Park Tourism Operators (RRRC & AMPTO, 2016) found that the majority of coral reefs visited by tourists in the Cairns area had escaped major damage. Unfortunately, the widespread media coverage of the event conveyed a picture of significant coral damage in this area. Neither the 2016 nor 2017 coral bleaching events were responded to effectively with post-disaster communication strategies.

The coral bleaching events were no surprise to the GBRMPA, which has recognised the potential for climate change events, including coral bleaching, for some time. The *2014 Outlook Report* (GBRMPA, 2014) stated that climate change remained the greatest threat to the long-term survival of the GBR. The challenge posed by climate change and strategies to respond to this threat are outlined in the 2050 Long-Term Sustainability Plan (Australian Government, 2015).

Coral protection and regeneration

In Australia considerable resources have been allocated to the protection of the GBR by the Queensland state government and the Australian government. Funding has been provided for a range of measures to improve the water quality of the GBR lagoon, including reducing agricultural run-off of fertilisers and pesticides, buy-back of commercial fishing licences and improving coastal sewerage discharge (Commonwealth of Australia, 2015). Considerable investment has also been made in supporting scientific research in the GBR.

A number of international destinations have invested in coral regeneration, including Florida (USA) and the Philippines. Considerable success has been reported in the Florida project (Florida Reef Resilience Program, http://frrp.org/coral-restoration/) which involves regenerating a range of corals and then replanting in affected areas. A similar project funded by the Reef Restoration Foundation commenced on Fitzroy Island near Cairns in early 2018 (https://www.reefrestoration foundation.org/). At the time this book was written the GBRMPA has yet to announce a policy on coral rehabilitation. Another strategy that has been suggested is cooling the surface water by mixing cooler deep water with warm shallow water. An experiment to test the potential of this system to protect limited areas of coral reef will be trialled in the GBR region in 2018.

Method

In late 2015 it became apparent that a coral bleaching event might occur in early 2016 and a survey was developed to identify tourists' reactions during and after the event. The survey was undertaken in Cairns by the authors, with surveys collected at the domestic terminal of the Cairns International Airport via a convenience sampling approach. Participants generally took about 10 minutes to complete the survey instrument. Due to the nature of the research, there are limitations that need to be considered when interpreting the findings. The survey was administered at the domestic terminal of the airport. The views of tourists departing via road, rail, sea or the international terminal are not represented; however, the sample did collect the views of international tourists travelling on domestic flights. The views of non-English speakers, such as the Chinese and Japanese, were not captured as the survey questionnaire was available in English only. Finally, while the potential for social desirability bias (Nederhof, 1985) is recognised, no specific steps were taken to reduce this potential bias.

Over the period of the survey (January 2016 to June 2017) 1,817 valid surveys were collected. Domestic tourists accounted for 45% of respondents. Just over two-thirds (69%) of all respondents visited the GBR. International respondents were more likely to visit the GBR (52%) than domestic respondents (48%). Respondents were asked about their motives for visiting Cairns using a 5-point Likert scale, where 5 indicated "very important" and 1 indicated "not at all important." Table 6.1 presents the ranking of the top ten motives for domestic and international respondents. As Table 6.1 indicates, domestic and international respondents reported very different motives for visiting Cairns. Domestic respondents did not nominate the GBR as a particularly important motive, while international respondents indicated that the GBR was the most important motive for visiting the destination.

Figure 6.2 tracks the ranking of the GBR as a motive for domestic and international respondents over the period of the two consecutive bleaching events. As Figure 6.2 indicates, the rank of the GBR as a motive for visiting Cairns fell quite dramatically after the first coral bleaching event from 3rd to 12th position in Quarter 3 2016, recovering to 7th position in Quarter 1 2017, before falling to 9th position in Quarter 2 2017. International respondents ranked the GBR as the top motive for visiting Cairns over the entire period of the survey.

As Figure 6.3 indicates, the decline in the number of international respondents visiting the GBR over the 18 months of the survey was relatively small. The pattern of domestic visitation was more volatile with a sharp decline in the rate of visitation between Quarter 2 (April to June 2016) and Quarter 3 (July to September 2016),

TABLE 6.1 Comparison of top ten motives for domestic and international respondents

Rank	Activity	Domestic N=804 (mean)		International N=985 (mean)
1	To have fun	4.44	Visit the Great Barrier Reef	4.66
2	Rest & relaxation	4.24	To have fun	4.56
3	Experience the natural environment	3.97	Go snorkelling/diving	4.12
4	Enjoy the tropical lifestyle	3.97	See Australian wildlife	4.11
5	Experience the climate	3.94	Experience the natural environment	4.06
6	Visit the Wet Tropics Rainforest	3.70	Visit the Wet Tropics Rainforest	3.94
7	Visit the Great Barrier Reef	3.67	Visit National Parks	3.75
8	The price matched my budget	3.67	Learn about the natural environment	3.75
9	Visit the beaches	3.67	Rest & relaxation	3.72
10	Visit National Parks	3.57	Visit World Heritage Area	3.66

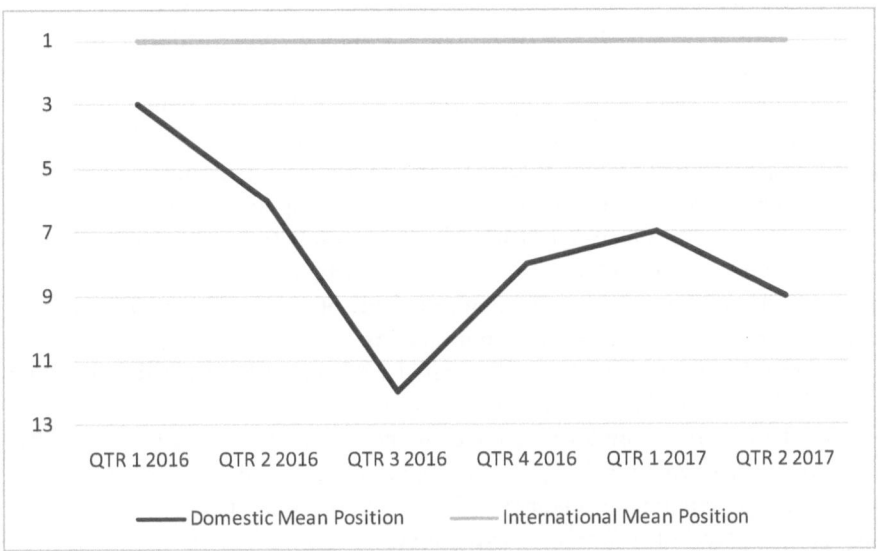

FIGURE 6.2 Ranking of the GBR as a travel motive for domestic and international respondents between Q1 2016 and Q2 2017

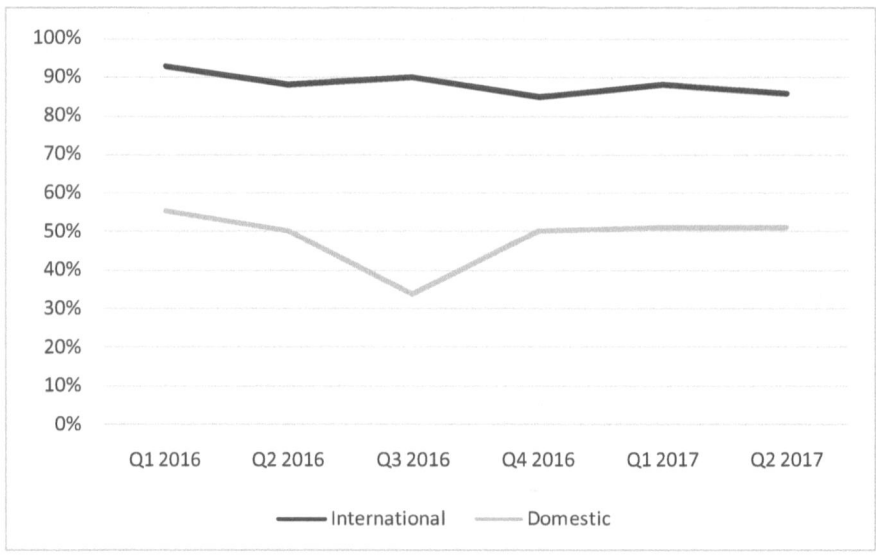

FIGURE 6.3   Pattern of reef visitation for domestic and international respondents over the period January 2016 to June 2017

indicating a response to negative media reports about coral bleaching. Surprisingly, there was no decline in the visitation rate for domestic visitors during the 2017 coral bleaching event.

Figure 6.4 compares the level of recall among domestic respondents who reported viewing media reports on the bleaching event with the level of concern about bleaching. In Quarter 2 2016 a little under 80% of respondents recalled seeing media reports on bleaching, with this level of response increasing over subsequent quarters. By Quarter 1 2017, the correlation between concern and recall of media had begun to fall. In Quarter 2 2017 this trend reversed and the level of concern increased significantly although the level of recall about media was only slightly higher than in the previous four quarters.

Figure 6.5 compares the level of reported viewing media about the bleaching event with concern about coral bleaching for international respondents. Over the period of the survey, the level of concern about coral bleaching grew by 34%. Overall, fewer international respondents reported viewing media about bleaching than domestic respondents. Figure 6.5 does, however, indicate a positive relationship between level of concern and exposure to media.

Respondents were also asked to describe their GBR experience on a 4-point scale from "good" to "awful." Combined, the scale of "poor" and "awful" never exceeded 5% for either domestic or international respondents. In most quarters, 80% of respondents ranked their experience as good.

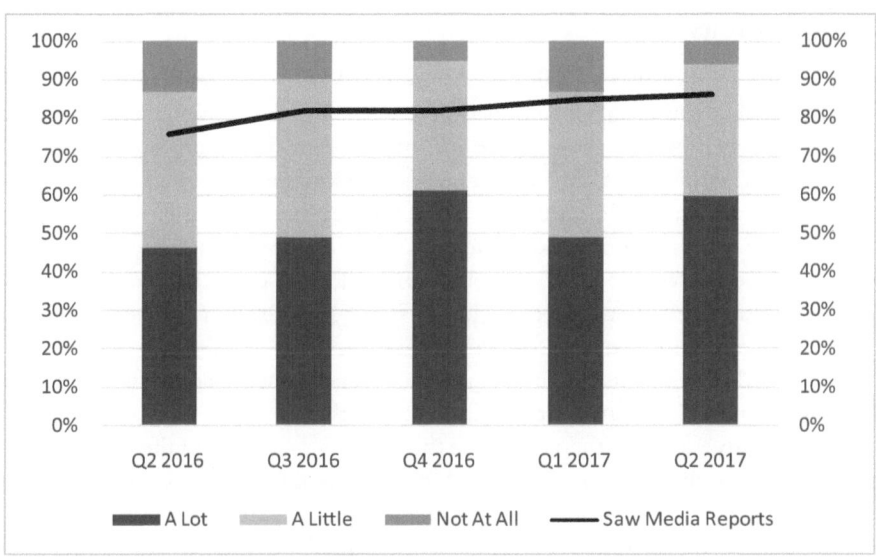

FIGURE 6.4 Comparison of concern by domestic respondents about coral bleaching with viewing of media reports

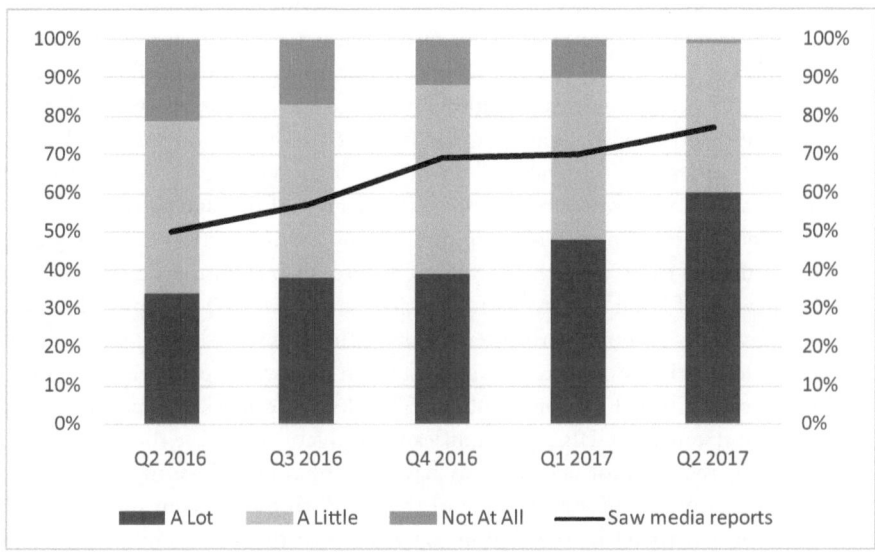

FIGURE 6.5 Comparison of concern by international respondents about coral bleaching with viewing of media reports

Survey results

The results of the survey indicates that respondents were generally aware that coral bleaching had occurred and that there was a high level of concern about the bleaching events. While the international market remained quite buoyant during the 2016 and 2017 events, the loss of one of the destination's key pull factors can be expected to eventually flow through to international visitor numbers in a manner similar to the response of the domestic market (see Figure 6.2). Figures 6.4 and 6.5 highlight the growing level of concern expressed by respondents over the effects of coral bleaching. Over the period of the survey, the level of concern expressed by domestic respondents rose by 15%, while the level of concern expressed by international respondents rose by 35%. If the level of concern continues to escalate there is a danger that some tourists will look for other destinations as a holiday site.

One positive result of the survey was that even at the height of the coral bleaching event, less than 5% of respondents rated their GBR experience as poor or awful. This positive message was never picked up in the media or used effectively by the destination to combat the generally poor media coverage.

Discussion

The objective of this chapter was to examine how tourists respond to coral bleaching events and consider the implications of long-term coral reef decline and how this may affect the sustainability of coral reef dependent tourism destinations. As the results of the Cairns Airport survey highlight, the 2016 and 2017 bleaching events had a negative effect on some segments of the domestic market but not the international market. On a broader scale, all destinations that promote coral reef tourism face a potential decline in visitors, although the extent will vary between destinations and be determined by a range of factors that are destination specific. Factors include the importance of coral reefs in the destination's suite of experiences, its location relative to major source markets, the degree of damage to the coral reef and the effectiveness of post-crisis communications and other recovery strategies.

The results of the Cairns Airport survey raise an interesting question about the importance of coral reefs to various visitor sectors. In the case of Cairns, the GBR is obviously important to international tourists and a decline in the quality of the destination's coral reef experience can be expected to lead to a sharp decline in international visitor numbers. The domestic market appears to have moved away from the GBR as a significant motivation to visit the destination, with lifestyle-related motives (such as having fun) now more important. This raises the question of the significance of the GBR to the tourism industry – do domestic tourists travel to the region to experience the GBR and while there take the opportunity to engage in other tourism activities, or are tourists visiting Cairns and while there take the opportunity to visit the GBR? If the latter is the case, the destination needs to revisit its suite of experiences and consider how it can build new non-GBR experiences.

Results of the Cairns study illustrate the danger faced by destinations when the appeal of a key tourism resource begins to decline because of a crisis event such as coral bleaching. In the case of Cairns, the 20% fall in the interstate markets in the period following the first bleaching event and the decline in the importance of the GBR as a travel motive for domestic tourists hints at what may happen if the quality of the GBR as a unique tourism experience continues to decline. It is not possible at the time of writing to determine if the decline in the interstate market is a short-term event or the beginning of a more long-term pattern. The danger for Cairns is if the decline in interest in the GBR as a key domestic travel motive migrates across to the international market. The failure of the destination to mount a more vigorous and effective post-crisis communication strategy, to counter exaggerated claims that the quality of the coral cover off Cairns had declined significantly, demonstrates the need for all destinations affected by crisis events to have a well-developed crisis management plan and to use post-crisis communication to counter any misleading coverage of the crisis event.

More broadly, the extent of the decline in visitors in all coral reef destinations that may follow future bleaching events will be determined by the role of coral reefs as the destinations' main tourist pull factors, the time lapse between coral bleaching events, the success or otherwise of crisis recovery strategies and the success of strategies to increase coral resilience. Of particular concern to the Cairns tourism industry is the ongoing failure to provide significant support for crown-of-thorns starfish (COTS) eradication (see Chapter 4 in this volume for a discussion on the problems caused by COTS) and past failure by the GBRMPA to support coral reef rehabilitation.

Estimating the strength of the GBR as a proximate motive for visiting Cairns, or any other coral reef dependent destination, is therefore important. The legal term "*sine qua non*" or the "but for" rule states that when an action is proximate, the absence of the factor to which the action is proximate would mean that an outcome would be impossible. In this case, Cairns is a popular destination because of its location adjacent the GBR. Under the "but for" rule, the absence of the GBR would mean that a visit to Cairns would not have been considered. The strength of the "but for" position of the GBR in the international market appears to be significant, as the results illustrate in Figure 6.2.

Repeated bleaching events may form a pattern that in the minds of consumers will create a negative image of the destination. Gurtner (2007) observed that crisis events often lead to a fall in visitor numbers but the impact is often short-lived if the crisis is not an ongoing or recurring event and that the destination's key attractions are restored quickly. Based on this view of the literature, the absence of further bleaching events in the next few years will provide Cairns/GBR and other destinations that have been similarly affected with some expectation that visitor numbers will quickly recover. However, the long-term view by the GBRMPA (2017) is that coral bleaching will continue to be a regular event as the global climate continues to warm. Corals are particularly sensitive to ocean heating (Queensland Museum, 2013) and failure to keep warming below the agreed 2°C warming target of the 2015 Paris Climate Change Accord will lead to further widespread damage to all coral reef systems.

Conclusion

All coral reef dependent destinations need to consider the potential impact of future coral bleaching events. The peripheral location of many of these destinations (Florida is an exception) and the degree of path dependence are key factors in determining the level of impact. Solutions will require new development pathways that will force destinations to reassess the basis of their current competitive position and re-examine opportunities for comparative resources to supplement or replace coral reef experiences.

Ideally, forward planning should adopt a series of assumptions about the probable level of global warming and develop destination response strategies on that basis. The Intergovernmental Panel on Climate Change (IPCC, 2014) has developed a number of models that predict the level of global temperature increase under a range of future greenhouse gas emission scenarios. Selecting a particular model is difficult unless there is a logical basis for selection. The 2015 Paris Climate Change Accord, which adopted a target of less than 2°C rise in global temperatures, appears to offer a logical base line for planning. Working within the parameters of the Paris Accord's target of global temperature rises being kept to below 2°C by the turn of the century, allied with sea level rises expected to be at least 1 metre (IPCC, 2014), destinations have a useful planning base line. To this should be added the best scientific estimates of the anticipated extent of coral decline within these parameters.

This chapter commenced with a statement that the long-term sustainability of coral reef dependent destinations may be affected by ongoing coral bleaching events that are expected to occur as sea temperatures increase in coming decades. In the short-term, effective post-crisis recovery marketing may help coral reef destinations retain visitors, but in the long-term, coral reefs are likely to continue to decline and unless alternative attractions are developed, destinations may face the prospect of long-term visitor loss. Path dependence theory provides useful insights into how adaptation strategies can be adopted by affected destinations. Destinations that remain locked into using coral reefs as a key selling point are at the greatest risk of decline. Escaping lock-in requires investment in new experiences, some of which may be found in the destination's stock of comparative resources.

Despite this gloomy picture, there remains an urgent need to continue the protection of all coral reefs with the hope that temperature increases will peak at a level below that which will lead to further global loss of coral reefs. There are many strategies that coral reef destinations can adopt to protect and even rejuvenate their reefs. From a tourism perspective, there is an urgent need to reduce destination stress on reefs. The 10-point scaffold outlined in Table 18.1 goes some way to providing a template to how destinations must respond to the decline in coral reefs.

References

Australian Bureau of Statistics (ABS). (2016). Cairns region data summary. Retrieved from http://stat.abs.gov.au/itt/r.jsp?databyregion

Australian Government. (2015). *Reef 2050 Long-Term Sustainability Plan*. Canberra: Commonwealth of Australia.

Brundtland, G., Khalid, M., Agnelli, S., Al-Athel, S.A., Chidzero, B., Fadika, L.M., et al. (1987). *Our Common Future: The World Commission on Environment and Development*. Oxford: Oxford University Press.

Cairns Airport Corporation. (2017). Year of passenger growth. Retrieved from https://www.cairnsairport.com.au/corporate/media/news/year-of-passenger-growth/

Carson, D.A., Carson, D.B., & Hodge, H. (2014). Understanding local innovation systems in peripheral tourism destinations. *Tourism Geographies, 16*(3), 457–473.

Commonwealth of Australia. (2015). *Reef Long Term Sustainability Plan*. Canberra: Commonwealth of Australia.

Deloitte Access Economics (DAE). (2017). *At What Price? The Economic, Social and Icon Value of the Great Barrier Reef*. Report to the Great Barrier Reef Foundation. Retrieved from https://www2.deloitte.com/content/dam/Deloitte/au/Documents/Economics/deloitte-au-economics-great-barrier-reef-230617.pdf

Doloreux, D., & Parto, S. (2005). Regional innovation systems: Current discourse and unresolved issues. *Technology in Society, 27*(2), 133–153.

Faulkner, B. (2001). Towards a framework for tourism disaster management. *Tourism Management, 22*, 135–147.

Grabher, G. (1993). The weakness of strong ties: The lock-in of regional development in the Ruhr area. In G. Grabher (Ed.), *The Embedded Firm: On the Socio-Economics of Industrial Networks* (pp. 255–277). London: Routledge.

Great Barrier Reef Marine Park Authority (GBRMPA). (2014). *Outlook Report 2014*. Townsville: GBRMPA.

Great Barrier Reef Marine Park Authority (GBRMPA). (2017). *Final Report 2016 Coral Bleaching Event on the Great Barrier Reef*. Townsville: GBRMPA.

Gurtner, Y. (2007). Crisis in Bali – lessons in tourism recovery. In E. Laws, B. Prideaux, & K. Chon (Eds.), *Crisis Management in Tourism* (pp. 81–97). Wallingford, UK: Centre for Agriculture and Bioscience International.

Harrison, D. (2015). Vanishing peripheries and shifting centres: Structured certainties or negotiated ambiguities. In T.V. Singh (Ed.), *Challenges in Tourism Research* (pp. 170–175). Bristol, UK: Channel View.

Hassink, R. (2010). Locked in decline? On the role of regional lock-ins in old industrial areas. In R. Boschma & R. Martin (Eds.), *Handbook of Evolutionary Economic Geography* (pp. 450–468). Cheltenham, UK: Edward Elgar.

Hendrikx, R., Zammit, C., Hreinsson, E., & Becken, S. (2013). A comparative assessment of the potential impact of climate change on the ski industry in New Zealand and Australia. *Climatic Change, 119*(3–4), 965–978.

Hoegh-Guldberg, O., & Ridgway, T. (2016, 21 March). Coral bleaching comes to the Great Barrier Reef as record-breaking global temperatures continue. Retrieved from The Conversation, https://theconversation.com/coral-bleaching-comes-to-the-great-barrier-reef-as-record-breaking-global-temperatures-continue-56570

Intergovernmental Panel on Climate Change (IPCC). (2014). *Climate Change 2014 Synthesis Report*. Geneva: Intergovernmental Panel on Climate Change. Retrieved from http://www.ipcc.ch/report/ar5/syr/

Martin, R. (2010). Roepke lecture in economic geography – rethinking regional path dependence: Beyond lock-in to evolution. *Economic Geography, 86*(1), 1–27.

Martin, R., & Sunley, P. (2006). Path dependence and regional economic evolution. *Journal of Economic Geography, 6*(4), 395–437.

National Oceanic and Atmospheric Administration Coral Reef Watch (NOAA). (2017). Global coral bleaching 2014–2017: Status and appeal for observations. Retrieved from https://coralreefwatch.noaa.gov/satellite/analyses_guidance/global_coral_bleaching_2014-17_status.php

Nederhof, A. (1985). Methods of coping with social desirability bias: A review. *European Journal of Social Psychology*, *15*, 263–280.

Prats, L., Guia, J., & Molina, F.X. (2008). How tourism destinations evolve: The notion of tourism local innovation system. *Tourism and Hospitality Research*, *8*(3), 178–191.

Prideaux, B. (2002). Creating visitor attractions in peripheral areas. In A. Fyall, A. Leask, & B. Garrod (Eds.), *Managing Visitor Attractions: New Directions* (pp. 58–72). Oxford: Butterworth Heinemann.

Prideaux, B. (2013). The importance of landscapes: Difference between first time visitors and returning visitors. *Tourism Tribune*, *28*(2), 9–12.

Queensland Museum. (2013). *The Great Barrier Reef: A Queensland Museum Discovery Guide*. Brisbane: Queensland Museum.

Reef and Rainforest Research Centre (RRC) & the Association of Marine Park Tourism Operators (AMPTO). (2016). *Coral Bleaching Assessment on Key Tourism Sites between Lizard Island and Cairns*. Cairns: RRRC & AMPTO.

Ritchie, B.W. (2009). *Crisis and Disaster Management for Tourism*. Bristol, UK: Channel View.

Ritchie, J.B., & Crouch, G.I. (2003). *The Competitive Destination: A Sustainable Tourism Perspective*. Wallingford, UK: Centre for Agriculture and Bioscience International.

Schofield, G. (2009). Cairns Chamber of Commerce 100 years of history. Retrieved from https://www.cairnschamber.com.au/uploads/media/Complete_CCoC_history.pdf

Scott, D., Dawson, J., & Jones, B. (2008). Climate change vulnerability of the US Northeast winter recreation–tourism sector. *Mitigation and Adaptation Strategies for Global Change*, *13*(5–6), 577–596.

Scott, N., Laws, E., & Prideaux, B. (2008) Tourism crises and marketing recovery strategies. *Journal of Tourism and Travel Marketing*, *23*(2), 1–13.

Smail, S. (2016, 20 April). Great Barrier Reef: Only 7 per cent not bleached survey finds. *ABC News* [online]. Retrieved 7 March 2017 from http://www.abc.net.au/news/2016–04–20/great-barrier-reef-bleaching/7340342

Tödtling, F., & Trippl, M. (2005). One size fits all? Towards a differentiated regional innovation policy approach *Research Policy*, *34*(8), 1203–1219.

Tourism Research Australia (TRA). (2017). International visitors in Australia. Tourism Research Australia: Sydney. Retrieved from https://www.tra.gov.au/ArticleDocuments/250/IVS_one_pager_June2017.pdf.aspx?Embed=Y

Tourism Tropical North Queensland (TTNQ). (2014). Tropical North Queensland destination tourism plan. Retrieved from http://media.ttnq.org.au/_documents/9-b5025ed792d34de7b56406a645d6703e.pdf)

United Nations World Tourism Organization (UNWTO). (2017). *UNWTO tourism highlights, 2017 edition*. Madrid: UNWTO. Retrieved from http://mkt.unwto.org/publication/unwto-tourism-highlights

Wall, G. (2015). Tourism in peripheries. In T.V. Singh (Ed.), *Challenges in Tourism Research* (pp. 180–185). Bristol, UK: Channel View.

Weaver, D. (2015). Moving in from the margins: Experience consumption and the pleasure core. In T.V. Singh (Ed.), *Challenges in Tourism Research* (pp. 176–179). Bristol, UK: Channel View.

Wismer, J. (2001). From amusement thrills to summertime chills. The rise and decline of the traditional American amusement park. Unpublished Masters thesis submitted to the Department of Sociology, Queen's University, Kingston, Ontario, Canada.

7

BELIZE BARRIER REEF SYSTEM

A threatened biodiversity hotspot

Leandra Cho-Ricketts

The Belize Barrier Reef

The aim of this chapter is to examine a range of issues that relate to the long-term sustainability of the Belize Barrier Reef System as a tourism experience. The Belize Barrier Reef System, spanning 220 km in length, is the world's second longest coral reef system after the Great Barrier Reef of Australia and comprises 80% of the Mesoamerican Barrier Reef. The Belize Barrier Reef is a world-renowned tourism destination, famous for its marine life and is the home of the Great Blue Hole. It is an outstanding natural system consisting of the longest barrier reef in the western hemisphere, three offshore atolls, hundreds of sand and mangrove cayes and coastal lagoons and was declared a World Heritage Site in 1996. The World Heritage Site comprises 12% of the entire Belize Barrier Reef, showcases its natural beauty and illustrates a classic example of the evolutionary history of reef development through seven unique marine protected areas: Bacalar Chico National Park and Marine Reserve, Blue Hole Natural Monument, Half Moon Caye Natural Monument, South Water Caye Marine Reserve, Glover's Reef Marine Reserve, Laughing Bird Caye National Park and Sapodilla Cayes Marine Reserve. The reef complex also contains fringing reefs, patch reefs, and pinnacle and faro reefs making the Belize Barrier Reef a truly unique and diverse representation of all the major reef types (Figure 7.1).

The Belize Barrier Reef is rich in diversity with 65 recorded coral species, over 500 species of fish, 350 mollusc species, along with a great diversity of squirts, sponges, crustaceans, echinoderms, other marine invertebrates, plants and birds creating a hotspot of biodiversity within the Caribbean region (UNESCO, 1996). It provides critical habitat for numerous threatened and endangered species, including hawksbill, loggerhead, leatherback and green sea turtles, the West Indian manatee, the American saltwater crocodile and several endemic and migratory birds that reproduce in the

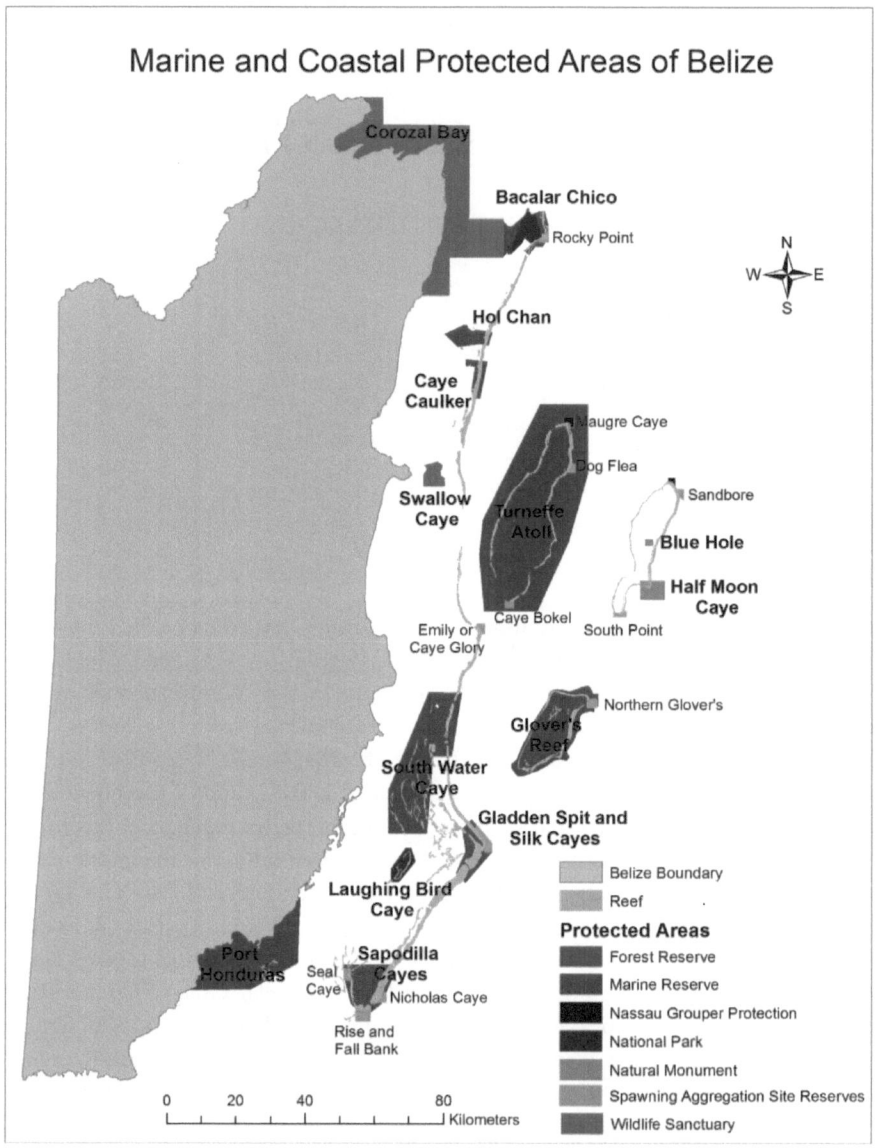

FIGURE 7.1 Map of the Belize Marine Protected Area Network including fish spawning aggregation sites

littoral forests of the cayes. Charles Darwin once referred to the Belize Barrier Reef as "the most remarkable reef in the West Indies" in his 1842 study of coral reefs. It is this wealth of biodiversity and reef development that attracts hundreds of thousands of visitors to Belize's reefs every year. In recognition of the biodiversity significance of the Belize Barrier Reef System, there is a national network of marine protected areas

(MPAs) from the north of the country at Rocky Point to the south at Sapodilla Cayes that protects and conserves the most significant, diverse and sensitive reef areas. This MPA network includes fish spawning sites, turtle and crocodile nesting sites, resilient reef sites, faroes, fringing, patch and atoll reefs, as well as unique, picturesque diving sites such as the Blue Hole.

Reef ecology

The Belize Barrier Reef complex is a well-developed barrier reef that runs parallel to the coast starting in the north at Rocky Point, where the reef meets the land, and ending in the hook shaped reefs of Sapodilla Cayes in the south. The main barrier reef is structurally divided into three provinces: the Northern, Central and Southern. Within the inner lagoon of the barrier reef the sheltered waters support extensive seagrass beds, patch reefs and shoals full of marine life that have enabled the establishment of sand and mangrove islands called cayes. The shelf lagoon of the southern province contains numerous rhomboid shoals called faroes or shelf atolls. These faroes are similar in structure to atolls but are much smaller and found only in the shallow barrier lagoon. Combined, they provide habitats for the West Indian manatee (*Trichechus manatus manatus*,) bottlenose dolphins, American crocodile (*Crocodylus acutus*) and sea turtles which attract many visitors to these waters.

In general, the barrier reef consists of three main ecological zones: the back reef, reef crest and fore reef (shallow and deep). The reef crest is a high-energy surf zone with shallow ramparts, often exposed and primarily made up of coral rubble and *Acropora palmata* (elkhorn coral). On the leeward side of the reef crest is the back reef zone, a shallow, fairly sheltered area containing a variety of sediment types and patches of corals near to the crest which gradually deepens towards the seagrass dominated inner lagoon. The fore reef is immediately seaward of the reef crest and is separated into a shallow fore reef up to 5 m, which is characterised by an *A. palmata* dominated zone often found in spurs or buttresses. Much of the *A. palmata* is now rubble due to white band disease which decimated this species in the 1990s. Further seaward is the deep fore reef (between 5–15 m in depth) characterised by a mixed coral zonation once dominated by *Orbicella spp.* complex. Beyond the deep fore reef is the fore reef slope that generally begins at the fore reef wall or drop off around 20 m and extends down to 100 m into the deep reef. The Laughing Bird Caye Faro is a unique and well-known reef formation within the main barrier reef of Belize (Figure 7.2). The island supports a nesting population of the laughing gull and is a perfect example of a faro found only within the shallow barrier reef shelf. It is also a popular tourism destination for snorkellers and divers. Gladden Spit, one of many reef promontories or "elbows" along the barrier reef, supports spawning aggregations of reef snappers and groupers (Figure 7.3). It is also a popular dive and snorkel location due to the annual migration of whale sharks during the snapper-spawning season when these gentle giants come to feed on fish eggs.

FIGURE 7.2   Laughing Bird Caye Faro

Source: Photo courtesy of Doug Perrine, CAVU.

FIGURE 7.3   Gladden Spit Marine Reserve

Source: Photo courtesy of Doug Perrine, CAVU.

The other major portion of the Belize Barrier Reef complex is made up of three oceanic atolls: Lighthouse Reef, Turneffe and Glover's Reef Atolls situated 45 km, 7 km and 20 km east of the barrier reef, respectively (Perkins & Carr, 1985). These atolls rest upon submarine limestone ridges with reefs that have similar zonation as that found on the main barrier reef. The eastern or windward side of the atolls generally have more well-developed reef formations while the western or leeward sides represent less developed reef zonation with deeper zones due in part to reduced wave energies. Lighthouse and Glover's are spectacular atolls closest in structure to the true volcanic atolls of the Pacific, with typical sandy cayes sitting atop the reef margins and unique features including the Great Blue Hole in Lighthouse and about 1,000 patch reefs in the inner lagoon of Glover's. The Great Blue Hole, located 100 km off the coast of Belize, is a 300 m wide and 125 m deep underwater limestone cave with numerous stalactites and stalagmites. Coral reefs encircle the shallow, light, turquoise waters of the lagoon and coral walls extend down to over 30 m. The Great Blue Hole is frequented by various species of sharks and is the most popular dive site in Belize (Figure 7.4). Half Moon Caye in Lighthouse has the only long-standing nesting colony for the red-footed booby (*Sula sula*). Unlike the other two atolls, Turneffe Atoll is dominated by extensive mangrove islands along the inner margins of the atoll, creating a large

FIGURE 7.4 Blue Hole, Lighthouse

Source: Photo courtesy of Belize Audubon Society.

FIGURE 7.5 Grassy Caye, Turneffe Atoll

Source: Photo courtesy of Turneffe Atoll Trust.

central lagoon which supports a healthy population of the American crocodile (*C. acutus*) and breeding grounds for the West Indian manatee (*T. manatus manatus*). Grassy Caye in Turneffe Atoll is one of numerous mangrove cayes characteristic of the atoll that support bird rookeries and provide critical fish nurseries, making Turneffe one of the most productive fishery areas in Belize (Figure 7.5).

Tourism

The tourism industry is one of the cornerstones of the Belizean economy. There is no definitive estimate of the sector's contribution to the gross domestic product (GDP) but most assessments indicate that tourism directly contributes anywhere from 15% to 25% of the GDP and accounts for about 28% of total employment (Belize Tourism Board, 2011). The total contribution from tourism in 2014 was BZD$1,311.8 million, 39% of the GDP (World Travel & Tourism Council [WTTC], 2015). In 2015 over 1 million cruise and overnight tourists visited Belize. The expected trend up to 2025 is for a gradual increase. Belize's tourism industry is eco-tourism and culture based and depends on the country's natural and cultural assets, in particular its marine resources including the Belize Barrier Reef.

The barrier reef complex with its high diversity of habitats and species is the backbone of the country's tourism industry. Studies show that over 50% of visitors come for the reef and marine life that it supports (Belize Tourism Board, 2012), with 67% of visitors engaging in snorkelling and 53% visiting marine protected areas such as the Half Moon Caye, Great Blue Hole and Hol Chan MPAs. While the majority of visitors snorkel and enjoy the vibrant and abundant marine life of Belize, a large proportion (25%) also engage in scuba diving. Belize has repeatedly been rated as one of the top diving destinations within the Caribbean and Central America due to locations such as the Great Blue Hole and the excellent dive sites across MPAs, including Hol Chan, Glover's Reef, Turneffe and Lighthouse Reef. Other marine activities include snorkelling along the main barrier reef and swimming with sharks and rays in the Hol Chan Marine Reserve. Whale shark tours at Gladden Spit and the Great Blue Hole dives are a special attraction for both divers and snorkellers (see Chapter 9 for further discussion of whale shark tourism). Aside from snorkelling and diving, approximately 23% of visitors to the reef come for sport fishing and general island activities such as sun bathing on the cayes. Given this high level of dependency on the reef, the Belize Barrier Reef is a priceless tourism resource for the country and without this extensive, unique and productive system, Belize's economy and the livelihood of its people would be significantly affected.

Ecosystem services

The Belize Barrier Reef contains a variety of coral reef types that support a wealth of biodiversity and, along with associated seagrass and mangrove ecosystems, provides critical ecosystem services (see Table 7.1). These ecosystem services include fisheries production, shoreline protection, waste and pollution regulation, water filtration, biodiversity, carbon storage, recreation and nursery functions (Hargreaves-Allen, 2011). An economic valuation study conducted by the World Resources Institute determined that Belize's barrier reef contributed benefits between US$135–176 million to tourism, US$13–14 million to fisheries and US$120–180 million to avoiding damages from storms (Cooper et al., 2009). These calculations were based on the nursery functions and fishery production that the coral reefs and associated systems contribute to sustain healthy and productive fisheries, such as the queen conch and spiny lobster export fisheries. It also looks at the huge biodiversity of life that catches the eye and touches the heart of people who visit, creating a vibrant tourism market for snorkelling, diving and sport fishing tours. Lastly, it includes the critical function of coastal protection through safeguarding urban settlements and tourism developments from hurricanes, storms and coastal erosion. The reefs, mangroves and seagrasses form an interconnected system that plays a crucial role in maintaining the health of habitats and species and has high conservation and cultural values.

The barrier reef directly supports local communities residing in coastal towns and villages and other users such as fishers, local tour operators, aquaculture farms, developers and hotels. Without these services provided by the reef, the well-being and livelihoods of these people would be significantly and negatively affected through coastal erosion, destruction of property from storms, flooding, loss of biodiversity, poor water quality, lack of productive fisheries and reduction in tourism visitors due to poor reef health. Besides use values such as goods and services, reefs also provide non-use values. These include existence values, which for the Belize Barrier Reef is the knowledge that this global treasure exists; option

TABLE 7.1 Ecosystem service values of Caribbean coral reefs

Ecosystem service values (US$ millions)

	Tourism	Fishing	Shoreline protection	Amenity (house prices)	Local recreation and culture
Tobago	100–130	0.8–1.3	18–33		
St Lucia	160–194	0.4–0.7	28–50		
Belize	135–176	13–14	120–180		
US Virgin Islands	103	3	7	37.1	51.1
Bermuda	405.9	4.9	265.9	6.8	36.5
Turks and Caicos	18.2	3.7	16.9		
Caribbean	2,100	310	700–2,200		

Source: Adapted from Mumby et al. (2014).

values that looks at having the reef in good health so that it is available for future use, including the potential use of marine species in pharmaceuticals; and bequest values that look at benefits derived from the barrier reef for this current generation while leaving it intact for future generations or other groups. However, these values have been very difficult to determine.

Threats and impacts

The health of the Belize Barrier Reef is a very good indicator of the level of impacts to the reef. The 2015 Mesoamerican Reef Report Card showed that Belize's reefs are in poor health, with a reef health index of 2.5 (Healthy Reefs, 2015). The reef health index looks at live coral cover, fleshy macroalgal cover, key herbivorous fish abundance and key commercial fish abundance. Although the overall health of the reef is poor, coral cover is reported as fair, with an average of 15% cover, and there are certain areas that stand out above others with the highest live coral cover, such as the offshore reefs within Glover's, Turneffe and Lighthouse, of greater than 30%. Herbivorous fish biomass was fair across the reef due largely in part to a 2009 legislation that protects parrotfish; however, commercial fish biomass remains poor suggesting continued overexploitation and inadequate legislation. This current health status of the reef is as a result of a series of anthropogenic impacts compounded by global climate change. The major threats to the reef include overfishing and destructive fishing, unsustainable coastal development, mass tourism, uncontrolled tourism activities, unsustainable land use practices and oil exploration. Belize's Barrier Reef has been classified as being at medium risk from these various threats (Burke et al., 2011). Despite having a system of 14 MPAs with approximately 13% of the reef under protection and management, the threats to the reef have not been successfully minimised, bringing into question whether these MPAs are being effectively managed.

Overfishing and destructive fishing

Overfishing of commercially important species such as spiny lobster, conch, snappers and groupers has resulted in declining populations evidenced through reduced catches and stocks. This is due to the open access nature of Belize's fisheries since the 1960s, where there have been little to no limits on fishing effort. In addition, destructive fishing practices such as the continued use of gill nets on the reef and along estuaries and fishing at spawning sites have most likely resulted in a decline in commercial fish species such as snappers and groupers (Heyman & Requena, 2002; Gale, 2012). The population of endangered Nassau grouper has been significantly reduced due to overfishing of the species during spawning at their aggregation sites on reef promontories. Recent regional and global demand for shark meat and fins has led to unsustainable harvesting of sharks causing declining populations (Chanona, 2015; San Pedro Sun, 2016).

Unsustainable coastal development

Belize, a developing country with a small population and wealth of natural resources, is ideally placed to attract investment and development particularly in its coastal environment. This coastal development is often not sustainable due to a lack of adequate policy and poorly regulated legislation. Between 1980 and 2010 there was very low mangrove habitat loss (Cherrington et al., 2010); however, since 2010 there has been an increase in the clearing of mangroves for coastal development. This unsustainable and unregulated development has led to the removal of mangroves from high diversity areas, resulting in loss of critical fish habitat and biodiversity and increase risk of erosion (McKee & Vervaeke, 2009), dredging and removal of seagrass habitat which results in the loss of fishery habitats and sedimentation of coral reefs, and increase in nutrient run-off causing eutrophication from improper sewage treatment. Unregulated development and clearing of mangrove islands have led to the Belize Barrier Reef being placed on the World Heritage Site's Danger List in 2009 (Krohn, 2007; UNESCO World Heritage Centre, 2009). The southern village of Placencia has also suffered impacts from large unsustainable coastal developments that have contributed to a decline in the health of the unique and important Placencia lagoon (Ortiz, 2015).

Mass tourism and uncontrolled tourism

The coral reefs of Belize have attracted numerous tourists since the 1990s with increased visitation in the twenty-first century due to its popularity and the promotion of Belize as a cruise destination (Diedrich, 2006; Belize Tourism Board, 2011).

This mass tourism has brought negative impacts to the reefs, including an increase in solid waste and a high volume of visitors to the reef which results in physical impacts including stepping on and breaking of corals. The overnight sector has also grown resulting in increased uncontrolled tourism activities. Most of the popular marine sites within and outside MPAs do not have limits of acceptable change. Therefore, tourism is uncontrolled in some areas, with numerous tour operator boats, snorkellers and divers in the water simultaneously, creating a high potential for negative physical impact to the reef. Sites such as the Hol Chan Marine Reserve receive one of the largest numbers of visitors on an annual basis – over 65,000 visitors (Belize Tourism Board, 2012). This level of visitation generates negative impacts on the health of the ecosystems.

Unsustainable land-use practices

One of the greatest threats to the Belize Barrier Reef is from unsustainable land-use practices on the mainland through banana, citrus and sugar farms that have traditionally utilised pesticides and fertilisers in their systems. A recent trend in the expansion of poultry, livestock and arable farms for corn, rice and beans has resulted in extensive land clearing and continued use of pesticides and fertilisers on these highly mechanised farms. The impacts arising from such unsustainable agricultural practices are loss of habitat and watershed degradation resulting in increased run-off into streams and rivers and soil erosion. These lead to run-off of fertilisers, pesticides and sediments into the coastal environment and into the reef, impacting water quality and marine life. This further leads to algal blooms such as the unprecedented bloom in 2011 which spread from the central barrier reef southward to Placencia and affected offshore reefs at Turneffe and Glover's. The shrimp industry (aquaculture), which is focused mainly in the southern plain of the country, has also contributed to nutrient run-off from farm effluents (Ledwin, 2010).

Oil exploration

Oil exploration is another threat to the barrier reef with the potential for significant and irreversible impact. Belize's land and sea areas have been divided into oil concession blocks that have been surveyed since the 1950s, but with the commercial oil find on the mainland in 2005, oil exploration on land has increased significantly with interest in exploring for oil extending to the sea. Based on the structure of Belize's reefs, oil exploration would target the shallow reef areas and offshore platforms. Given the sensitive and vulnerable nature of the barrier reef, any oil spill or related industry work would have a significant negative impact on the health of the barrier reef and its marine life. Any impacts from oil exploration would quickly spread and affect large tracts of the reef. Belize would stand to lose not only the coral reef but also over BZD$690 million in total income from tourism and fisheries, negatively affecting its GDP and economic standing (WTTC, 2015; Belize Fisheries Department, 2014).

Global climate change

Besides these localised impacts, the coral reefs of Belize are also under threat from the impacts of climate change. Globally, sea surface temperature is increasing. Warmer sea surface temperatures stress corals and result in mass bleaching events, which leads to further loss and degradation of coral reefs. Belize's reefs attract thousands of tourists annually for snorkelling and diving activities. The barrier reef has already suffered significant impacts from mass bleaching events in 1995 and 1998, which resulted in 52% of corals showing bleaching and 25% of colonies showing partial mortality (McField, 1999). In addition to warming, the world's oceans have also been slowly but steadily increasing, with more rapid increases predicted in the next three decades. Sea level rise in the Caribbean is following the global trend at around 1.5–3 mm a year (Bindoff et al., 2007), with studies predicting sea level rise of 0.5–2.15 m by 2100 (Rahmstorf, 2007). This poses a significant risk of flooding, salt-water intrusion and erosion. Warmer sea surface temperatures are also associated with the increasing frequency and intensity of tropical cyclones. Hurricanes and tropical storms in the North Atlantic seem to have increased in intensity over the last 30 years, with larger peak wind speeds and heavier precipitation associated with increased sea surface temperatures. It is likely that tropical cyclones will become more intense and this will pose significant threat to coastal populations due to flooding, erosion and destruction of infrastructure. Belize's reefs have been impacted in the last 15 years by at least four tropical storms and hurricanes that have caused varying degrees of damage to the reef and associated ecosystems. The most recent storms were Hurricane Richard in 2010 and Hurricane Earl in 2016, both of which affected the central barrier reef and Turneffe Atoll.

Impacts on tourism

If these threats are not addressed in the near future, Belize's tourism industry will suffer major impacts that will have far reaching economic implications. The National Sustainable Tourism Master Plan of Belize (Belize Tourism Board, 2011) highlighted several challenges to the tourism industry, some of which are further exacerbated by climate change. Inadequate natural asset management, mainly due to a lack of awareness, knowledge and financial limitations, will lead to degradation of the natural resources and in turn the destination. Insufficient waste disposal and sewage systems result in pollution, which has a negative effect on marine resources and the quality of the tourism product. A lack of proper land-use planning and land-use regulation has led to haphazard and inadequate urban development, the results of which include beach erosion, nutrient pollution and land-use conflict. Finally, a lack of public awareness of the importance of the Belize Barrier Reef to the livelihoods and well-being of the citizens has led to unsustainable use of natural resources including overfishing, destruction of the reef and degradation of MPAs.

Management strategies

Despite the numerous challenges facing the Belize Barrier Reef, all is not lost. There have been excellent efforts at conserving this global treasure since the late 1990s, with significant achievements made in the last 15 years. In 1998 the Coastal Zone Management Act was passed establishing a coastal zone management framework. In 2016 the government finally adopted an integrated Coastal Zone Management Plan that sets the blueprint for achieving sustainable use of its marine resources. The fishing industry joined efforts with the conservation community to lobby for protection of fish spawning aggregations, and in 2003 legislation was passed to protect 13 spawning aggregations. Thirteen years later, sites that have had effective enforcement and management are showing stabilisation and recovery of numbers, such as Sandbore in Lighthouse Reef, North East Point at Glover's Reef and Caye Glory on the main barrier reef. Combined with the protection of spawning aggregation sites, a size limit and closed season was implemented for Nassau grouper in 2009 to provide further protection for the species. Also in 2009, a ban was placed on spear fishing within MPAs and on the fishing of herbivorous species. Shrimp trawling can be very destructive and was banned in Belize's waters in 2011. In 2012 the MPA system coverage was increased with the addition of the country's largest MPA, the Turneffe Atoll Marine Reserve (Healthy Reefs, 2015).

Fast forward to 2015, the Belizean government approved the move from an open access fishery to a managed access structure which uses catch shares to empower fishers to become better stewards of their resources. This approach gives fishers access to their fishery through designated fishing areas, thereby allowing them better fishing returns and reducing overall fishing effort. This new fisheries management approach is being implemented countrywide in 2016 following a 4-year pilot of the programme in two MPAs – Glover's Reef and Port Honduras Marine Reserve. Another management success has been the accreditation of some of the major shrimp farms in 2015 into the Aquaculture Stewardship Council (ASC) through a certification programme led by the World Wildlife Fund in partnership with the shrimp industry. This certification holds the farms to maintain environmental standards to treat and regulate effluent levels and reduce habitat loss, among others. In addressing the concerns about oil exploration and its impact to the Belize Barrier Reef, which contributed to the WHS Danger Listing, the government announced a ban on oil exploration in the entire World Heritage Site in December 2015. This will enable the country to assess its ability to effectively and safely manage oil exploration activities while safeguarding the barrier reef. In 2015 the National Climate Change Policy, Strategy and Action Plan was adopted. This plan aims to mainstream climate change policy across all sectors of the country and implement mitigation and adaptation measures to help address the negative impacts of climate change, which are already affecting the vulnerable coastal areas and coral reefs of Belize.

While there are good management strategies being implemented for the conservation and sustainable management of the Belize Barrier Reef, there are still major gaps remaining to achieve a comprehensive management framework. Key among these is the National Land Use Policy and Plan that is crucial in defining and regulating land use through various zonation and prescriptions for sustainable land-use practices. This policy will inform the development of revised land-use legislation, which is expected to address the removal of mangrove islands as land. Another critical need is the approval of the updated Fisheries Bill, a modern and innovative legislation that will ensure reef fisheries are more sustainably managed. This draft bill has been awaiting the political will to approve it since 2012. The revised mangrove legislation, which grants greater protection for mangroves and stiffer penalties for violations, is also associated with the land-use policy and plan. This piece of legislation has also been awaiting approval and adoption since 2010. Finally, through country-wide consultations with stakeholders, an expansion of MPA replenishment zones has been finalised and, when approved, will enable the National Protected Area System to be more effective in conserving critical habitats and allowing fish stocks to recover.

The vision for the Belize Barrier Reef moving forward is for an effective system of MPAs that is properly managed and supported through the integrated Coastal Zone Management Plan, a modern fisheries bill, and an updated and sustainable land-use plan and legislation ensuring the removal of the Belize Barrier Reef World Heritage Site from the UNESCO Danger List and the sustainable use of this globally significant natural wonder. In achieving this vision, the country is continuously challenged with limited financial and human resources, as is the case with all small developing countries. There are large gaps in scientific capacity and data to inform decisions. There are a few non-governmental organisations that have some scientific capacity and that help contribute data. The University of Belize, through its Environmental Research Institute, has also contributed significantly to building scientific capacity through research and training since 2010. Despite this, there still remains significant limitations in accessing funding and information to make informed decisions regarding the sustainable use and conservation of the Belize Barrier Reef. It is anticipated that with time the government, along with the national university and other partners, will work to address these problems.

References

Belize Fisheries Department. (2014). *Fisheries Statistics 2014*. Belize City: Belize Fisheries Department.

Belize Tourism Board. (2011). *National Sustainable Tourism Master Plan for Belize 2030*. Belize City: Belize Tourism Board.

Belize Tourism Board. (2012). *Travel and Tourism Statistical Digest*. Belize City: Belize Tourism Board.

Bindoff, N.L., Willebrand, J., Artale, V., Cazenave, A., Gregory, J., Gulev, S., & Hanawa, K. (2007). Observations: Oceanic climate change and sea level. In *Climate Change 2007: The Physical Science Basis*. Contribution of Working Group I to the Fourth Assessment Report of the Intergovernmental Panel on Climate Change. Cambridge: Cambridge University Press.

Burke, L., Reytar, K., Spalding, M., & Perry, A. (2011). *Reef at Risk Revisited*. Washington, DC: World Resources Institute.

Chanona, J. (2015, February). Death of rare shark highlights need for gillnet ban: Scalloped hammerhead drowns in Hopkins fisherman's gillnet. *Oceana Press Release*. Retrieved from http://belize.oceana.org/press-center/press-releases/death-rare-shark-highlights-need-gillnet-ban

Cherrington, E.A., Hernandez, B.E., Trejos, N.A., Smith, O.A., Anderson, E.R., Flores, A.I., & Garcia, B.C. (2010). *Technical report: Identification of threatened and resilient mangroves in the Belize Barrier Reef System*. Panamá: Water Center for the Humid Tropics of Latin America and the Caribbean (CATHALAC).

Cooper, E., Burke, L., & Bood, N. (2009). *Coastal capital Belize: The economic contribution of Belize's coral reefs and mangroves*. World Resources Institute Working Paper. Washington, DC: World Resources Institute.

Diedrich, A. (2006). Assessment of the Impacts of Tourism Development in Coastal Communities in Belize. PhD Dissertation. University of Rhode Island.

Gale, K. (2012, August 17). Stop the destruction of Belize's fisheries – ban gill netting! *Amandala Newspaper*. Retrieved from http://amandala.com.bz/news/stop-destruction-belizes-fisheries-ban-gill-netting/

Hargreaves-Allen, V. (2011). The economic value of ecosystem services in the Exumas Cays: Threats and opportunities for conservation. Conservation Strategy Fund, June 2011.

Healthy Reefs. (2015). Healthy reefs for healthy people – Mesoamerican Reef: An evaluation of ecosystem health. 2015 Report Card.

Heyman, W., & Requena, N. (2002). *Status of multi-species spawning aggregations in Belize*. Belize: The Nature Conservancy.

Krohn, S. (2007, December 13). Dredging in Pelican Cayes threatens World Heritage Site. *Channel Five Belize News*. Retrieved from http://edition.channel5belize.com/archives/5822

Ledwin, S. (2010). Assessment of the ecological impacts of two shrimp farms in Southern Belize. MSc Dissertation. School of Natural Resources and Environment, University of Michigan.

McField, M.D. (1999). Coral response during and after mass bleaching in Belize. *Bulletin of Marine Science, 64*(1), 155–172.

McKee, K.L., & Vervaeke, W.C. (2009). Impacts of human disturbance on soil erosion and habitat stability of mangrove-dominated islands in the Pelican Cays and Twin Cays ranges, Belize. *Smithsonian Contributions to the Marine Sciences, 33*, 415–428.

Mumby, P.J., Flower, J., Chollett, I., Box, S.J., Bozec, Y., Fitzsimmons, C., . . ., & Williams, S.M. (2014). *Towards reef resilience and sustainable livelihoods: A handbook for Caribbean coral reef managers*. University of Exeter, Exeter.

Ortiz, D. (2015, September 24). Placencia Lagoon under sustained sewage pressure. *Channel Five Belize News*. Retrieved from http://www.7newsbelize.com/sstory.php?nid=33803

Perkins, J.S., & Carr, A.C. (1985). The Belize Barrier Reef: Status and prospects for conservation management. *Biological Conservation, 31*, 291–301.

Rahmstorf, S. (2007). A semi-empirical approach to projecting future sea-level rise. *Science, 315*(5810), 368–370.

San Pedro Sun. (2016, February 24). Unsustainable shark fishing is killing the Belize marine ecosystem. *San Pedro Sun Newspaper*. Retrieved from http://www.sanpedrosun.com/conservation/2016/02/24/unsustainable-shark-fishing-is-killing-the-belize-marine-ecosystem/

UNESCO. (1996). *Belize Barrier Reef Reserve System*. UNESCO World Heritage Centre. Retrieved from http://whc.unesco.org/en/list/764/

UNESCO World Heritage Centre. (2009, June 27). Belize Barrier Reef Reserve System and Colombia's Los Katios National Park enter UNESCO's danger list. *UNESCO World Heritage Centre News*. Retrieved from http://whc.unesco.org/en/news/530/

World Travel & Tourism Council (WTTC). (2015). *Travel and tourism economic impact 2015 – Belize*. London: World Travel & Tourism Council.

8

CORAL REEFS OF THE FLORIDA KEYS

The threats of a changing sea

Roberta Atzori and Alan Fyall

Introduction

Often compared to tropical rainforests for their biological diversity, the Florida Keys' coral reefs have traditionally been a highly productive ecosystem. They provide habitat for a large variety of fish, crustaceans and other sea creatures. The reefs are a vital part of the Florida Keys' tourism industry, supporting recreational fishing, diving and snorkelling businesses, and have an estimated asset value of US$8.5 billion (Johns, Leeworthy, Bell, & Bonn, 2001). Despite their great economic and recreational value, however, the reefs today are severely threatened by numerous factors. Once damaged, coral reefs are more vulnerable to disease and less able to support the marine life and the local community and businesses that depend on them. As a result, an impacted reef loses value as a tourism resource. Increased sea surface temperatures, aggravated by ocean acidification, are reinforcing constant human-caused stresses to the Keys' coral reefs, such as land-based pollution, unsustainable fishing practices and direct physical damage from boating, recreational fishing and diving. The objective of this chapter is to examine the factors that are currently threatening the Keys' coral reefs ecosystem and to provide an overview of how these challenges have been addressed by the local government and community so far. The health of the coral reefs in the Florida Keys is clearly in danger with there being a greater need than ever before for a solution that ensures their long-term health and sustainability.

The Florida Keys' coral reefs

The Florida Keys lie between the Gulf of Mexico and the Atlantic Ocean, consisting of about 102 square miles of land. They form an elongated chain of about 1,700 low-lying islands that extends for more than 220 miles in length, from the south-eastern tip of the Florida peninsula (about 15 miles south of Miami) to the Dry Tortugas (about 67 miles west of Key West). The highway that connects all the islands with the mainland – known locally as the Overseas Highway – is the southern-most stretch of US 1 that runs along the entire Atlantic coast from Maine south (Craig, 2003). The Florida Keys, the exposed portion of an ancient lime-stone or coral reef, are rocky islands. Although some sandy beaches exist, they are not common.

Just about 6 miles offshore on the Atlantic side of the Keys are the coral reefs of the Florida Keys, starting near Miami and extending south-west to the Dry Tortugas. Popularly considered the third longest coral reef in the world – after the Great Barrier Reef in Australia and the Yucatan/Belize reef system – the Florida Keys' reef is the only living coral reef barrier in the continental United States. It started to form during the Pleistocene Period, 100,000–125,000 years ago, resulting in an almost continuous tract of outer reefs and patch reefs. The warm, shallow waters of the Florida Keys, ideal for coral reef growth, are typically 15–30 feet deep (Hoffmeister & Multer, 1968). The climate of the Keys, due to the Gulf Stream and the tempering effects of the Gulf of Mexico, has a mild, trop-ical-maritime climate, defined as tropical savanna. No record of frost or snow has ever been reported in Key West or in the Lower and Middle Keys, although ice has been reported in the Upper Keys. There are two main seasons in the Florida Keys: the dry season from November through May, which usually receives no more than 25% of annual rainfall, and the wet season from June through October in which showers and thunderstorms tend to concentrate.

The importance of coral reefs in the Florida Keys

Coral polyps of the Florida Keys are enjoyed by millions of snorkellers and divers, but they are also of vital importance for the whole coral reef ecosystem. The three main types of habitats that interact in the Florida Keys' reef ecosystem are mangrove forests, seagrass beds and coral reefs. Mangroves, growing along the shoreline, protect the coast and create a perfect nursery area for young fish and invertebrates. Seagrass beds grow in the shallows and provide food, shelter and breeding grounds to a variety of fish and invertebrates. The coral reefs promote habitat formation, increasing the biodiversity of the ecosystem (Craig, 2003). The Florida Keys reef ecosystem is home to over 50 species of corals and over 150 spe-cies of fish (Hoffmeister & Multer, 1968). Overall, the reef environment supports more than 6,000 species of plants, fish and invertebrates, including several threat-ened and endangered species.

Coral colonies are made up of multitudes of genetically identical tiny living ani-mals called polyps. These polyps have microscopic algae called zooxanthellae that live within their tissue and give corals their typical rainbow colours. The zooxanthellae, working like an internal symbiotic vegetable garden, carry out photosynthesis and provide energy that help corals create reef structures. The corals are the main build-ing block of coral reefs, building these enormous structures through the construction of their skeletons made of calcium drawn from seawater. When subject to stress, the corals expel their zooxanthellae, making the corals more vulnerable to disease. Since the stunning colours of corals are due to zooxanthellae, when zooxanthellae are lost, coral appears white or, as commonly known, "bleached". Bleaching is a stress response of corals that occurs when the coral–algae relationship breaks down. Unless the stress on the coral is reduced and the zooxanthellae can be replaced, the coral will die. Bleached corals are still alive, and if the environmental conditions return to normal in less than about 6 weeks, the corals can regain their zooxanthellae and return to a healthy state. However, if the stressors are severe or protracted in time, bleaching can lead to the death of corals. Bleached corals are more vulnerable to disease, predation and death because they are deprived of their principal energy source (Hughes et al., 2003). If corals cannot continue living and multiply, reefs will disappear due to several factors, including wave erosion and predators. The death of corals and the subsequent loss of the reefs would result in a chain reaction affecting the entire Florida Keys ecosystem.

The health of its coral reefs is critical for the Florida Keys. First, they provide coastal protection by reducing wave energy from storms and hurricanes, a particularly important factor during Florida's hurricane season. By reducing the strength of waves, the reefs protect people living near the coast from damage, erosion and flooding. Second, the fishing and tourism industries rely on the reefs to sustain their activities. Millions of tourists and local residents enjoy scuba diving, snorkelling and fishing on the reefs. The loss of the region's coral reefs would result in major losses of revenue and resources from tourism and fishing in the Florida Keys.

Tourism and the economy of the Florida Keys

Tourism is the economic engine of the Florida Keys. Coral reefs support local jobs and fisheries by providing opportunities for tourism and recreation. Activities based on boating, diving and recreational fishing bring in billions of dollars into the economy every year (Park et al., 2002). A socioeconomic study commissioned by the National Oceanic and Atmospheric Administration ([NOAA], Johns et al., 2001), estimated that Southeast Florida reefs have an asset value of about US$8.5 billion, generating US$4.4 billion per annum in local sales, US$2 billion in local income and sustaining over 71,000 local jobs. The Florida Keys host several million snorkellers, divers, recreational fishers and boaters every year. Approximately 4.5 million tourists visit the Keys annually. Tourism is the largest sector in the islands, with 54% of jobs and 60% of all spending attributed directly and indirectly to tourism (Monroe County Tourist Development Council, 2017). So many people have become charmed by its coral reefs that Florida is regarded as one of the top dive destinations in the world. The total reef-adjacent tourism value is US$850.6 million, which is nearly double that of Australia's reef-adjacent tourism value (US$473.1 million; Spalding et al., 2017).

In addition to diving and snorkelling, recreational fishing is one of the most popular leisure activities in and around the Florida Keys' coral reefs. The Florida

Keys are called "the sport fishing capital of the world". Recreational fishers catch a variety of crustaceans, such as spiny lobster, Tortuga pink shrimp and stone crab, as well as big game fish such as marlin, tarpon and bonefish. At the same time, the Florida Keys' coral reefs support a very important commercial fishing industry. The area is one of the richest fishing grounds in the Gulf of Mexico, and the National Marine Fisheries Service estimates the commercial value of US fisheries from coral reefs to be over US$100 million (NOAA, 2017).

What is threatening the Florida Keys' coral reefs?

The combined effects of numerous stressors, however, are threatening the coral reefs of the Florida Keys. The consequences of inconsiderate tourism, the pressure of overfishing, the damage of large-scale development and the more recent impacts of changed ocean composition and temperatures are contributing to the depletion of the reefs. This combination of stressors is causing corals to be infected by diseases such as black, white and yellow band disease, along with the Aspergillosis fungus, which attacks purple sea fans.

According to Reef Relief, the biggest threat to Florida Keys' coral reefs is the decline in water quality. Corals need clean, nutrient-free waters to prosper. A healthy coral reef has 30–40% live coral coverage. However, in the Florida Keys, coral coverage is now reduced to a dangerous 3% (Reef Relief, 2017). Coral proliferation has declined due to the lack of healthy coral colonies and clean water. The Keys coral reefs have undergone what has been defined a "phase shift", in which hard corals have been mostly replaced by soft corals and sponges that tolerate poor water quality. The physical structure of the reef is changing as the rocky coral formations are replaced by dead boulders, encrusting sponges and algae, leaving reduced habitat for all the sea life that lives upon the reef structure. Nuisance algae have now replaced areas where corals once flourished. Visibility has plunged. It is rare nowadays that visibility exceeds 80 feet at Key West-area reefs, whereas before visibility used to be over 100 feet and locals referred to it as "gin clear". For instance, the new shipwreck in the Florida Keys – the *Vandenberg* – can only be viewed with underwater lights when immediately upon it due to poor visibility at 140 feet (Reef Relief, 2017).

The impacts of climate change

The causes behind coral bleaching may be attributed to the effects of climate change on the ocean system. Strong evidence has pointed to increased sea surface temperatures as one of the main causes (Glynn, 1993). Moreover, ocean acidification resulting from the accumulation of carbon dioxide is today imposing additional stress on the coral reefs (Karl et al., 2009). The combination of warmer and more acidic waters makes coral ecosystems among the most threatened marine habitats in Florida and the world (Hoegh-Guldberg et al., 2007).

Increased sea surface temperatures

Increased sea surface temperatures stress corals, causing coral bleaching and favouring a variety of marine diseases (Hoegh-Guldberg, 2010). Two types of heat-related stress can generate bleaching – the first is a short-term, acute temperature stress (i.e. several days of particularly high water temperatures between 25°C and 32°C) and the second is a cumulative temperature stress (weeks of consistent moderately high water temperatures). Corals are very sensitive to small temperature changes. Coral bleaching events that have occurred around the world have been attributed to sea surface temperatures rising as little as 1°C higher than the normal average monthly maximum temperature of the hottest months of the year (Goreau & Hayes, 1994). Coral communities worldwide have different heat stress thresholds that usually generate bleaching in response to heat stress. The Atlantic Oceanographic and Meteorological Laboratories (AOML) scientists identified specific patterns of increased water temperatures on coral reefs that tend to precede bleaching events. The index that is the most reliable indicator for bleaching events in the Florida Reef Tract is the maximum monthly sea surface temperature >30.5°C (86.9°F). When sea surface temperatures rise and stay above this temperature threshold, bleaching events are likely to occur. For instance, in the summer of 2014 several local divers noticed that below the ocean surface corals were bleaching. In August, the Coral Bleaching Early Warning Network, a community-based monitoring programme, received 34 reports describing paling or partial bleaching and a further 19 reports indicating significant bleaching (Jones, 2006). Data gathered from the Molasses Reef C-MAN station (located southeast of Key Largo) showed that the winter of 2014 was the warmest on record since the station started recording data in 1988. Comparably, the second warmest winter recorded in the Keys was the winter of 1996/97, which preceded the worst bleaching events ever documented in the Florida Keys – the summers of 1997 and 1998 (Jones, 2006).

Ocean acidification

Within a few years, ocean acidification – labelled "the evil twin of global warming" – has emerged as another major threat to coral reefs. The term "twins" refers to the fact that both issues are the result of greenhouse gas emissions. Ocean acidification is aggravating the impact of increased water temperatures. The process of ocean acidification occurs as the oceans absorb carbon dioxide (CO_2) from the atmosphere, activating a chemical reaction that reduces the ocean water pH. As more fossil fuels are burned and the concentration of CO_2 in the atmosphere increases, the uptake of CO_2 becomes excessive for the ocean to process as it used to do, causing the water to become more acidic. More acidic waters affect the calcareous structure of the coral skeletons, reducing their strength and making them more fragile and prone to damage.

Until recently, projections based largely on laboratory studies predicted that ocean pH would not decrease enough to cause coral reefs to start dissolving until

2050–2060 (Muehllehner et al., 2016). However, the results of the study conducted by Muehllehner et al. (2016) show that the limestone that forms the basis of coral reefs along the Florida Reef Tract is already dissolving during the autumn and winter months on several reefs in the Florida Keys. As explained by Chris Langdon, one of the study's authors, in the natural scheme of things, during spring and summer months the ocean's environmental conditions such as water temperature, light and seagrass growth are favourable for the formation of coral limestone. On the other hand, during autumn and winter months, low light, temperature conditions and annual decomposition of seagrass result in a slower pace or minor loss of reef growth. However, with the reef dissolution resulting from ocean acidification, the natural summer growth cycle of coral is no longer able to offset the winter cycle, and the loss of limestone is exceeding the amount of limestone that corals are able to produce annually.

The study findings showed that the upper Florida Keys – closer to Miami's polluted coast – were the most impacted, with reefs gradually improving heading south. For instance, Fowey Rocks, a popular dive spot in Biscayne National Park off Key Biscayne, is already disappearing. Since the data for the study was collected in 2009–2010, and the worst bleaching years recorded in the Florida Keys occurred in 2014–2015, a new analysis would be able to assess whether current conditions have worsened.

The impacts of tourism

Over the years, tourism activities have harmed the Florida Keys' coral reefs. The physical impact of many boats, divers and fishermen represent another major stressor for the health of corals (Bruckner et al., 2005). Just a slightest touch is enough to crush the fragile living coral polyps on the surface of the reefs, making the whole coral head exposed to infection and disease. Coral heads take hundreds of years to grow, and reef visitors touching, standing and scraping the corals with hands, fins or other equipment can easily damage corals and introduce bacteria that may lead to the death of the entire coral head. Inconsiderate anchoring by recreational boat owners can also destroy wide areas of reefs by crushing living polyps. The cumulative effect of boat groundings and propeller damage is equally as damaging. Every year the Florida Keys National Marine Sanctuary officials report thousands of small boat groundings in the Keys. There are more registered boats in the Keys than any other area of Florida, in addition to other visiting and transiting vessels (VisitFlorida, 2011). Similarly, when inexperienced boaters enter the shallow waters of the Florida Keys, their boat propellers stir up the bottom sediments causing "prop dredging". Not only does this create a milky white trail of calcium carbonate sediment that blocks sunlight and suffocates bottom-dwelling organisms that are part of the reefs ecosystem, but propellers also often displace and shred seagrass beds that may never re-grow. Additionally, oil pollution from the many recreational boats contributes to polluting the water in which the corals live, as well as introducing hydrocarbons into the food chain.

The combined impacts of numerous stressors

The pressure of a number of stressors combined has frayed the delicate ecosystem of the Florida Keys coral reefs. In addition to the impacts from ocean warming and acidification and inconsiderate visitors' behaviours, the impacts from unsustainable fishing pressure and land-based pollution have intensified the problem. As discussed in previous sections of the chapter, the Florida Keys coral reef ecosystems support significant commercial, recreational and subsistence fishery resources for the local and state economy. The reefs fisheries, even if relatively small in scale, may have extremely large impacts on the ecosystems if conducted unsustainably. The impacts from unsustainable fishing on coral reef areas can lead to the depletion of reef habitats. Such losses often have a domino effect, not just on the coral reef habitats themselves, but also on the local economies that depend on them. Moreover, certain types of fishing equipment often cause serious physical damage to coral reefs, seagrass beds and other vital marine habitats.

In addition, it is also necessary to consider the impact from land-based pollution sources such as toxicants, sediments and nutrients flowing into the ocean due to coastal development, deforestation, agricultural run-off and oil and chemical spills. These sources of pollution inhibit coral growth and reproduction and cause disease and mortality in delicate species (Hoegh-Guldberg et al., 2007; Pastorok & Bilyard, 1985; Rogers, 1990; Smith & Buddemeier, 1992). The excess of nutrients (organic and inorganic materials originating from sewage, fertilisers and other pollutants) that are discarded into the oceans promotes the growth of harmful algae which compete with corals for habitat. Algal blooms can cause oxygen levels to become so low that fish and other marine life cannot survive. Reef Relief (2017) reported that every year, 700 tonnes of nutrients are discarded into Keys waters from agricultural run-off from the Everglades. As reported in a recent article from CNN (Sanchez, 2016), algae have already become the dominant species on the Florida Reef Tract, covering the ocean floor and suffocating the reef by blocking sunlight. As reported by the same source, the coastal outflows in Florida were scheduled to be closed in 2017, after a push from the state's Department of Environmental Protection (DEP). However, the plan was deferred, allowing them to remain running until 2025 despite an escalating controversy about the damage to Florida Keys' coral reefs (Kleypas et al., 1999; Pendleton et al., 2016; *The Guardian*, 2016).

Action taken to support conservation and protection of the reefs

Over the last decades, recognising the threats coral reefs are facing, the State of Florida has been taking integrated action to protect the vulnerable ecosystem of the Keys reefs. In 1980 the State of Florida designated the Florida Keys as Areas of Critical State Concern. The legislation has been enacted to protect the unique ecosystem of the Keys by regulating land development and other activities considered detrimental to the environment.

Florida Keys National Marine Sanctuary

Recognising the value of the Florida Keys coral reefs to the entire country, the United States Congress in 1990 established the Florida Keys National Marine Sanctuary (FKNMS) through the Florida Keys National Marine Sanctuary and Protection Act. The Keys' marine environment has been described as the marine equivalent of tropical rainforests for supporting high levels of biological diversity, being fragile and vulnerable to damage from human activities, and possessing high value to human beings if properly conserved. Managed by NOAA, the FKNMS covers 2,900 square nautical miles surrounding the Florida Keys, from south of Miami westbound to the Dry Tortugas (excluding the Dry Tortugas National Park). Visitors to the sanctuary can engage in several activities that include diving, swimming, snorkelling and fishing. However, strict rules and regulations are in place and communicated to visitors to prevent these activities causing injury to the sanctuary ecosystem. According to Craig (2003), in the management plan for the FKNMS, NOAA de-emphasised the role of fishing in degrading the Florida Keys' coral reefs, while emphasising instead the role of tourist use and abuse, land-based pollution and vessel groundings as the sources of reef degradation. However, Craig argues that overfishing did play a large role in degrading the Florida Keys coral reef ecosystem (see Craig, 2003, for details).

Florida's Coral Reef Protection Act

The Florida's Coral Reef Protection Act introduced in 2009 was designed to increase protection of Florida's endangered coral reefs by raising awareness of the damage associated with vessel groundings and anchoring on coral reefs (Florida Department of Environmental Protection, 2017). The law affects both commercial and recreational vessels that transit South Florida's waters, holding those that injure reefs responsible for causing damage to coral reefs. However, the law is enforced only if a vessel owner or operator self-reports an injury, leaving unreported damage unpunished.

That same year, after a presentation led by the Southeast Regional Director of the National Marine Sanctuary Program, the Florida Keys Sanctuary Advisory passed a resolution recognising the threat from ocean acidification. What can be noticed is that when the sanctuary management plan was developed, the perceived threats to coral reefs were associated with direct human impacts rather than with climate change impact. Following the presentation, this has unquestionably changed.

The Florida Reef Resilience Program

The Florida Reef Resilience Program (FRRP) is a public and private partnership that brings together the collaborative effort of scientists, reef managers, conservation organisations and reef users with the aim of developing strategies to improve the health of Florida reefs and improve the economic sustainability of commercial enterprises depending on the reefs (see http://frrp.org/). The programme has applied the concept of resilience since 2005 to maximise protection of resilient reefs and increase the viability of those that are less resilient (Bergh, 2009). Increased ocean temperatures and ocean acidification obviously cannot be effectively addressed at the local level alone. However, the FRRP has been established to enable reef managers and reef users to help make coral reefs, and the local communities that depend on them, more resilient to climate change related stresses.

Florida Keys BleachWatch Program

In 2005, modelled on the Great Barrier Reef's BleachWatch Program (created in 2002 during a mass bleaching event), the FKNMS and Mote Marine Laboratory established the Florida Keys BleachWatch Program to work as an early warning network for coral bleaching. The BleachWatch Program combines climate and sea surface temperature data collected by NOAA's Coral Reef Watch Program with field observations on the reef condition collected by a trained Observer Network, and aims to detect conditions that may lead to mass bleaching events.

The programme helps in assessing the full impact of coral bleaching by monitoring when bleaching events occur and their duration and severity, as well as identifying the coral species that are most vulnerable and the recovery and resilience potential of the coral reefs ecosystem. A team of trained recreational, commercial and scientific divers helps monitor the reef by collecting field observations and producing reports on the current conditions of the reefs. Divers are alerted when weather conditions such as calm winds and clear sunny days are favourable for bleaching in order that field data can be gathered before, during and after a bleaching event. Divers report reef conditions whether they observe coral bleaching or not, and after each visit they fill out a report on their observations which is sent to a BleachWatch coordinator. After combining all available data collected, including pertinent weather information and NOAA Coral Reef Watch analysis, together with BleachWatch Observer Network reports, a Current Conditions Report is generated to provide an overview of reef conditions. The Mote Marine Laboratory makes the report available to the public on its website (https://mote.org/).

Conclusions

The issues discussed in this chapter indicate the extent of the challenges that face the Florida Keys and its tourism industry. Although governmental organisations and the local community have been working in partnership in an attempt to manage the situation, stressors from ocean warming and acidification continue to aggravate an already fragile ecosystem weakened by unsustainable fishing practices, inconsiderate or inattentive visitor behaviour and numerous sources of pollution (Brander et al., 2007; Cesar et al., 2003). Although difficult to manage, some control is possible over reef health, fisheries and visitor behaviour.

Florida faces enormous threats from climate change. The projected impacts on the state's coastal and marine systems alone will have destructive consequences for the state's economy and the tourism industry. For the Keys' coral reefs, short-term actions such as transplanting corals or finding more resistant species cannot be a solution if the reef structure is dissolving under them because of acidification. Developing strategies to increase resilience of the reef ecosystems is the key. To decrease the impacts of higher ocean temperatures and acidification it is necessary to reduce the negative impacts of the many human-induced stressors on the reef ecosystems.

Several recommendations can be made to enhance the management of the reefs and increase resilience. For example, to cope with the multitude of stressors affecting the reefs, greater emphasis needs to be given to habitat protection, focusing on education and outreach, regulation and enforcement. Additionally, it is essential to continue to closely monitor ocean water temperatures and develop effective management responses to deal with bleaching events and disease outbreaks. For education and outreach efforts to be managed effectively it will be necessary to educate locals as well as visiting boaters, divers and snorkellers on sustainable behaviour to limit reef damage. Communication and engagement efforts among locals and visitors need to include awareness and communication of coral reef resilience and climate change issues. Regulations that are already in place need to be enforced. The presence of undercover inspectors would be an incentive to control destructive fishing practices and boaters' damaging anchoring and groundings.

The Florida Keys community has an important responsibility to protect the fragile ecosystem that their economy and quality of life depends upon. By implementing appropriate management strategies and focusing on resilience they can contribute to the alleviation of the effects of a changing ocean and help preserve the reef ecosystem for generations to come.

References

Bergh, C. (2009). Initial estimates of the ecological and economic consequences of sea-level rise on the Florida Keys through the Year 2100. The Nature Conservancy, Sugarloaf Key, FL. Retrieved from http://www.frrp.org/SLR.htm

Brander, L.M., Van Beukering, P., & Cesar, H.S. (2007). The recreational value of coral reefs: A meta-analysis. *Ecological Economics, 63*(1), 209–218.

Bruckner, A., Buja, K., Fairey, L., Gleason, K., Harmon, M., Scott, H., . . ., & Wiley, P. (2005). *Threats and stressors to US coral reef ecosystems*. The State of Coral Reef Ecosystems of the United States and Pacific Freely Associated States, pp. 12–44.

Cesar, H., Burke, L., & Pet-Soede, L. (2003). *The economics of worldwide coral reef degradation*. Arnhem, The Netherlands: Cesar Environmental Economics Consulting (CEEC).

Craig, R.K. (2003). Taking steps toward marine wilderness protection? Fishing and coral reef marine reserves in Florida and Hawaii. *McGeorge Law Review, 34*(2), 155–266.

Florida Department of Environmental Protection. (2017). Coral Reef Protection Act. Retrieved 20 February 2017 from https://floridadep.gov/ogc/ogc/documents/coral-reef-protection-act

Glynn, P.W. (1993). Coral reef bleaching: Ecological perspectives. *Coral Reefs, 12*(1), 1–17.

Goreau, T.J., & Hayes, R.L. (1994). Coral bleaching and ocean 'hot spots'. *Ambio-Journal of Human Environment Research and Management, 23*(3), 176–180.

Hoegh-Guldberg, H. (2010). Climate change and the Florida Keys. Report retrieved 20 February 2017 from http://sanctuaries.noaa.gov/science/socioeconomic/floridakeys/pdfs/climateflkeys_main.pdf

Hoegh-Guldberg, O., Mumby, P.J., Hooten, A.J., Steneck, R.S., Greenfield, P., Gomez, F., . . . , & Knowlton, N. (2007). Coral reefs under rapid climate change and ocean acidification. *Science, 318*(5857), 1737–1742.

Hoffmeister, J.E., & Multer, H.G. (1968). Geology and origin of the Florida Keys. *Geological Society of America Bulletin, 79*(11), 1487–1502.

Hughes, T.P., Baird, A.H., Bellwood, D.R., Card, M., Connolly, S.R., Folke, C., . . ., & Lough, J.M. (2003). Climate change, human impacts, and the resilience of coral reefs. *Science, 301*(5635), 929–933.

Johns, G.M., Leeworthy, V.R., Bell, F.W., & Bonn, M.A. (2001). Socioeconomic study of reefs in southeast Florida: Final report. Hazen and Sawyer Environmental Engineers & Scientists, Hollywood, Florida.

Jones, S. (2006). The science behind coral bleaching in the Florida Keys. Retrieved 20 February 2017 from http://www.aoml.noaa.gov/keynotes/keynotes_0914_coral_bleaching.html

Karl, T.R., Melillo, J.M., & Peterson, T.C. (Eds.). (2009). Global climate change impacts in the United States: A state of knowledge report from the US Global Change Research Program. New York: Cambridge University Press. http://www.globalchange.gov/usimpacts

Kleypas, J.A., Buddemeier, R.W., Archer, D., Gattuso, J.P., Langdon, C., & Opdyke, B.N. (1999). Geochemical consequences of increased atmospheric carbon dioxide on coral reefs. *Science, 284*(5411), 118–120.

Monroe County Tourist Development Council. (2017). Key West and Monroe County demographics and economy. Retrieved 20 February 2017 from http://www.keywestchamber.org/uploads/4/6/5/2/46520599/demographics_and_economy.pdf

Muehllehner, N., Langdon, C., Venti, A., & Kadko, D. (2016). Dynamics of carbonate chemistry, production, and calcification of the Florida Reef Tract (2009–2010): Evidence for seasonal dissolution. *Global Biogeochemical Cycles, 30*(5), 661–688.

National Oceanic and Atmospheric Administration (NOAA). (2017). Coral reefs support jobs, tourism, and fisheries. Retrieved 20 February 2017 from https://floridakeys.noaa.gov/corals/economy.html

Park, T., Bowker, J.M., & Leeworthy, V.R. (2002). Valuing snorkeling visits to the Florida Keys with stated and revealed preference models. *Journal of Environmental Management, 65*(3), 301–312.

Pastorok, R.A., & Bilyard, G.R. (1985). Effects of sewage pollution on coral-reef communities. *Marine Ecology Progress Series Oldendorf, 21*(1), 175–189.

Pendleton, L.H., Hoegh-Guldberg, O., Langdon, C., & Comte, A. (2016). Multiple stressors and ecological complexity require a new approach to coral reef research. *Frontiers in Marine Science*, March.

Reef Relief (2017). Threats to coral reefs. Retrieved 20 February 2017 from https://www.reefrelief.org/

Rogers, C.S. (1990). Responses of coral reefs and reef organisms to sedimentation. *Marine Ecology Progress Series, Oldendorf, 62*(1), 185–202.

Sanchez, B. (2016, June 27). Florida's coral reef system in rapid decay, scientists say. CNN. Retrieved 20 February 2017 from http://www.cnn.com/2016/06/27/us/florida-coral-reefs/

Smith, S.V., & Buddemeier, R.W. (1992). Global change and coral reef ecosystems. *Annual Review of Ecology and Systematics, 23*(1), 89–118.

Spalding, M., Burke, L., Wood, S.A., Ashpole, J., Hutchison, J., & Ermgassen, P. (2017). Mapping the global value and distribution of coral reef tourism. *Marine Policy, 82*, 104–113.

The Guardian. (2016). Florida's coral reefs rapidly 'wasting away' under stress of climate change. Retrieved 20 February 2017 from https://www.theguardian.com/environment/2016/may/04/florida-coral-reefs-disintegrating-climate-change

VisitFlorida. (2011). 5 Great boating destinations in Florida. Retrieved 20 February 2017 from http://www.visitflorida.com/en-us/articles/2011/august/1848-floridas-top-5-boating-destinations.html

9

WHALE SHARK TOURISM AT NINGALOO REEF

Successes, challenges and what's next?

Sarah Duffy, Roger Layton and Larry Dwyer

Introduction

Rarely does one look at the news without seeing a story about a natural resource under threat, from destruction of the Amazon to coral bleaching at the Great Barrier Reef to the impact of climate change in the Arctic. Some of these natural locations are tourist destinations, alongside other, sometimes competing industries and face challenges to their continued attractiveness and overall sustainability. We will use Ningaloo Reef and their whale shark tourism industry as an illustrative example to explore issues of sustainable resource management. Ningaloo Reef is Australia's largest fringing coral reef, skirting for 300 kilometres alongside the west coast. The region is a striking colour palate of turquoise Indian Ocean, hemmed by white sand, edged by the vast red desert. The reef may be visited from two towns, Exmouth and Coral Bay, which are located over 1,200 kilometres north of the state's capital city, Perth (Figure 9.1). In the past, the reef's key defence from being "loved to death" was its low profile and isolation. This is changing as the region has received global recognition and accessibility has improved.

Whale shark tourism is a key source of local pride and point of difference for the region. Whale sharks are a protected species and a popular drawcard for tourists. Whale sharks, similar to coral reefs, fisheries and forests, fit the criteria to be classified as a common pool resource (CPR). A commons refers to a resource held in common by a collective who have access and gain benefit from increasing their appropriation of the resource (Burger & Gochfeld, 1998). CPRs are finite and, as a result, growth is often a serious pressure, signalling a need to temper growth and increasing the potential for conflict as issues of cooperation and competition arise (Duffy et al., 2017). Whale sharks may be depleted in that too much, or particular types of, interaction with tourists can be distressing for them, or because they swim so close to the surface they can be seriously injured by collisions with boats.

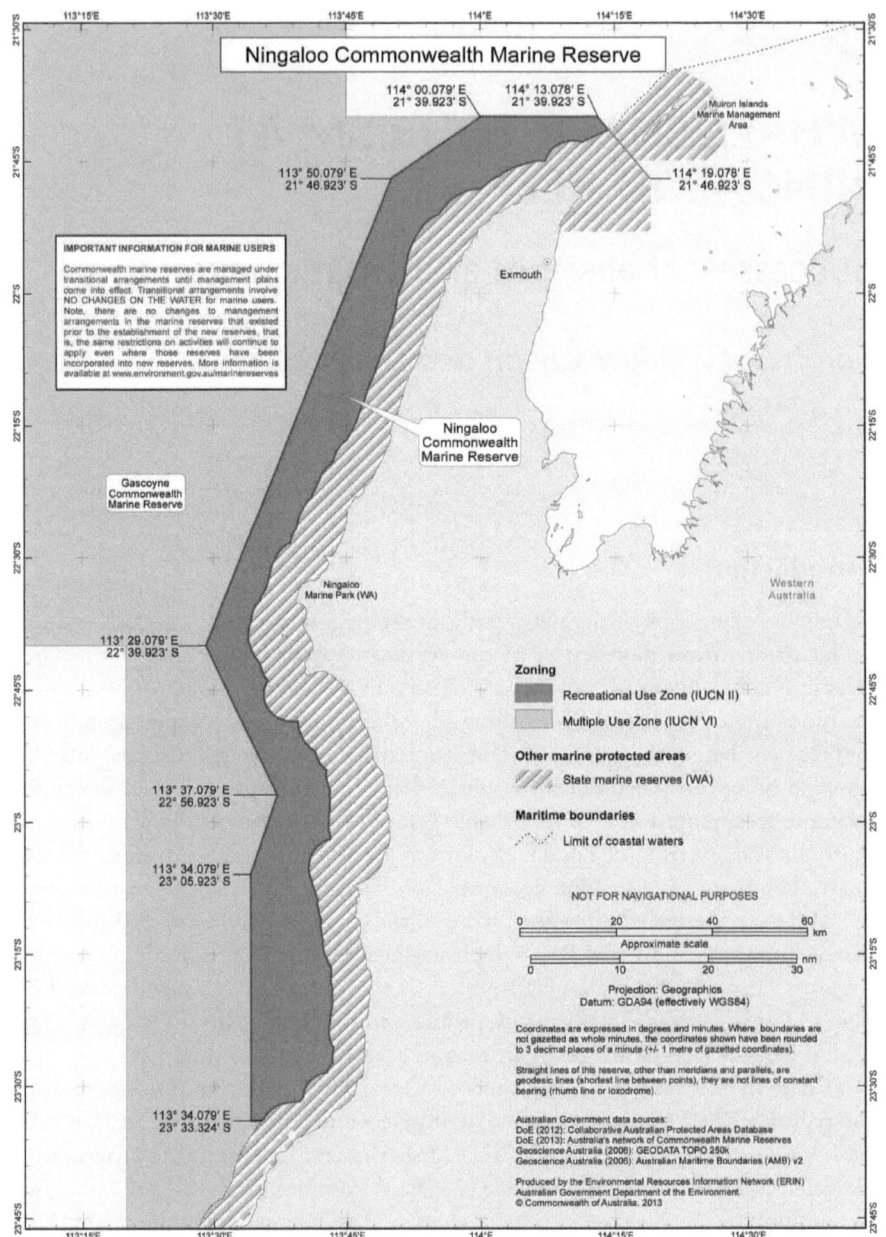

FIGURE 9.1 Map of the Ningaloo Coastal Region of Western Australia

Source: Commonwealth of Australia, http://www.environment.gov.au/topics/marine/marine-reserves/north-west/ningaloo-maps.

A tourism industry has developed in Exmouth and Coral Bay offering the opportunity to swim with whale sharks under particular conditions. The state government saw the potential for the industry to have negative consequences for the whale sharks and its importance as an asset for the state, and so they intervened and regulated the industry. The regulations will be explained in more detail; however, they involve limiting the number of swimmers and boats that are allowed to be in the water with the whale sharks and a minimum distance to be maintained from the animals. This is also to protect the swimmers from any accidental harm caused by overcrowding or getting too close to a whale shark and accidentally colliding with their powerful tail. If whale sharks left the region, or the whale shark industry was no longer functioning to its present levels, there would be a significant impact on the local community because the tourism industry is the largest contributor to the gross domestic product (GDP). Striking a balance between conservation and growth is therefore a critical issue for the local community and the region's tourism industry.

This chapter provides a brief overview of tourism at Ningaloo Reef, focusing on the whale shark industry. The chapter then discusses the meaning of CPR and identifies some of the unique challenges faced by the region. This is followed by a more detailed examination of the whale shark tourism industry from this perspective. Finally, the chapter considers a range of solutions that may help to ensure sustainable CPR management.

Overview of tourism at Ningaloo Reef

Tourism is the largest component of Exmouth's economy. There are two main settlements alongside the Ningaloo Reef: Exmouth and Coral Bay. Exmouth is the larger settlement and is closer to the airport; however, Coral Bay offers easier access to Ningaloo Reef itself. Both Exmouth and Coral Bay are destinations that rely on natural, water-based attractions. There are a variety of ways visitors can enjoy the reef and its rich wildlife, most notably whale sharks, dolphins, humpback whales, killer whales, manta rays, stunning coral varieties, turtles and a variety of sharks. Visitors can enjoy fishing, snorkelling, scuba diving (one of the top ten pier dives in the world is at Exmouth), sailing, kayaking, aerial flights over the reef and glass-bottom boat tours. In addition to the water-based attractions, visitors can tour or hike in the adjacent national park that offers the opportunity to see native animals such as wallabies and a variety of reptiles against stunning scenery characteristic of the Australian Outback. Tourism Australia (TA) estimates that the area receives over 200,000 visitor nights a year (Tourism Research Australia [TRA], 2014). This is significant given that the permanent population of Exmouth is 2,393 residents (Australian Bureau of Statistics [ABS], 2011) while Coral Bay has 221 residents (ABS, 2011). The tourist season runs from March to October, with significant spikes during school holidays. Summer is the low season as the average daily temperature is scorching, exceeding 40 degrees Celsius. Summertime is also cyclone season.

Visitor profile Exmouth

Most of the visitors to Exmouth are from Western Australia, although the number of international tourists is increasing. International tourists make up twice the number of tourists travelling from other states within Australia (TRA, 2014). International visitors stay the longest, while interstate visitors stay the shortest. Domestic visitors predominantly travel as family units (35%), whereas international visitors mostly travel unaccompanied (48%). The domestic market also has a significant number of tourists older than 65 years (17%). The international market includes a significant backpacker sector, with 17% of international tourists aged between 15 and 24.

What is a common pool resource?

A CPR has two defining characteristics. The first is that the exclusion of users through physical barriers or legal means is difficult or prohibitively costly. In the case of whale sharks, they are wild, migratory animals. The second characteristic of a CPR is that consumption can deplete the resource or is rivalrous (Dietz et al., 2003). Depletability resulting from use means that open-access arrangements can lead to a situation where no individual bears the full cost of resource degradation, however, they derive short-term benefits from their use (Hardin, 1968). Rivalrous means that the resource is not endless and that one person's use diminishes or prevents another's use. This can be demonstrated in whale shark tourism when only a limited number of tourists can experience swimming with a whale shark before the experience is diminished from overcrowding or, in more extreme cases, the whale sharks retreat. Given that nature-based tourism often centres on resources such as forests, coral reefs, fisheries or groups of wild animals (all CPRs), understanding the characteristics of CPRs and how this impacts their management is important and is a useful guide for determining resource sustainability.

Hardin (1968) suggested that without coercion or privatisation, the ruin of our CPRs is inevitable. In other words, Hardin (1968) believed that if there is no-one controlling resource use, the resource will be "used up." Hardin (1978) claimed there are only two solutions to the commons dilemma: privatisation or government intervention. We now know that the future need not be as bleak as Hardin feared. Commons scholars have provided useful insights into how resource use is affected by management decisions and use patterns and how this can be sustainably managed (Acheson, 2003; Carlsson & Sandström, 2007; Dietz et al., 2003; Healy, 2006; Ostrom, 1990). One of the most important implications of this body of work is that individuals, communities and agencies have the capacity to plan, cooperate and communicate with one another to maximise a particular resources capacity and they can do this successfully without government intervention (Burger & Gochfeld, 1998).

All commons are not created equal. They are unique and vary depending on their social context. Control of the commons and the mechanisms that allow this (e.g. formal regulation or informal rules) can be democratic or otherwise, and access can be equal or unfair (Burger & Gochfeld, 1998). These mechanisms influence how the CPR is managed, how the rules are constructed and applied, as well as the alignment between activities, incentives and sustainable outcomes, the level of investment and type of infrastructure. Ultimately these mechanics are likely to influence if the resource is maximised or depleted or somewhere in between.

Challenges of common pool resource management

The challenge of CPR management is how to reduce or prevent externalities (Agrawal, 2001) which, in the case of tourism destinations, are often resource degradation, rivalry and overcrowding. Externalities can be a positive or negative consequence of a transaction experienced by third parties. In this situation we are interested in reducing negative externalities. Free riders are regularly at the heart of these types of issues because they benefit from using the resource without investing in infrastructure to maintain or enhance the resource. For example, if one were to hire a boat and set out in search of whale sharks they are "free riding" (if they can find a whale shark!) because they are avoiding paying the fee to swim with whale sharks, which contributes to research and their management. Additionally, they may contravene the guidelines that regulate the number of swimmers, boats in the water and minimum distances from the whale sharks. If users free ride, the collective benefit will suffer (Ostrom, 1990). Further to this, those do who invest in a CPR might not achieve as high a return due to free riders or if the actions of others diminish or destroy the resource. For example, if visitors to Ningaloo Reef overfish their quota, stress or collide with the whale sharks or stand on the coral when snorkelling, these activities will be diminished for others in the future.

Table 9.1 shows the nature of the CPR and lists the type of externality that threatens the resource. It is essential to identify the threat to assess the risks and consequences. If the resource can be enhanced, this is also indicated and has implications for investment. The use is classified, which has consequences for the health of the resource and the likelihood of government intervention. The type of management regime is specified, followed by the possibility of exclusion. Type of use and the level of demand are listed because these influence planning. The existence of a shared future among users is also shown. Finally, in recognition that humans will behave in ways that are rewarded, the alignment of rewards and conservation of the resource is considered.

We will discuss whale shark tourism in greater depth before analysing the example using CPR theory.

TABLE 9.1 Common pool resources classification

Example	Whale shark industry (Duffy, 2016)	Lobster Fisherman Maine (Acheson, 2003)	Sustainable tourism (Healy, 2006)	Fish stocks, Ningaloo (Duffy, 2016)
Externality	Congestion, rivalry & resource degradation	Resource degradation, rivalry	Resource degradation, congestion	Resource degradation
Access rights of users	Access & withdrawal	Access, withdrawal, informal management, exclusion	Locals – access; Tourists – access	Access & withdrawal
Rule-making power	Some input – limited	Informal – mutually agreed upon & enforced	Locals – limited opportunity via the commission that manages the falls; Tourists – none	Limited to non-existent
Resource can be enhanced	Yes (the facilitating infrastructure)	Yes	Yes	Yes
Specified use	Non-extractive	Extractive	Non-extractive	Extractive
Management	State government – limited entry	Self-governance	State government	State government – licence required
Exclusion	Impossible – migratory	Possible through territorialism	Impossible	Unfeasible
Transferability	Not without sanction from management regime	Yes	No	No
Type of use & level of demand	Commercial – increasing	Commercial & recreational – increasing	Recreational – increasing	Recreational – increasing
Shared future among users	No	Yes	No	Locals – yes; Tourists – no
Incentives aligned with sustainable outcomes	Incentives to invest in value-added facilities are weakened by open access. Tour operators have personal incentive to conserve the resource.	Yes, individuals have strong incentive to preserve the resource to ensure future benefit.	The efforts of management are ultimately to increase visitation to Niagara Falls; however, they do focus on mitigating impacts.	Incentives to invest in value-added facilities are weakened by open access (free riding tourists & adjacent industries). Individuals have strong incentives to stretch or break the rules for personal benefit.

Source: Duffy et al. (2017).

Whale shark tourism

Ningaloo Reef is one of the best places in the world to encounter the remarkable whale shark, the world's largest fish, which gathers in greater numbers at Ningaloo than have been recorded anywhere else. The average size of an adult whale shark is approximately 10 metres in length and 9 tonne in weight. From 1828 to 1987 there were just 320 sightings of the whale shark worldwide recorded in the scientific literature. They are a filter feeder and have a mouth that may be 1.5 metres wide, but pose no danger to humans. Whale sharks are grey-blue in colour with a white belly. They are covered in white spots that form a unique constellation on the body of each individual shark.

Whale sharks were seen by fisherman and others frequently out on the reef, but it was not until local doctor Geoff Taylor began to film whale shark sightings in the early 1980s that word spread that they visited regularly (Taylor, 1994). In 1992 a documentary film was made using the footage Taylor had taken, sparking worldwide interest in whale sharks and Ningaloo Reef. Camera crews were among the first visitors, followed later by recreational swimmers seeking this sublime experience. Whale shark tours first began in the region in 1989. In response to this demand, local fishing or scuba diving tour operators diversified their business to offer whale shark tours. Slowly a tourism industry began to form as more visitors arrived in the area wanting to swim with the whale sharks. The state governing body, the Department of Parks and Wildlife (hereafter, DPaW), established the first licensing agreement in 1993 (Colman, 1997a). The state recognised the potential negative externalities of the whale shark tours and intervened as part of a broader initiative to ensure that the state's natural assets were effectively managed. In 2004, 5,247 people swam with whale sharks at Ningaloo. Ten years later, this number had increased to almost 20,000 people.

It is important to note that the whale shark is recognised as the most important stakeholder in this activity. Clearly, there are challenges to overcome to be able to "know" their perspective; however, it is important an effort is made to do so. The work of Safina (2015) presents a convincing argument that animals, like humans, are capable of thought, forward planning and emotion. Animals lack the legal protection available to humans and sometimes there is a lack of regard for their welfare. It has been argued that ecotourism is morally acceptable if local communities control and benefit from the industry and the animals themselves are not disturbed (Bordas, 2004). Although we acknowledge the importance of this debate, we will not enter into it here, but instead wish to focus on what the whale shark ecotourism industry reveals about the reactions of whale sharks to interactions with tourists.

The ethics of non-consumptive ecotourism, for example swimming with whale sharks, can be difficult to balance alongside economic imperatives or public policy encouraging the development of a tourism industry. Since the animals are not being killed, touched or visibly hurt, it could be easy to dismiss any concerns. However, it is worth noting that a study of the impact of a similar style of ecotourism on dolphin populations found that dolphins changed their behaviour significantly when dolphin-watching vessels approached them (Steckenreuter et al., 2011). The industry was regulated by the state government and its practitioners had received special dispensation to approach the dolphins more closely. The study observed that dolphins spent less time feeding and resting, which may have adverse impacts on their long-term health and reproductive success (Steckenreuter et al., 2011).

It is impossible to directly know the feelings of whale sharks towards tourists; however, we can deduce that, as is the case with dolphins, they find particular behaviours objectionable. Studies led by Quiros (2007) and Norman (1999, 2002) have documented the whale sharks' reactions to the tourism industry. Both studies found close proximity, touching and obstruction of its path by a swimmer were among the conditions found to trigger reactions in the whale shark. While the whale shark is peaceful, too many boats and too many swimmers can be problematic. If swimmers are too close to the whale sharks, they may be injured if hit by a shark's powerful tail. The whale sharks may also suffer. There are numerous videos and images online of tourists in other parts of the world interacting with whale sharks in ways that marine scientists believe have an adverse impact on the sharks' health. However, a lack of baseline data concerning whale shark behaviour makes it difficult to definitively attribute causality to tourists because these reactions have also been observed in whale sharks when swimmers and vessels are not present (Stevens et al., 1998).

Overcrowding may also disturb the whale sharks and could impact how likely they are to return, causing changes to migratory pathways (Norman, 2002). Whale sharks swim very close to the surface and are vulnerable to boat strikes; increased boat traffic therefore increases the potential for collisions to occur (DPaW, 2013). Additionally, as migratory animals, whale sharks are more complex from a governance perspective because their protected status varies across their migratory pathway. For example, within Australia the whale shark is listed as "Vulnerable" under the Environment Protection and Biodiversity Conservation Act 1999. Globally, the World Conservation Union has included the whale shark in the Red List of Threatened Species, marking it as "Vulnerable" because the population is believed to be dwindling. The Taiwanese hunt and consume whale sharks, presenting a threat to the species.

Research undertaken by Norman (1999) at Ningaloo Reef from 1995 to 1997 specifically studied the interaction between whale sharks and aspects of the tourism industry. It found that, for the most part, the tourism industry was respectful of the whale sharks, and the conditions governing the industry and its customers seemed to be ensuring minimal impact on the whale sharks. Thus, it appears that the industry and governing body are mindful of the feelings and experience of the whale sharks. Since the research was conducted almost 20 years ago, it would be worth investigating whether the growth of the industry has had an impact on the whale sharks.

If the whale sharks were to depart the region, or the whale shark tourism industry was no longer functioning to its present levels, there would be a significant impact on the local community. Swimming with whale sharks is a key drawcard for the region and forms the basis of the area's unique selling proposition. There is the possibility of cruise ships stopping at Exmouth for day trips, which would mean a large influx of passengers who might wish to experience a whale shark tour. There are numerous murals, statues and representations suggesting that whale sharks form an integral part of the town's identity and are a source of local pride. Their presence was an important justification for the area's listing as a World Heritage Area. Whale sharks have also become increasingly significant for the state of Western Australia. They feature on the state's emblem and are a unique drawcard for visitors. Ningaloo is the state's fourth most popular destination, according to the most recent visitor survey (Tourism Western Australia [TWA], 2013). In 2005 the sector's contribution to the local economy was AU$12 million annually. Since then the number of tours and passenger numbers has more than doubled, suggesting the financial loss for the town would be substantial (Catlin et al., 2012). The whale shark industry contributes to the state economy directly through attracting visitors, taxes collected and the bureaucracy that manages the industry. Without consideration of the welfare of the whale sharks and subsequent management of the risks to them, substantial consequences may result.

The whale shark sector is not currently facing a crisis; however, it could in the future. There are a number of pressure points on the event horizon which could create problems, including increased boat traffic threatening the health of whale sharks and the temptation to increase the number of visitors who swim with the whale sharks. Most people of the Ningaloo region will experience some of the consequences of the tug-of-war between growth and conservation because whale sharks are not the only CPR under strain. Fish stocks, the coral reef and local infrastructure connected to the whale shark sector are all subject to pressures of overuse. Visitors to the region have been steadily increasing and there is no reason to suggest that this trend will not continue. Furthermore, it is easy to be lulled into a false sense of security because the first stages of over-exploitation or stress are difficult to identify (Acheson, 2006).

Assessing whale shark tourism against CPR theory

Extensive study of CPRs has led to the development of the set of design principles shown in Table 9.2. Each principle has been examined as to how they relate to the whale shark industry to identify which constructive structural elements are present and which are lacking (Ostrom, 1990). These principles have been found to be pivotal in the successful management of a CPR. However, it is critical to note that there are very few situations where all the principles will be operating in a positive way.

The challenges to CPR management outlined in Table 9.2 are largely internal to the industry. It is also important to consider external threats. The Exmouth tourism industry is both threatened by and would suffer without the oil, gas and mining industries (Duffy, 2016). Significant oil and gas production occurs offshore in the Carnarvon Basin and in the Exmouth Gulf. The oil and gas industry

TABLE 9.2 Design principles applied to the whale shark marketing system

1A. User Boundaries	Clear and locally understood boundaries between legitimate users and non-users are present.
Supply	*MET.* The industry is managed by a state government body, the Department of Parks and Wildlife (DPaW). There are 15 licences conditionally dispensed by the DPaW for a duration of 5 years (Mau, 2008).
Demand	*MET.* A tourist may participate without a tour; however, barriers are high to circumvent this system.
1B. Resource Boundaries	Clear boundaries that separate a specific common-pool resource from a larger social–ecological system are present.
Supply & Demand	*MET.* Resource boundaries apply equally to the demand and supply side of the industry. The tours are conducted within the confines of Ningaloo Marine Park (NMP) as specified by state government legislation. However, from the supply side, there are other locations where whale sharks may be seen.
2A. Congruence with Local Conditions	Appropriation and provision rules are congruent with local social and environmental conditions.
Supply	*UNMET.* For reasons outlined previously, there is conflict between the regulator (DPaW) and the tour operator, suggesting that there is some degree of incongruence with the local social conditions.
Demand	*MET.* Based on interviews conducted with both domestic and international tourists, the rules were well received and no causes of contention were expressed.
2B. Appropriation and Provision	Appropriation rules are congruent with provision rules; the distribution of costs is proportional to the distribution of benefits.
Supply & Demand	*MET.* 100% of the cost of the management programme is recovered from the tour operators. This cost is passed on to tourists. Based on the interviews conducted, tourists felt the cost was high, but many expressed a sense of satisfaction from supporting research and a sustainable industry.
3. Collective-Choice Arrangements	Most individuals affected by a resource regime are authorised to participate in making and modifying its rules.
Supply	*UNMET.* There is a strong feeling among the tour operators that although they can make suggestions their suggestions are easily disregarded by DPaW if they disagree.
Demand	*UNMET.* Tourists are not involved in making or modifying rules. However, if they do not wish to comply with them, they may swim with whale sharks at other locations.
4A. Monitoring Users	Individuals who are accountable to or are the users monitor the appropriation and provision levels of the users.

Demand	*MET.* Tour operators are formally monitored by DPaW. This cost is borne by the tourists consistent with the licensing conditions. Informal monitoring also takes place, that is, the tour operators monitor one another on the water and will informally deal with contraventions, similar to the lobster fishermen in Maine (Acheson, 2003).
Supply	*MET.* Tour operators are required to communicate to tourists the rules relevant to them and monitor their compliance.
4B. Monitoring the Resource	Individuals who are accountable to or are the users monitor the condition of the resource.
Demand	*MET.* There are two resources, Ningaloo Reef and the whale sharks. Due to the high mobility of the whale sharks and their vulnerability it is difficult to monitor them and isolate the impacts of tourism. However, all whale sharks swum with are photographed and this information is supplied to DPaW. The condition of Ningaloo Reef is monitored, and due to the nature of this resource it is a more complex undertaking. The monitoring is conducted by a variety of research bodies rather than users. It is difficult to distinctly separate the effects of adjacent industries such as recreational fishing and oil and gas exploration for example.
5. Graduated Sanctions	Sanctions for rule violations start very low but become stronger if a user repeatedly violates a rule.
Supply & Demand	*MET.* Both formal and informal sanctions graduate in severity depending on the violation and if it has been repeated.
6. Conflict-Resolution Mechanisms	Rapid, low-cost, local arenas exist for resolving conflicts among users or with officials.
Supply & Demand	*UNMET.* There is no independent, swift conflict-resolution mechanism. The options are formal legal procedures or lobbying the minister.
7. Minimal Recognition of Rights	The rights of local users to make their own rules are recognised by the government.
Supply	*UNMET.* Interviews suggested that the locals ability to make their own rules were often not recognised and were a significant cause of resentment and contention.
8. Nested Enterprises	When a common pool resource is closely connected to a larger social–ecological system, governance activities are organised in multiple nested layers.
Supply	*MET.* It was found that local, regional and state-level staff were associated with the administration of the region; it was noted the Federal Government had no direct involvement, although they issued a mandate to the state government.

Source: Duffy et al. (2017).

provides consistent, well-paying employment opportunities (unlike the seasonal tourism industry) and has funded small-scale community infrastructure projects. The need to transport fly-in fly-out workers employed in the oil, gas and mining sectors has resulted in Exmouth being serviced by two daily jet flights, whereas Monkey Mia (also a World Heritage Area), which also relies on tourism, is only serviced by small regional commuter aircraft a few times a week.

An oil spill of the type that occurred in the Gulf of Mexico in 2010 would be disastrous for the tourism industry; however, the oil, gas and mining industry have measures in place to mitigate risks of this type. They frequently consult with the local community and run training drills to hone the response effort. However, as long as they continue their operations, spillages are a possibility. The town remains divided on the presence of oil and gas exploration in the region and has actively rejected some projects from going ahead. There is tension between development and growth for the town and preservation of the established way of life (Duffy, 2016).

The commons issue: managing whale shark tourism

The management goals for tourism in Ningaloo Marine Park (NMP) are to offer low impact commercial tourism activities which add to the recreational and edu-cational experience of the NMP users and to ensure that tourist operations do not negatively impact on the ecological or cultural heritage values (Department for Planning and Infrastructure, 2004). The boundaries of NMP have facilitated the development of a set of social arrangements for regulating the preservation, main-tenance and consumption of its resources (Duffy, 2016). In general, management of human–whale shark interaction has focused on licensing, education, research and monitoring, executed by DPaW in consultation with commercial operators (Colman, 1997b; Mau, 2008). These regulations limit the number of commercial whale shark operators to 15. Twelve boats are based at Exmouth, with the remain-ing three boats based in Coral Bay. The licence conditions allow ten swimmers in the water with a whale shark at a time. Swimmers are guided by a person called a "spotter" who observes the whale shark and ensures that the group of swimmers observe the minimum distance from the whale shark (4 metres from the tail and 3 metres from the body). A photographer is also in the water with the dual purpose of taking photographs of the whale shark for research purposes and to capture the tourists' experience. DPaW train all of staff who work on the commercial tour operator boats to ensure they understand the code of conduct. DPaW also advise the industry on the latest research and maintain the library of images of whale sharks. DPaW monitors tourism operators and conducts checks from the air and in the water to ensure the rules are being adhered to and complaints are investigated. Sanctions are applied when rules are contravened.

In a tourism context, the users with the greater level of market demand for the resource are usually external to the region. This influences how the rules of use are developed, institutionalised and enforced. A large proportion of external users with high levels of demand is a challenge because studies have found that a shared future among resource users is a key determinant of sustainable resource management (Dietz et al., 2003). Some studies have suggested that successful resource management is more likely if "outsiders" can be excluded from use (Ostrom, 1990). However, this is counter to the logic of the tourism industry and, with a comprehensive management programme in place, it seems that the industry may avoid these complications. A shared future may not always be the pivotal factor it currently is, given that Western Australian tourists have been identified locally as the whale shark sector's most significant environmental threat (Jones et al., 2009).

Tour operators, despite their understanding of the need to protect the resources, are also concerned with economic gain. There are also many opportunities for unsupervised interactions to take place in such an expansive location, or for pressures of market demand to take priority. As the demands on whale shark tour operators have increased, there has been shift away from largely local operators towards more experienced tour operators from the capital, Perth, which may have the unintended effect of eroding the local connection (Dietz et al., 2003). The strict regulation of the industry strives to prevent such occurrences. Only those holding a licence may conduct whale shark tours and there are strict specifications. This includes having a photographer on hand to collect data on the whale shark and to act as an observer (meeting design principle 4A and 4B). The design principles provide a yardstick against which to appraise the management, as shown in Table 9.2. However, they are static and capture only a snapshot in time. This can be problematic and, unless regularly updated, misleading. It also fails to explain why some conditions exist and why others do not. The design principles are an important tool for understanding; however, it is critical that they are updated regularly, thoroughly investigated and that attention is given to external influences that may impact the system.

Issues of fairness

While the reasons for developing public policies are often complex, one of their chief functions is to police markets. In this sense, policing refers to the creation modification and, where necessary, the control of behaviour and outcomes in a market exchange or to substitute acts of authority for acts of exchange (Hurst, 1982). Efforts to control behaviour are evident in the whale shark industry as public policy intervention attempts to resolve the allocation of a scarce resource (Duffy, 2016). Fairness relates not only to the distribution of costs, but also to ensuring that the benefits of a scarce resource are equally shared. The cost of the regulation needs to be considered in light of fairness. The government will incur costs implementing public policy, often so too will those who must comply with the regulation (the whale shark tour operator). The high cost of swimming with a whale shark (approximately US$320–420) can be prohibitive for some tourists. The allocation of both costs and perceived benefits of regulation may contribute to polarisation of those involved. It is useful to compare the management of whale sharks with local fish stocks (also a CPR) to consider this issue within a similar context (see Table 9.3).

TABLE 9.3 Comparison of common pool resource management, Ningaloo Marine Park

	Whale sharks	*Fish stocks*
State government managed	Y	Y
Licence required for use	Y	Y
100% cost recovery	Y	N
Powerful lobby	N	Y
Information provided to users about resource health	Y	N
Extractive experience	N	Y

Source: Duffy (2016).

In comparison to the whale shark industry, the licence fees for recreational fishing recover only 35% of the costs of the management programme (Duffy, 2016). The state government absorbs the deficit. In contrast to the whale shark tour operators, recreational fishermen are represented by a powerful lobby group. The disparity in the cost of the management programme is interesting considering swimming with a whale shark is a non-extractive experience, whereas recreational fishing is extractive.

Conclusion

In a world experiencing a diminishing supply of pristine environments, it is critical to understand the nature of a tourism CPR, how this impacts on growth and the management strategies required for coping with adverse effects arising from their use (Duffy, 2016). Governance of CPRs is a tricky balancing act because not only is there potential to deplete the resource but excessive regulation can also severely stunt economic growth (Techera & Klein, 2011). The design principles suggest that for a resource to be sustainably managed, certain conditions must exist. They are clear boundaries, fair rules, a balance in private and public funding, thorough monitoring programs, graduated sanctions, accessible and low-cost conflict resolution, the right to self-organise, collectively devise rules and finally consideration of related systems. Not all of the ideal conditions are present in the whale shark tourism industry. Conspicuously absent are low-cost conflict resolution and contribution to rule making. Tension can be attributed to these missing conditions. Additionally, it is important to consider threats external to the industry. Overall, however, the whale shark sector appears to be successfully negotiating the precarious balance between development and growth and protecting the local environment and way of life (Duffy, 2016).

References

Acheson, J.M. (2003). *Capturing the commons: Devising institutions to manage the Maine lobster industry*. Lebanon, NH: University Press of New England.

Acheson, J.M. (2006). Institutional failure in resource management. *Annual Review of Anthropology, 35*, 117–134.

Agrawal, A. (2001). Common property institutions and sustainable governance of resources. *World Development, 29*(10), 1649–1672.

Australian Bureau of Statistics (ABS). (2011). *Census. Exmouth, Coral Bay, Cape Range National Park (C) population data.* Retrieved 10 November 2013 from http://www.abs.gov.au/

Bordas, M.I.S. (2004). Animal rights? No, human responsibility. *Human Ecology Special Issue, 12*, 149–160.

Burger, J., & Gochfeld, M. (1998). The tragedy of the commons 30 years later. *Environment: Science and Policy for Sustainable Development, 40*(10), 4–13.

Carlsson, L.G., & Sandström, A.C. (2007). Network governance of the commons. *International Journal of the Commons, 2*(1), 33–54.

Catlin, J., Jones, T., & Jones, R. (2012). Balancing commercial and environmental needs: Licensing as a means of managing whale shark tourism on Ningaloo reef. *Journal of Sustainable Tourism, 20*(2), 163–178.

Colman, J. (1997a). *Western Australian Wildlife Management Program No. 27: Whale shark interaction management, with particular reference to Ningaloo Marine Park.* Fremantle, WA: Western Australian Department of Conservation and Land Management, pp. 1–39.

Colman, J. (1997b). *Whale shark interaction management with particular reference to Ningaloo Marine Park.* Marine Conservation Branch, Western Australia Department of Conservation and Land Management, p. 63.

Cox, M., Arnold, G., & Villamayor Tomás, S. (2010). A review of design principles for community-based natural resource management. *Ecology and Society, 15*(4), 38.

Department for Planning and Infrastructure. (2004). Ningaloo coast regional strategy Carnarvon to Exmouth (pp. 1–6). Retrieved 6 December 2014 from https://www.planning.wa.gov.au/dop_pub_pdf/Foreword(1).pdf

Dietz, T., Ostrom, E., & Stern, P.C. (2003). The struggle to govern the commons. *Science, 302*(5652), 1907–1912.

Duffy, S. (2016). *New perspectives on marketing systems: An investigation of growth, power, social mechanisms, structure and history.* Doctor of Philosophy, University of New South Wales.

Duffy, S., Layton, R., & Dwyer, L. (2017). When the commons call "enough", does marketing have an answer? *Journal of Macromarketing, 37*(3), 268–285.

Hardin, G. (1968). The tragedy of the commons. *Science, 162*(3859), 1243–1248.

Hardin, G. (1978). Political requirements for preserving our common heritage. *Wildlife and America, 31*, 1017.

Healy, R.G. (2006). The commons problem and Canada's Niagara falls. *Annals of Tourism Research, 33*(2), 525–544.

Hurst, J.W. (1982). *Dealing with statutes.* New York: Columbia University Press.

Jones, T., Hughes, M., Wood, D., Lewis, A., & Chandler, P. (2009). *Ningaloo coast region visitor statistics: Collected for the Ningaloo destination modelling project.* Queensland, Australia: Sustainable Tourism Cooperative Research Centre, pp. 1–61.

Mau, R. (2008). Managing for conservation and recreation: The Ningaloo whale shark experience. *Journal of Ecotourism, 7*(2–3), 213–225.

Norman, B. (1999). *Aspects of the biology and ecotourism industry of the whale shark Rhincodon typus in north-western Australia.* Masters dissertation, Murdoch University, Perth, Australia.

Norman, B. (2002). *Review of current and historical research on the ecology of whale sharks* (Rhincodon typus), *and applications to conservation through management of the species.* Department of Conservation and Land Management, Perth, Australia.

Ostrom, E. (1990). *Governing the commons: The evolution of institutions for collective action.* Cambridge: Cambridge University Press.

Quiros, A.L. (2007). Tourist compliance to a Code of Conduct and the resulting effects on whale shark (*Rhincodon typus*) behavior in Donsol, Philippines. *Fisheries Research*, *84*(1), 102–108.

Safina, C. (2015). *Beyond words: What animals think and feel*. New York: Macmillan.

Steckenreuter, A., Harcourt, R., & Möller, L. (2011). Distance does matter: Close approaches by boats impede feeding and resting behaviour of Indo-Pacific bottlenose dolphins. *Wildlife Research*, *38*(6), 455–463.

Stevens, J., Norman, B., Gunn, J., & Davis, T. (1998). *Movement and behavioural patterns of whale sharks at Ningaloo Reef: The implications for tourism*. National Ecotourism Grant Final Report. Canberra: CSIRO.

Taylor, G. (1994). *Whale sharks: The giants of Ningaloo Reef*. Sydney: Angus & Robertson.

Techera, E.J., & Klein, N. (2011). Fragmented governance: Reconciling legal strategies for shark conservation and management. *Marine Policy*, *35*(1), 73–78.

Tourism Research Australia (TRA). (2014). *Exmouth overnight visitor fact sheet*. Tourism Western Australia. Retrieved 15 March 2015 from http://www.tourism.wa.gov.au/Research-Reports/Facts-Profiles/Pages/Region-Fact.aspx

Tourism Western Australia (TWA). (2013). *Western Australia overnight visitor fact sheet Dec 10/11/12*. Retrieved 11 February 2014 from http://www.tourism.wa.gov.au/Research_and_Reports/Regional_Fact_Sheets/Pages/Regional_Fact_Sheets.aspx

PART III

Development and management of coral reef tourism activities

10

ECOTOURISM AND CORAL REEF RESTORATION

Case studies from Thailand and the Maldives

Margaux Y. Hein, Fanny Couture and Chad M. Scott

Introduction

Coral restoration is increasingly becoming a popular reef management strategy in response to the consortium of threats currently affecting the world's coral reefs. The theories and techniques of mainstream coral restoration are often derived from terrestrially based experience and knowledge (Delbeek, 2001), and corals are planted on the reefs similarly to the way forests are planted on land.

Climate change and localised anthropogenic disturbances are reducing the structural and biological complexity of the world's reef ecosystems, leading to serious declines in coral coverage and diversity, and in turn threatening fish biomass and the balance of marine ecosystems (Pandolfi, 2003; Hoegh-Guldberg & Bruno, 2010). Coral restoration is used to mitigate disturbances, increase reef size or complexity, and enhance or restore the physical and biological integrity of coral reefs. Physical restoration uses specially designed and carefully selected materials such as concrete or steel to create or replace reef structure and habitats (Edwards & Gomez, 2007). Examples include artificial reef techniques such as concrete Reef Balls™, coral frames and even steel structures connected to a weak electric current to stimulate calcium carbonate accretion through electrolysis (e.g. Goreau & Hilbertz, 2012). Biological restoration refers to the attachment of coral fragments on these structures or directly on the reef substrata (Edwards & Gomez, 2007). Coral fragments can be broken off healthy "donor" coral colonies or collected as loose coral fragments naturally present on the reef due to storms, fish feeding or tourism activities. The latter are commonly referred to as "corals of opportunity" (Monty et al., 2006). Coral ramets (asexually produced fragments genetically identical to the mother colony) created through propagation can also be grown *in situ* or *ex situ* coral nurseries following the "coral gardening" concept developed by Baruch Rinkevich (Rinkevich, 1995; Epstein et al., 2001) and then transplanted to the reef. Recent advances are also using coral genets (genetically distinct corals produced sexually) specifically bred and cultivated for restoration purposes.

Planting corals is advocated as a means to promote both the resistance and recovery of reef ecosystems and is thus an important tool to manage coral reef resilience. Unfortunately, the costs of coral restoration in terms of monetary costs and time and labour often prevent coral restoration efforts being undertaken on a scale that is meaningful compared to the scale of disturbances. A recent valuation by Bayraktarov et al. (2015) revealed that coral restoration costs ranged from US$11,717/ha to US$2,879,773/ha. At this price, the area of reefs that can be restored is severely limited and the scale of potential benefits meagre.

One way in which costs can be reduced is through the use of volunteer workforce in ecotourism settings. Engaging the general public in scientific projects is commonly referred to as "citizen science" and is increasingly used in all fields of research for its potential to generate vast amounts of data at minimum expense, and to more effectively reach out and communicate to the public (Silvertown, 2009; Theobald et al., 2015). A key strength of coral restoration efforts is the hands-on and interactive nature of the tasks that are easily integrated in recreational activities such as snorkelling and scuba diving. Many coral restoration programmes have embraced ecotourism as a way to finance the restoration efforts. Getting free labour to participate in the coral planting exercise not only reduces the costs of coral restoration, but also spreads potential benefits to a socio-cultural dimension

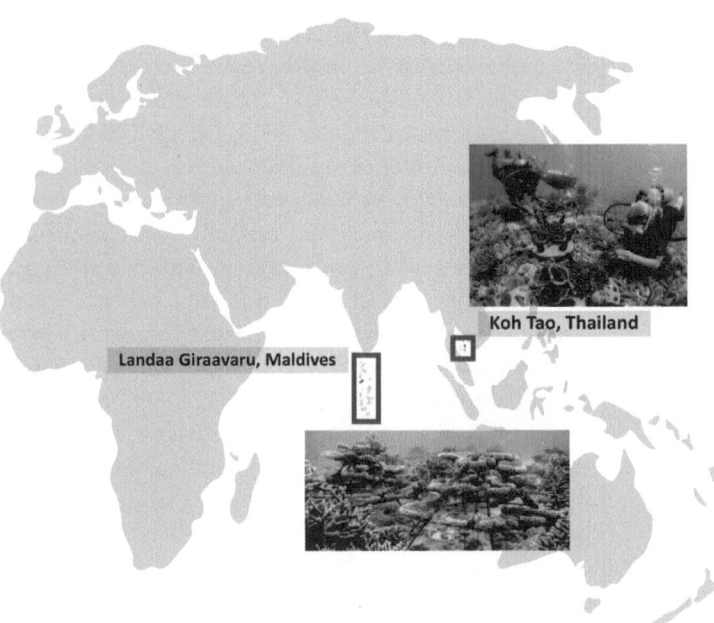

FIGURE 10.1 Map of case study locations

through increased education opportunities, conservation awareness and steward-ship (Scott & Phillips, 2010; Dela Cruz et al., 2014). Some programmes manage to have tourists pay to be involved in the restoration efforts, therefore creating busi-nesses that are profitable to both local economies and to the reef ecosystems. In this chapter, we present two restoration programmes – Reefscapers in the Maldives and New Heaven Reef Conservation Program (NHRCP) in Thailand – that have successfully integrated coral restoration as ecotourism activities (see Figure 10.1).

Reefscapers programme in Landaa Giraavaru, Baa Atoll, Maldives

The Maldives have the eighth largest coral reef ecosystem worldwide, covering an area of about 21,300 km^2 (Jaleel, 2013). These reefs are also some of the most biodiverse reefs in the Western Indian Ocean, with an estimated 241 species of Scleractinian corals spread among 60 genera (Food and Agriculture Organization [FAO], 1997). Unfortunately, this unique environment is currently under threat from anthropogenic stressors such as resort activities, coral mining, waste disposal, pollution and heavy fishing. More importantly, the monsoon-dominated climate of the Maldives makes coral reefs very sensitive to environmental hazards such as coral bleaching, crown-of-thorns (*Acanthaster planci*) outbreaks and storms (McClanahan, 2000). Coral bleaching was first reported on these reefs in the late 1980s, and a dramatic large-scale event occurred during the 1998 El Niño Southern Oscillation period. Bleaching followed the low-pressure, dry north-east monsoon and water temperatures increased by 1–3°C above the historic average during peak warming periods (Wilkinson et al., 1999). Repeated censuses over the Maldives estimated that 90% of the Maldivian coral reefs died, with a mortality reaching nearly 99% around some islands (Goreau et al., 2001; McClanahan, 2000; Edwards et al., 2001). The post-bleaching live coral cover was the lowest recorded in 40 years, with live coral cover dropping to 2% around some islands (Sheppard, 1999). The loss of reef structures critically impacted coral species richness (McClanahan, 2000), as well as invertebrates and fish populations (Goreau et al., 2001). After 1998, the recovery of the Maldivian coral reefs was slow and uneven, possibly because of limited coral recruitment (Bianchi et al., 2006), rarity of fast-growing corals (Bozec & Mumby, 2015), decline in fish biomass and diversity (McClanahan, 2011), and particularities in oceanographic patterns along Maldivian geographical gradients (McClanahan & Muthiga, 2014).

The future potential for the recovery of Maldivian coral reefs is looking rather bleak. According to a recent Intergovernmental Panel on Climate Change (IPCC) study, sea surface temperatures are predicted to increase by up to 2.61°C by 2050 and by up to 4°C before the end of the century (IPCC, 2014), thereby increasing the frequency and intensity of bleaching events. Reef recovery is also hindered by other multiple and escalating threats associated with global climate change and anthropogenic activities. Coral restoration efforts may help increase the poten-tial for reef recovery by increasing coral cover and reef structural complexity thus providing a more complex and diverse habitat for coral reef organisms to repopulate the area. Implementing efficient and large-scale coral reef restoration protocols in the Maldives could help re-establish functioning coral reef ecosystems and sustain Maldivian livelihoods and economy.

Reefscapers is a coral propagation project developed by Seamarc, a Maldivian Systems Engineering and Marine Consulting company. Seamarc's mission is to develop solutions for environmental challenges, including beach erosion, waste disposal and coral propagation (Seamarc, 2015). One of their coral propagation programmes is located in Landaa Giraavaru, Baa Atoll, home of the luxury Four Seasons Resort. The programme was the first of its kind in the Maldives when it started in 2001, and it has now been replicated in many other Maldivian atolls. Coral propagation efforts consist in attaching coral fragments on small artificial reef structures (referred to as "coral frames"). Coral fragments are collected from healthy coral colonies on the nearby reef. Only small portions of these "donor" colonies are broken off to produce coral ramets in order to minimise potential negative effects to the natural reef. The frames are produced by a local cooperative, and are sponsored either by resort guests or the resorts themselves. The main goal of the coral propagation effort is to create a whole new artificial coral reef system on the depleted reef around Landaa Giraavaru. Other goals include the establishment of new dedicated snorkelling trails, increasing scientific knowledge regarding coral propagation, and the creation of links between local communities, visitors and hotel managers, while increasing awareness and conservation stewardship.

The restoration effort on Landaa Giraavaru

As of 2017, 2,800 frames have been deployed around Landaa Giraavaru, featuring about 40 different species of corals at around 19 specific reef sites and covering about 5,585 m² (Seamarc, unpublished data). The rate of frame deployment averages 35 coral frames per month with an estimated 263 frames sponsored by hotel guests annually between 2010 and 2015. Hotel guests sponsor the frames either through direct sponsoring or as parts of special "packages" (e.g. wedding packages) sold by the resort. Three sizes of frames are available for purchase by guests: small (US$150, 1.31 m², 41 coral fragments), medium (US$300, 2.41 m², 65 coral fragments) and large (US$500, 4.2 m², 100 coral fragments). Guests also actively participate in attachment of coral fragments to the frames under the supervision of Reefscapers' coral biologists. Attachment of the coral fragments is done on land to allow guests and biologists to better interact. The frames are then dropped underwater on the damaged reef areas and their exact location is logged into a central Geographical Information System database.

Each frame is regularly maintained by on-site coral biologists and photographed every 6 months to assess coral growth and mortality. Most importantly, these pictures are uploaded on the Reefscapers website so that guests can follow up the evolution of their frames (Figure 10.2; Marine Savers, 2016). Educational newsletters about specific coral species or environmental events (e.g. coral spawning, bleaching) are regularly sent to the hotel guests and provide a constant reminder of the efforts and awareness of potential restoration challenges. The 19 artificial coral reefs around Landaa Giraavaru are now major snorkelling sites, with approximately 250 hotel guests visiting them daily (Seamarc, unpublished data).

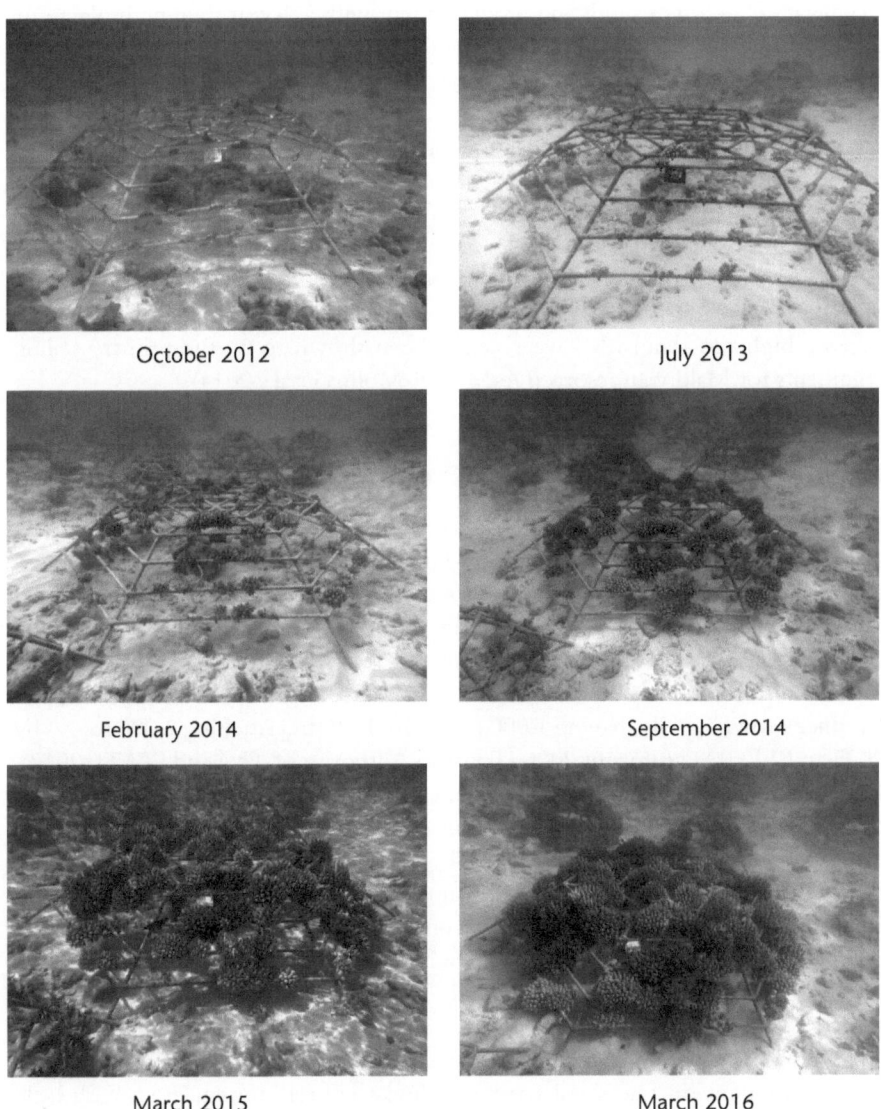

October 2012   July 2013

February 2014   September 2014

March 2015   March 2016

FIGURE 10.2 Photography time series showing the evolution of one of the large frames of the Reefscapers programme around Landaa Giraavaru. The frame was transplanted in October 2012

Source: Reefscapers http://reefscapers.com/

Restoration benefits beyond hotel guests

The Reefscapers programme also strives to simultaneously integrate local communities, tourists and luxury area managers. Two main connections have been made with local communities, including the Apprenticeship Program and the Educational Partnership with local schools. The Apprenticeship Program recruits young Maldivians (17–20 years old) and provides them with a 2-year training programme in hospitality. On Landaa Giraavaru, training in marine biology is also available. Apprentices enrolled in the Marine Biology Program have weekly classes on subjects that include coral reefs and coral restoration. This programme provides a government accredited curriculum and nationally recognised competency qualifications (Four Seasons, 2018). Local schools also visit regularly to learn about the programme, the importance of coral reefs and the making of frames (sponsored by the resort).

Seamarc's coral propagation programme is a prime example of how coral restoration efforts can be integrated in the context of luxury tourism. This connection between luxury tourism and environmental protection is essential in the Maldives because high eco-standards have recently been shown to be the primary selling arguments for Maldivian resorts (De-Miguel-Molina et al., 2014).

New Heaven Reef Conservation Program, Koh Tao, Thailand

Koh Tao, which literally translates to "Turtle Island," is a small but economically important island located in the Western Gulf of Thailand. The picturesque island is surrounded by dense coral reefs and submerged granite pinnacles, with some of the highest coral coverage and biodiversity in the Gulf of Thailand (Yeemin et al., 2006; Phongsuwan et al., 2013). Prior to the 1980s, the island was almost completely uninhabited, with the first rooms and resorts only built in 1986. The island has since grown rapidly, having 1,000 rooms by 1995 and reaching 2,000 rooms by 2005 and 140,000 visitors in 2006 (Tourism Authority of Thailand [TAT], 2017). In recent years, the island has become one of Asia's most popular tourist destinations, and is second in the world in terms of the number of scuba divers certified each year (Wongthong & Harvey, 2014). As of 2016, the 19 km² island has 67 dive schools and is the main point of interest for snorkellers and divers visiting from neighbouring islands or coming from the mainland on one-day trips. The burgeoning diving industry has led to rapid development of support infrastructure, such as restaurants, bars and resorts, and the island now receives more than 500,000 visitors per year (Wongthong & Harvey, 2014) – in fact the Tourism Authority of Thailand received 700,000 visitors to the island in 2016.

Development on the island has been almost entirely unregulated, with little to no governance or adherence to planning guidelines such as those published by the Thailand Institute for Scientific and Technical Research in the early 1990s (Szuster & Dietrich, 2014). By 2014, most of the primary forest on the island had been altered (Weterings, 2011; Szuster and Dietrich, 2014) and road construction or development projects have led to large amounts of sedimentation and eutrophication in the island's bays and coral reefs (Larpnun et al., 2011). These pressures are further exacerbated by a complete lack of governance, effective polices or management measures in place to protect the marine resources of Koh Tao (Flumerfelt, 1999; Yeemin et al., 2006).

Several recent studies have identified and thoroughly documented the five most salient reef threats on Koh Tao as: the mass coral bleaching events of 1998, 2010 and 2014 (Yeemin et al., 2006; Chavanich et al., 2012; Hocksema et al., 2013; Phongsuwan et al., 2013); coral disease (Hein et al., 2014); development, run-off, erosion/sedimentation, water pollution and other land-based threats (Larpnun et al., 2011; Weterings, 2011; Szuster & Dietrich, 2014; Romeo, 2014); coral predation (Hoeksema et al., 2013; Scott et al., 2014; Moerland et al., 2016); and over-use by scuba diving, snorkelling and other marine-based tourism (Weterings, 2011; Nichols, 2013, Wongthong & Harvey, 2014; Hein et al., 2014).

The NHRCP was founded in 2007 under the belief that by training divers in reef ecology, research and restoration it is possible to not only reduce the threats to reef health, but to provide widespread positive impacts to local and global coral reefs. Since then, the NHRCP has become the primary group responsible for the monitoring, protection and restoration of the reefs and marine resources around the island. To create a holistic reef management programme, the NHRCP has a wide range of ongoing projects, including reef monitoring and research/publications, coral nursery maintenance, artificial reef construction, coral spawning and larval culturing, clean-ups, coral predator removal, shark monitoring, seahorse studies and much more. The NHRCP also works closely with local stakeholders, universities and government groups on many projects, including installing and maintaining the island's 130+ mooring buoys, rehabilitating and head-starting sea turtles, rearing and releasing giant clams (*Tridacna squamosa*), designing and implementing marine regulations, and zoning.

The NHRCP receives no direct outside funding for the ongoing work it has undertaken for the last 9 years and operates as a Social Enterprise Organization – meaning that it operates in the free market like a normal business but re-invests the majority share of profits back into the programme. Revenue is derived from the sales of marine conservation training courses and certifications, supplying both the operating capital and manpower for the programme's reef management activities. Students and volunteers come from the island and all over the world to receive hands-on and in-depth training in marine conservation theory and skills enabling them to give back to the ecosystems they love to dive in. Participants come from a wide range of backgrounds, including young-gap year travellers looking to do something new and add value to their time abroad; experienced divers looking to try something new and learn about the coral reefs; Masters or PhD students looking to get more dive training and assistance on research and thesis work; and people wanting to work on or even start their own marine management programmes in other areas.

The NHRCP offers courses ranging from 1 day to 4 weeks, or longer internships. During these courses, students are taught about coral reef ecosystems, threats and solutions. They are given time to practice new skills and are taken out to the reef to collect data and contribute to the ongoing projects undertaken by the programme. In this way, the NHRCP not only educates and trains students, but is also able to operate a financially sustainable marine resource management programme that does not rely upon grants, funding partners or governments. This allows the NHRCP to operate more efficiently than many non-profits or non-governmental organisations (NGOs), be able to rapidly respond to reports of new threats or problems, and reallocate resources to where they are most needed on a day-by-day basis.

After the initial success of the NHRCP, the directors of the programme were asked to lead a marine branch of the local community group, Save Koh Tao. This voluntary coalition of many of the island's dive schools lasted from 2008–2014 and led to improved management of the island's coastal areas over the period (Scott & Phillips, 2010). The local community and dive businesses effectively cooperated to manage their own resources with monthly meetings, clean-ups, educational workshops and other activities. This further supports the ideal that free market based conservation programmes that utilise voluntourism and other creative means of income generation can not only support and contribute to conventional management structures (government, NGOs, non-profits, etc.), but also in some cases may be preferable to them (Scott & Phillips, 2010).

NHRCP is unique in its commitment to long-term increases in coral reef abundance and resilience through a focus on preserving and increasing genetic diversity (see Baums, 2008). Unlike traditional and mainstream restoration programmes, the NHRCP never uses so-called "donor corals" for coral propagation. Instead, the NHRCP only uses "corals of opportunity", or those that have naturally been broken or are under threat for other reasons (e.g. burial, predation, etc.) and would most likely not survive on their own. Additionally, the NHRCP has been working since 2010 to utilise coral larval capturing and culturing to produce a genetically distinct and diverse feedstock of corals for restoration. Since 2012, the NHRCP has also been involved in the selective breeding of corals, carefully choosing parent corals based on long-term health and survivorship and using intra-specific hybrids to create more resilient coral larvae and colonies for restoration of marginalised areas.

Between 2007 and 2016, the NHRCP has trained and certified over 900 scuba divers through their marine conservation programme, plus hosted over 1,500 participants as part of 1–3-day group or school trips (ranging from elementary school through college age). They maintain a database of nearly 1,000 monthly reef surveys along permanent transect lines that span from 2006 to 2016. From that data and other research, the NHRCP has published 18 papers in peer-reviewed journals. In addition, over 20 theses or project papers have been written by students in the NHRCP programme. They have released hundreds of sea turtles and thousands of giant clams, collected tens of thousands of corallivorous *Drupella spp.* snails and destroyed over 1,000 crown-of-thorns starfish. The NHRCP is continually constructing and deploying artificial reefs in marginalised or disturbed areas, and also using artificial reefs and sculptures to create new reefs in sandy areas. The group has also collected, rehabilitated and transplanted hundreds of thousands of corals using nurseries and transplanting to artificial reefs or natural substrates (see Figure 10.3).

FIGURE 10.3 Examples of volunteers' involvement in coral restoration efforts with the New Heaven Reef Restoration Program. A: Transplanting corals in bottle nurseries; B: Moving concrete structures to restoration site; C: Maintaining established metal structures; D: Building new metal structures

Source: Photo courtesy of Chad Scott.

Koh Tao is an example of the environmental and social problems that can arise from the rapid growth and development brought about by tourism and how local businesses can act to offset their impacts and even help to mitigate problems using creative business models and voluntourism. The lack of effective governance and failure of non-profits and NGOs to respond effectively and timely is not unique to Koh Tao, and is also a common trend seen in the region. The NHRCP has proven to be an effective and sustainable programme over its 9 years on the island. The same business model is adaptable to a wide variety of conservation needs and ecosystem types, and provides an example of an alternative to traditional environmental protection structures. This model not only allows for a holistic coral reef management and restoration programme, but also creates jobs, contributes to the local economy and encourages so called "green" or responsible tourism. By educating and training students and divers in real-world conservation techniques, the programme is helping to dissolve the line between the public and professional resource managers to hopefully help communities stand up for themselves and take responsibility for protecting their own resources.

Concluding thoughts

Both case studies show a growing interest by tourists in coral restoration at both ends of the marine tourism industry spectrum from backpacking-style tourism in Thailand to a five-star resort in the Maldives. These two projects work towards a common goal but with very different styles. The luxury setting in the Maldives encourages tourists to invest more money than time in the restoration efforts. Resort guests pay to spend a few precious hours with local marine biologists to attach corals to a coral frame. In Thailand, tourists invest more time than money, as they are involved for a period ranging from 1 day to 3 months. The differences in the type of coral restoration performed also stems from the different type of threats affecting the two locations. Koh Tao's corals are strongly impacted by localised anthropogenic factors, and NHRCP thus focuses on the education of locals and tourists. Their restoration efforts have also led to a myriad of other conservation actions, such as setting up mooring lines or facilitating rules and regulations for the island's reef resources. In the Maldives, coral reefs are more strongly impacted by climate change related threats like rising sea temperatures. Reefscapers' restoration efforts are thus more focused on increasing the quantity and quality of the corals that survive to improve both the reef resilience and the reef aesthetics for the tourists.

The involvement of citizen science tourists in the coral restoration efforts convey many advantages to the restoration programmes:

1. More hands are available, resulting in more coral fragments planted in minimum time and over larger areas. Scale limitations, both temporally and spatially, are one of the main criticism of coral restoration (Yap, 2003; Precht et al., 2005; Edwards & Gomez, 2007), and increasing the size of the workforce through ecotourism appears as an efficient way to address some of these challenges.
2. Direct opportunities for restoration managers to educate a large crowd to the current threats to coral reef ecosystems. Both of these programmes have strong focus on education. Reefscapers offers lectures on coral restoration and hotel guests are able to have one-on-one experiences with local marine biologists; NHRCP offers daily lectures on a range of marine biology topics from marine invertebrate ecology to reef monitoring techniques. Moreover, in both locations, tourists physically transplant coral fragments either on artificial structures or on the reef itself. These hands-on approaches to conservation actions have significant potential for fostering increased conservation stewardship. The two projects also offer regular and comprehensive follow-up information about the restoration efforts in the form of emails, newsletter and strong presence on social media.
3. Reduce the costs of restoration actions. In both case studies, guests, students and interns are directly financing the restoration efforts. These economic contributions not only allow further scaling of the restoration actions but are also

paramount to the long-term sustainability of these restoration programmes. Corals are long-lived organisms with slow life-history strategies, and restoring reef areas is not just about planting corals but also about making sure these transplants survive and contribute to strengthening overall reef resilience. Such endeavours require long-term thinking with regular maintenance of the restored areas (e.g. predator control, reattachment of fallen fragments) and long-term monitoring for enhanced learning and adaptive management. A business approach to restoration programmes of the kind outlined in this chapter can help secure long-term funding and thus improve the prospect of coral restoration success.

Coral restoration science is still in its early stages and much is yet to be learned in how to maximise overall effectiveness. The long-term ecological success of these projects is uncertain given the rate of warming ocean temperatures and the devastating effects of bleaching on coral reefs. The increasing number of active coral restoration actions across the world's reef regions is a prime example of the will to step up and actively attempt to preserve these precious ecosystems. The projects presented in this chapter are only two of many examples of tourism-driven coral restoration efforts around the world's reefs, and yet they are evident demonstrations that ecotourism can have a critical role to play in the facilitation of coral restoration action.

References

Baums, I.B. (2008). A restoration genetics guide for coral reef conservation. *Molecular Ecology*, 17(12), 2796–2811.

Bayraktarov, E., Saunders, M.I., Abdullah, S., Mills, M., Beher, J., Possingham, H.P., . . . & Lovelock, C.E. (2015). The cost and feasibility of marine coastal restoration. *Ecological Applications*, 26, 1055–1074.

Bianchi, C.N., Morri, C., Pichon, M., Benzoni, F., Colantoni, P., Baldelli, G., & Sandrini, M. (2006). Dynamics and pattern of coral recolonization following the 1998 bleaching event in the reefs of the Maldives. In: Suzuki, Y., Nakamori, T., Hidaka, M., Kayanne, H., Casareto, B. E., Nadaoka, K., . . ., & Tsuchiya, M. (Eds.), *Proceedings of 10th International Coral Reef Symposium*. Okinawa, Japan: Tokyo Japanese Coral Reef Society, pp. 30–37.

Bozec, Y.M., & Mumby, P. J. (2015). Synergistic impacts of global warming on the resilience of coral reefs. *Philosophical Transactions of the Royal Society B*, 370(213), 267.

Chavanich, S., Viyakarn, V., Adams, P., Klammer, J., & Cook, N. (2012). Reef Communities after the 2010 Mass Coral Bleaching at Racha Yai Island in the Andaman Sea and Koh Tao in the Gulf of Thailand. *Phuket Marine Biology Centre Research Bulletin*, 71, 103–110.

De-Miguel-Molina, B., De-Miguel-Molina, M., & Rumiche-Sosa, M.E. (2014). Luxury sustainable tourism in small island developing states surrounded by coral reefs. *Ocean & Coastal Management*, 98, 86–94.

Dela Cruz, D.W., Villanueva, R.D., & Baria, M.V.B. (2014). Community-based, low-tech method of restoring a lost thicket of *Acropora* corals. *ICES Journal of Marine Science*, 71(7), 1866–1875.

Delbeek, J. (2001). Coral farming: Past, present and future trends. *Aquarium Sciences and Conservation*, 3, 171–181.

Edwards, A.J., & Gomez, E.D. (2007). *Reef restoration concepts & guidelines: Making sensible management choices in the face of uncertainty*. Coral Reef Targeted Research & Capacity Building for Management Program, St. Lucia, Queensland, Australia.

Edwards, A.J., Clark, S., Zahir, H., Rajasuriya, A., Naseer, A., & Rubens, J. (2001). Coral bleaching and mortality on artificial and natural reefs in Maldives in 1998, sea surface temperature anomalies and initial recovery. *Marine Pollution Bulletin*, 42, 7–15.

Epstein, N., Bak, R.P.M., & Rinkevich, B. (2001). Strategies for gardening denuded coral reef areas: The applicability of using different types of coral material for reef restoration. *Restoration Ecology*, 9(4), 432–442. http://doi.org/10.1046/j.1526-100X.2001.94012.x

Flumerfelt, S.L. (1999). Dive tourism on Koh Tao, Thailand: Community heterogeneity and environmental responsibility (Thesis). Department of Sociology & Anthropology, University of Guelph, ON.

Food and Agriculture Organization (FAO). (1997). Regional workshop on the conservation and sustainable management of coral reefs. 'Profile and status of coral reefs in Maldives and Approaches to its management' by Abdulla Naseer. Marine Research Section, Ministry of Fisheries and Agriculture, Male, Republic of Maldives.

Four Seasons. (2018). Training young Maldivians to become tomorrow's hoteliers. Retrieved from http://www.fourseasons.com/maldiveslg/my_four_seasons/training_young_maldivian s_to_become_tomorrows_hoteliers/

Goreau, T.J., & Hilbertz, W. (2012). Reef restoration using seawater electrolysis in Jamaica. In: Goreau, T.J., & Trench, R.K. (Eds.), *Innovative Methods of Marine Ecosystem Restoration*. Boca Raton, FL: CRC Press, pp. 35–45.

Goreau, T.J., McClanahan, T., Hayes, R.L., & Strong, A. (2001). Conservation of coral reefs after the 1998 global bleaching event. *Conservation Biology*, 14(1), 1–18.

Hein, M.Y., Lamb, J.B., Scott, C., & Willis, B.L. (2014). Assessing baseline levels of coral health in a newly established marine protected area in a global scuba diving hotspot. *Marine Environmental Research*. doi: 10.1016/j.marenvres.2014.11.008

Hoegh-Guldberg, O., & Bruno, J.F. (2010). The impact of climate change on the world's marine ecosystems. *Science*, 328(June), 1523–1528.

Hoeksema, B.W., Scott, C.M., & True, J.D. (2013). Dietary shift in coralivorous *Drupella* snails following a major bleaching event at Koh Tao, Gulf of Thailand. *Coral Reefs*, 32(2), 423–428.

Intergovernmental Panel on Climate Change (IPCC). (2014). *Climate Change 2014: Synthesis Report*. Contribution of Working Groups I, II and III to the Fifth Assessment Report of the Intergovernmental Panel on Climate Change [Core Writing Team, R.K. Pachauri & L.A. Meyer (Eds.)]. Geneva: IPCC.

Jaleel, A. (2013). The status of the coral reefs and the management approaches: The case of the Maldives. *Ocean & Coastal Management*, 82, 104–118.

Larpnun, R., Scott, C.M., & Surasawadi, P. (2011). Practical coral reef management on a small island: Controlling sediment on Koh Tao, Thailand. In: Wilkinson C., & Brodie, J. (Eds.), *Catchment Management and Coral Reef Conservation*. Townsville, Australia: Global Coral Monitoring Network and Reef and Rainforest Research Centre, pp. 94–95.

Marine Savers. (2016). Coral Propagation Program. Retrieved from marinesavers.com/coral-propagation

McClanahan, T.R. (2000). Bleaching damage and recovery potential of Maldivian coral reefs. *Marine Pollution Bulletin*, 40(7), 587–597.

McClanahan, T.R. (2011). Coral reef fish communities in management systems with unregulated fishing and small fisheries closures compared with lightly fished reefs – Maldives vs. Kenya. *Aquatic Conservation*, 21, 186–198.

McClanahan, T.R., & Muthiga, N.A. (2014). Community change and evidence for variable warm-water temperature adaptation of corals in Northern Male Atoll, Maldives. *Marine Pollution Bulletin*, 80, 107–113.

Moerland, M.S., Scott, C.M., & Hoeksema, B.W. (2016). Prey selection of corallivorous muricids at Koh Tao (Gulf of Thailand) four years after a major coral bleaching event. *Contributions to Zoology*, 85(3), 291–309.

Monty, J.A., Gilliam, D.S., Banks, K.W., Stout, D.K., & Dodge, R.E. (2006). Coral of opportunity survivorship and the use of coral nurseries in coral reef restoration. *Proceedings of the 10th International Coral Reef Symposium*, 1165–1673.

Nichols, R. (2013). Effectiveness of artificial reefs as alternative dive sites to reduce diving pressure on natural reefs, a case study of Koh Tao, Thailand. BSc Thesis in Conservation Biology, University of Cumbria, Cumbria, UK.

Pandolfi, J.M. (2003). Global trajectories of the long-term decline of coral reef ecosystems. *Science*, 301(5635), 955–58. doi:10.1126/science.1085706

Phongsuwan, N., Chankong, A., Yamarunpatthana, C., Chansang, H., Boonprakob, R., Petchkumnerd, P., . . ., & Bundit, O. (2013). Status and changing patterns on coral reefs in Thailand during the last two decades. *Deep-Sea Research II*, 19–24.

Precht, W.F., Aronson, R.B., Miller, S.L., Keller, B.D., & Causey, B. (2005). The folly of coral restoration programs following natural disturbances in the Florida Keys National Marine Sanctuary. *Ecological Restoration*, 23(1), 24–28. doi:10.3368/er.23.1.24

Rinkevich, B. (1995). Restoration strategies for coral reefs damaged by recreational activities: The use of sexual and asexual recruits. *Restoration Ecology*, 241–251.

Romeo, L. (2014). Tracing anthropogenic nutrient inputs using δ15N levels in algae tissue Koh Tao, Thailand. Masters Thesis, MAS Marine Biodiversity and Conservation, CMBC, Scripps Institute of Oceanography, UCSD.

Scott, C.M., & Phillips, W.N. (2010). A sustainable model for resource management and protection achievable through empowering local communities and businesses. *Proceedings of Ramkhamhaeng University International Research Conference 2010*, 13–14 January, Bangkok, Thailand.

Scott, C.M., Mehrotra, R., & Urgell, P. (2014). Spawning observation of *Acanthaster planci* in the Gulf of Thailand. *Journal of Marine Biodiversity*. doi: 10.1007/s12526–014–0300-x

Seamarc. (2015). Systems engineering and marine consulting in the Maldives. Retrieved from http://seamarc.com

Sheppard, C.R.C. (1999). Coral decline and weather patterns over 20 years in the Chagos Archipelago, Central Indian Ocean. *Ambio*, 28, 472–478.

Silvertown, J. (2009). A new dawn for citizen science. *Trends in Ecology and Evolution*, 24, 467–471.

Szuster, B.W., & Dietrich, J. (2014). Small island tourism development plan implementation: The case of Koh Tao, Thailand. *Environment Asia*, 7(2), 124–132.

Theobald, E.J., Ettinger, A.K., Burgess, H.K., DeBey, L. B., Schmidt, N.R., Froehlich, H.E., . . ., & Parrish, J.K. (2015). Global change and local solutions: Tapping the unrealized potential of citizen science for biodiversity research. *Biological Conservation*, 181, 236–244. doi:10.1016/j.biocon.2014.10.021

Tourism Authority of Thailand (TAT). (2017). About Thailand. Retrieved from https://www.tourismthailand.org/About-Thailand/Destination/Ko-Tao

Weterings, R. (2011). A GIS-based assessment to the threats to the natural environment on Koh Tao, Thailand. *Kasetsart Journal of Natural Sciences*, 45, 743–755.

Wilkinson, C., Linden, O., Cesar, H., Hodgson, G., Rubens, J., & Strong, A.E. (1999). Ecological and socioeconomic impacts of 1998 coral mortality in the Indian Ocean: An ENSO impact and a warning of future change? *Ambio*, 28, 188–196.

Wongthong, P., & Harvey, N. (2014). Integrated coastal management and sustainable tourism: A case study of the reef-base SCUBA dive industry from Thailand. *Ocean & Coastal Management*, 95, 138–146.

Yap, H.T. (2003). Coral reef 'restoration' and coral transplantation. *Marine Pollution Bulletin*, 46(5), 529. doi:10.1016/S0025–326X(03)00040–7

Yeemin, T., Sutthacheep, M., & Pettongma, R. (2006). Coral restoration projects in Thailand. *Ocean and Coastal Management*, 49, 562–575.

11

TOURISM AND FISHING IN PARADISE

A case study of the Maldives

Kelsey Miller

Introduction to the Maldives

The Republic of the Maldives is fortunate to have tropical waters teeming with colourful fish and coral reefs, with abundant turtles, charismatic megafauna, invertebrates and other marine fauna. At least 1,200 species of fish, 173 species of scleractinian corals and many more marine organisms inhabit the Maldivian reef systems. Rich in marine natural resources but with few other natural resources, the country's top two industries of tourism and fisheries revolve around marine life.

The Maldives is among the world's top luxury tourist destinations, famous for its exclusive five- and six-star resorts that compete for extravagance. The country has been considered a success story for its tourism development and sustainability (Domroes, 2001), a description that is frequently repeated by resort advertisements and the government's stated goal to be the leader in sustainable tourism. It is among the world's top destinations for honeymooning, recreational diving and other water activities. In 2014, over 1.2 million tourists visited, approximately three times the country's population (Ministry of Tourism [MoT], 2015a). However, as discussed by Hein et al. in Chapter 10, the country faces serious problems related to climate change and coral bleaching in particular.

Fishing greatly contributes to food security, gross domestic product (GDP) and employment. Roughly 96% of animal protein comes from fish and the Maldives has the highest per capita fish consumption in the world (Food and Agriculture Organization [FAO], 2014). Fishing was the largest industry prior to the rise of tourism. The country's main fishery is pole-and-line skipjack tuna, which was certified as sustainable by the Marine Stewardship Council (MSC) in 2012 (MSC, 2016). In comparison, reef fish fisheries, bait fisheries and the former shark fisheries draw from coral reef resources. In many parts of the world, fisheries and tourism are frequently in conflict with each other because tourism depends on fish remaining in the water, and fishing extracts fish, potentially impacts on marine habitat and can be considered unsightly. However, these two industries have coexisted rather peacefully in the Maldives.

This chapter examines the factors that have led to the success of tourism and fishing in the Maldives. It describes how the Maldives has established itself as an elite tourist destination while sustaining a large fishing industry, explains how it manages its marine ecosystems and suggests potential strategies for improving and mitigating current and future threats that may emerge. This chapter evaluates the tourism and fishing industries and discusses their impact on and relationship to each other.

Country background

The Maldives is located in the central Indian Ocean and is comprised of 1,190 small islands across 26 coral atolls (Figure 11.1). The country covers almost 90,000 km² of water, with approximately 300 km² of land, roughly 99.5% is ocean (National Bureau of Statistics [NBS], 2015b). Over 80% of the country is less than 1 m above sea level and the highest natural point is only 2 m above sea level. It is thus considered to be among the most vulnerable countries to climate change effects, such as sea level rise and storm surges (Becken et al., 2011).

The country had an estimated population of 402,071 in 2014, with over a third of the population living in the nation's capital, Malé (NBS, 2015a), which has an area of 5.8 km². Over one-third of the country's population lives on small islands with less than 1,000 residents (NBS, 2015a). These islands typically have little infrastructure and few employment opportunities. Services such as paved roads, running water, schools and health care are expensive and rare. Many of these islands are disproportionally dependent on fishing for food security and income. The Maldives is only one of four countries that has graduated from a least developed country to a developing country in 2011, yet 24% of the population still lives below the international poverty line of US$2/day (NBS, 2015b).

Tourism in the Maldives

Tourism is the largest industry in the Maldives, accounting for over 25% of GDP, and tourist receipts total US$2.6 billion (MoT, 2015b; NBS, 2015b). The industry has experienced rapid growth since its inception in 1972 and is largely responsible for the escalating GDP. The world's wealthy come to the 111 resorts, with some charging over US$30,000 per night. Almost all resorts follow the standard of "one island, one resort", and each resort is on its own private island, which is only for resort guests and employees (Figure 11.2). The Maldives has won numerous awards, such as World's Leading Island Destination, World's Most Romantic Destination from World Travel Awards, "Seven Star Destination Winner" from Seven Star Global Luxury Awards, and numerous diving destination awards.

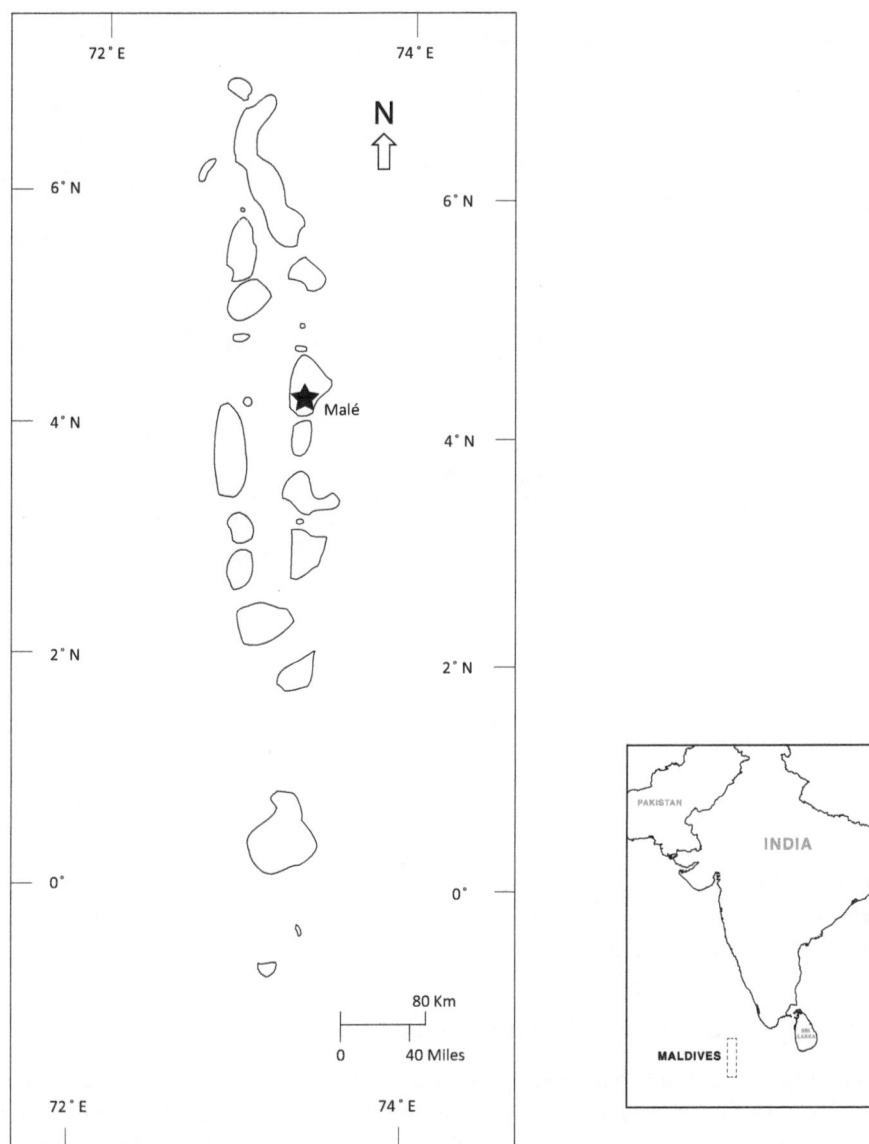

FIGURE 11.1 Map of the Republic of the Maldives

Since the opening of two resorts in 1972, the industry has experienced sustained rapid growth, with a recent 5-year average annual increase of 13% (MoT, 2015b). Until 2008, hotels and guest houses were not allowed on inhabited local islands, which separated local people from access to tourists and discouraged budget travel. Resorts account for over 93% of all tourist nights in the Maldives and "safari boats" – luxury vessels that can visit multiple areas – constitute about 2% of tourist nights (MoT, 2015b).

FIGURE 11.2 Resort island and tourism in the Republic of the Maldives

Source: Photo courtesy of Paola Mattana Lamperti.

The Maldives tourism is described as sophisticated and the country is among the world's most expensive tourist destinations (Zubair & Bouchon, 2014). Tourists come to the Maldives primarily for rest and relaxation (63%), snorkelling and diving (17%) and honeymooning (15%), choosing Maldives for its natural beauty (MoT, 2015a). While diving and snorkelling are still top activities, historically this may have been higher, with up to half of all tourists visiting for that reason (Anderson & Waheed, 2001).

Developing a sustainable tourism industry

The tourism industry has been praised for its approach to tourism development and has been commended for its planning and environmental considerations (Domroes, 2001, 2008; UN World Tourism Organization [UNWTO], 2004). Many small island developing states (SIDS) rely heavily on tourism and the Maldives is no exception. There were few controls on tourism development in the early years of development and the first law concerning tourism was enacted about 7 years after its inception (Maldives Foreign Investment Act in 1979). The first law specific to tourism was introduced 20 years later (Maldives Tourism Act in 1999).

Since 1980, tourism development has been guided by national tourism strategies which have focused on a Quality Tourism Strategy that aimed at catering to the luxury market (Henderson, 2008). However, these strategies have no regulatory authority and require further legislation or voluntary actions to be effective. The First Tourism Master Plan (TMP) in 1983 was contracted by Danish consultants to help guide the industry. It focused tourism in the central atolls for logistical reasons and highlighted conservation to ensure viable tourism. The Second TMP (1996–2005) encouraged more widespread development of tourism and planned to position the Maldives as a "premium marine eco-destination". The Third TMP (2007–2011) encouraged more sustainable and equitable development, focused on sophistication and responsibility and saw the country as a model of environmentally and socially sustainable tourism. It was also the first to acknowledge the importance of fisheries and shared management of resources. The Fourth TMP (2013–2017) included goals regarding the environment and highlighted waste management and management plans for sensitive marine habitats. It also acknowledged potential resource conflict with the fishing industry.

Tourism is managed under the Ministry of Tourism and the Ministry of Housing, Transport and Environment. The primary environmental regulatory laws are the Maldives Tourism Act 1999 and the Maldives Preservation Act 1993. Regulations for resorts govern water production, coral mining, waste treatment and limits on development (Becken et al., 2011; Domroes, 2001). Environmental impact assessments (EIA) are required for developments or improvements that could impact on waterways, reefs or terrestrial environments. While EIA thoroughness has been questioned (Scheyvens, 2011; Zubair et al., 2011), regulations are increasingly strict. Furthermore, many resorts go far beyond government regulations; the first completely solar powered resort opened in the Maldives in 2016 and many resorts have received national and international environmental awards.

Dive tourism has generated many of the conservation incentives for healthy aquatic resources. The value of shark dive tourism was first evaluated in 1993 and estimated at US$2.3 million (Anderson & Ahmed, 1993). Diving and snorkelling with manta rays was valued at US$8.1 million in 2011 in direct revenue (Anderson et al., 2010) and whale sharks were estimated to generate between US$7.6–9.4 million from one marine protected area alone (Cagua et al., 2014). Total revenue derived from this tourism (e.g. accommodation, food, transport) was not estimated, but is likely to be much higher.

Tourism negatives yet to be addressed

Environmental standards for tourism

Some researchers have pointed out that some resort aspirations of sustainability have not been met (e.g. Becken et al., 2011; Zubair et al., 2011). Similarly, the tourism industry has not addressed climate change impacts of luxury tourism and large carbon footprints associated with long-haul travel and imports for construction and development. Many resorts work independently on projects such as coral restoration, solar power, waste management and turtle rehabilitation, but country-wide standards could be improved.

Education of tourists can help to promote conservation and is often an essential part of ecotourism. Some dive companies brief tourists about proper dive procedures, including environmentally friendly techniques and behaviours around sensitive organisms; however, not all do (Anderson et al., 2010). Educating tourists about the fragility of coral reef environments and encouraging them to participate and donate to conservation projects could provide much needed resources and attention.

Socio-economic sustainability

High-end tourism has brought exceptional growth to Maldivian GDP, and with such a small population, the per capita GDP is substantial. However, the benefits of tourism are not distributed evenly (Rasheed, 2014; Scheyvens, 2011). While the Maldives has a gross net income four times higher than the South Asian average (World Bank Group, 2016), the Maldives also has greater inequality than nearby countries, generating strong criticism (Rasheed, 2014, 2015). Due to the all-inclusive nature and isolation of resorts, many other businesses, such as crafts, restaurants and transportation, do not directly benefit from tourism.

Many resort managers, owners and employees are foreigners, further reducing local benefits. In 2014, 42.3% of resorts were managed by foreigners, with an additional 17.1% operated as joint ventures and a high proportion of leaseholders (32.4%) and operators (56.7%) are foreign or joint ventures (MoT, 2015b). This creates economic leakages as benefits from tourism are exported to other countries (Henderson, 2008; Scheyvens, 2011; Shakeela & Cooper, 2009). While regulations require at least 50% of resort workers to be Maldivian, foreigners are hired if properly trained locals are unavailable (Shakeela & Cooper, 2009). Up to US$30 million or 5% of GDP was estimated to be sent overseas by foreign workers in 2001 (Shakeela & Cooper, 2009). Less than 2% of resort employees are Maldivian women, and some authors have suggested alternative housing/working arrangements would encourage them (Scheyvens, 2011). Understanding the causes of low local employment can help develop strategies to recruit and retain qualified Maldivians.

Promoting environmental research and sustainability

Several researchers have highlighted that although a large amount of revenue is derived from marine ecotourism, very little revenue is used for conservation (Bhat et al., 2014). However, a Green Tax was introduced in November 2015 as an amendment to the Maldives Tourism Act. Tourists are charged $6 per night with the stated goal being to assist in waste management. Over half of all conservation funding is from overseas sources and finding more stable income sources is required (Bhat et al., 2014). Less than 1% of national public sector budget allocations go to environmental areas, and only 3% of all donor support to the country (Emerton et al., 2009). Some money generated from tourism, especially dive tourism, could be used to support scientific research and conservation efforts (Anderson et al., 2010;

Topelko & Dearden, 2005). Similarly, tourism donations, either through volunteer programmes or financial contributions, could also support the marine resources that benefit this industry, such as coral restoration, clean-ups and waste management.

Fishing in the Maldives

Fishing was the lifeblood of the Maldives for at least a thousand years until 1978 when it was surpassed by tourism (Sinan, 2011). It is still significant for the economy, employment and food. Fishing techniques and practices have changed little over the centuries, although the boats are now powered by diesel rather than by sail, and are frequently larger. Approximately 1,000 fishing boats catch tuna by pole-and-line (Figure 11.3) and handline, targeting the prized skipjack (*Katsuwonus pelamis*) and yellowfin tuna (*Thunnus albacares*; Adam et al., 2014).

Tunas and tuna-related species account for roughly 90% of fisheries landings; other marine fishes (e.g. reef fishes) account for only about 10% of the country's landings (Adam, 2006). Reef fisheries developed in the late 1990s, largely to provide for the expanding tourism sector (Sinan, 2011). The reef fishery is a small mixed fishery of "white fish", including jacks (Carangidae), snappers (Lutjanidae) and emperors (Lethrinidae; Anderson, 2006), with estimates of roughly 10,000–29,000 metric tonnes (Sattar et al., 2014). Reef fisheries are primarily driven by tourism, as local Maldivians traditionally prefer tuna. Sattar et al. (2012) estimated 7,100 tonnes

FIGURE 11.3 Pole-and-line fishing in the Maldives

(round weight, i.e. the weight of the fish that is caught, not the portion that is just meat) were consumed in 2006 by tourists – a three-fold increase since 1988. Shark fishing used to be a minor fishery of roughly 1,440 tonnes/year (Sinan, 2011) and giant clam was harvested for a couple of years, but ceased due to overexploitation (Ahmed et al., 1997). Aquarium fish, bêche-de-mer (much more extensively in former years) and grouper are also fished (Sinan, 2011).

Baitfish harvested for the pole-and-line and handline tuna fisheries are significant at approximately 15,000 tonnes/year (Adam, 2006), although estimates have not been updated in a decade. These fish, caught inside the atolls, are rarely landed and so they are not included in national statistics. Live baitfish species used in tuna fisheries include small pelagic and reef-associated species, comprised primarily of silver sprat (*Spratelloides gracilis*), as well as blue sprat (*Spratelloides delicatulus*), anchovy (*Encrasicholina heteroloba*), several species of cardinalfishes (Apogonidae), fusiliers (Caesionidae) and damselfishes (*Chromis* spp.; Jauharee et al., 2015).

Developing sustainable fisheries

The pole-and-line skipjack fishery was certified as sustainable by the MSC in November 2012. The success of Maldivian fisheries is due in part to ecological factors. The pelagic nature of the main fishery means impacts to the reef environment have been minimal, and little bycatch is caught (Miller et al., 2017). The relatively small human population and abundant reef systems, especially in comparison to many south-east Asian reefs, has also limited the risk of overfishing.

Some of the strongest environmental fisheries regulations were put in place for cultural reasons, not conservation. Most types of net fishing (pelagic gillnetting, trawling and purse seining) were banned to protect the interests of local pole-and-line tuna fishermen and to retain the employment of approximately 10,000 Maldivian fishermen (Anderson et al., 2012). Restrictions on foreign workers have also helped to maintain the traditional Maldivian culture of fishing. In 1998, shark fishing was banned in the central tourist zone, with a full ban throughout the country 10 years later (see later). Pre-emptive restrictions on exports of baitfish, corals, etc., were also introduced.

Certification by the MSC came with conditions that required improved research and fishery monitoring, along with stronger management plans, both of which are in progress. Tracking of fish purchases for commercial sale and mandatory vessel registration and logbooks have been implemented. The Maldives became a full member of the Indian Ocean Tuna Commission (IOTC) in 2011.

Fishing negatives yet to be addressed

Fisheries science research

All reef fisheries and some aspects of tuna fisheries require substantial research to better determine appropriate harvest levels and assess current harvest rates. Priority needs to be given to species that may have been regionally overexploited, such as grouper, bêche-de-mer and spiny lobster (Anderson, 2006; Gillett, 2004). Specifically, reef fish and baitfish require stock assessments and accurate land-ings statistics. Baitfishing, essential to pole-and-line and handline tuna fisheries, is poorly understood. Because baitfish are rarely landed, only coarse estimates of catches and species composition have been conducted. There are some indica-tions of regional overexploitation (Anderson, 2006) and research into improving baitfish utilisation would be helpful. Similarly, accurate metrics of reef fish sold directly to islands or resorts are needed. Fisheries science is conducted by the Marine Research Centre (MRC) and Ministry of Fisheries and Agriculture (MoFA) with limited funds and staff. Few resources are available to manage fisheries research throughout the country, and new funding sources need to be identified. For example, these might be partnering with foreign research organi-sations or channelling more national resources to the MRC to ensure adequate harvest limits are determined and not exceeded.

Fisheries management and regulation

Refining national harvest limits, gear use and marine protected areas (MPAs) may be useful, depending on research findings. Creating effective national management plans for all fisheries, especially those of concern, should be a top priority. Tuna fisheries rely on the cooperation of other countries to set and adhere to fishery targets and regulations and, as such, the Maldives should continue its membership and support within the IOTC.

Historically, Maldivian fishermen have been unfamiliar with the concept of overfishing. As tunas have always been plentiful, fishermen have not learned painful lessons of overfishing, leading to a view that fish cannot be overfished (Anderson, 2006). The fishing culture has therefore not prioritised conservation, and there are few fishing regulations, restrictions, MPAs or taboos that have developed in Oceania and other regions (Johannes, 1978). Thus, in addition to developing regu-lations, education will be necessary to increase awareness.

Enforcement of new and existing regulations

Developing proper management plans and regulations are critical to reef fish fish-eries, but enforcement may be equally significant. Little enforcement is available and the dispersed geography of the country purports that costs will be significant. Local stakeholders were rarely engaged in planning marine regulations (Anderson, 2006; Hameed & Ali, 2001), potentially reducing willingness to comply. Poaching has been noted for bêche-de-mer and sharks, using illegal fishing gear and fishing within restricted areas (Anderson, 2006, personal observation). Despite bans on shark fishing, shark numbers have continued to decline (Anderson et al., 2010). However, many of the most destructive methods are not used (e.g. dynamite, trawl, muro-ami). In addition to enforcing regulation on Maldivian fishermen, enforce-ment on foreign fleets may be necessary. Fishermen claim that large amounts of fish are taken illegally from Maldivian waters by unlicensed foreign fleets, but the country does not have monitoring capacity for the expansive waters. Enforcement may require assistance from other countries.

Decline of fishery stocks may have other consequences. As shallow-water marine resources become scarce, fishermen began using scuba to access deeper dwelling stocks. Many of these fishermen have little dive training or knowledge of decompression illnesses and dive deeper and longer than recommended for safe diving in the live baitfish fishery (personal observation) and bêche-de-mer fishery (Anderson, 2006), with potential for serious injury or death.

Industry comparison

Financial value and contribution to GDP are only two ways marine resources can be valued (Table 11.1). Tourism and fishing can also be evaluated through the number of jobs (or more specifically, jobs for Maldivians), distribution of economic benefits, potential future values for tourism or fisheries, food security, food for tourists, recreation (for tourists or locals) and ecosystem values and services (see Anderson, 2006), although many of these are difficult to quantify. Although

TABLE 11.1 Comparison of tourism and fisheries in the Maldives

	Tourism	Fishing
Dollar value	Receipts of $2.6 billion in 2014	Valued at US$118 million in 2013
GDP contribution	25% of GDP	2.8% GDP from fisheries and agriculture
Former GDP contribution	Exponential growth since 1972	18% in 1980s
Employment	15,892 Maldivian; 15,216 (foreigners)	9,940 (all Maldivian); not including processing jobs
Culture	Recent expansion (1972)	Long heritage (centuries)
Marine resource use	Non-extractive	Extractive
Export	N/A	98% of all export value is marine products
Socio-economic	Contributes majority of funds to government budget	Provides income and food to areas most needed
Security of industry	Dependent on world economy, politics, marine health	Dependent on fish stocks and marine health
Environmental impacts	Energy and resource use for transport, imports, construction & development and luxury lifestyle	Extraction of resources, fuel use and construction and development resources
Benefits to other industry	Alternative jobs, support of fisheries through purchases	Provides food resources for tourism, can be a draw for tourists

Source: Compiled from multiple sources.

TABLE 11.2 Timeline of relevant fisheries, tourism and conservation events in the Maldives

	1970s	1980s	1990s
Tourism	1972 Tourism begins with opening first two resorts	1980 Introduction of tourism, airport departure and import taxes 1981 Hulhule Island International airport opens	1990 Coral mining is banned in some areas; further regulations are established 1990 National Environmental Action Plan I: Concerns environmental conservation and reducing impacts of development and includes a regulatory framework 1999 Tourism Act passed: Concerns tourism development, leases for resorts, hotels and marinas, operations and standards 1999 National Environmental Action Plan II is passed: Concerns sustainable tourism, including environmental and socio-economic development
Fishing	1970s Rapid expansion of shark fisheries in Maldives	1981 Shark fishing is prohibited during daytime in tuna fishing areas 1987 Fisheries Law #5/87 passed: Concerns all aspects of fisheries within Maldives waters	1991 Export of giant clams is prohibited; their harvest closes in 1995 1992 Shark fishing with livebait is prohibited near tuna schools while other vessels are tuna fishing 1993 Ban of cetacean fishing is passed 1995 Export of rays is prohibited 1995 Ban of fishing for whale sharks is passed 1996 Ban of shark fishing or any other fishing detrimental to pole-and-line tuna fishing within 3 miles radius of any fish aggregating device (FAD) 1998 10-year moratorium on shark fishing in central tourism zone
Conservation			1992 Maldives ratifies the Convention on Biological Diversity 1993 First valuation of shark fisheries and reef shark diving tourism 1995 First 15 marine protected areas (MPAs) are declared 1999 Ten more MPAs are established
Other			1993 Environmental Protection and Preservation Act of Maldives: Includes guidelines for protection and preservation, requirements on Environmental Impact Assessments and plans for harmful substances

(continued)

TABLE 11.2 (*continued*)

2000s	2010s
Tourism	
2006 Maldives Tourism Development Corporation is created to allow wider tourism investment	2010 Second amendment to the Maldives Tourism Act regulates development of marinas and changes lease duration for resorts
2006 Regulation and Protection and Conservation of the Environment in the Tourism Industry is passed	2012 Fishing is restricted within resort marine areas
2007 Environmental Impact Assessment Regulation passes	2015 Maldives introduces Green Tax for tourists
Fishing	
2009 Shark fishing is banned within 12 nautical miles of any atoll	2010 Ban on shark fishing throughout Maldives
	2011 Ban on capture, keeping, trade or harming sharks
	2011 Maldives becomes a full member of the Indian Ocean Tuna Commission
	2012 Maldivian skipjack tuna pole-and-line fishery is Marine Stewardship Council (MSC) certified
Conservation	
2002 The National Biodiversity Strategy and Action Plan of the Maldives passed: Concerns value, conservation and sustainable resource use	2011 Baa Atoll is declared UNESCO World Biosphere Reserve
2004 Additional MPA is declared in Addu Atoll	2013 Standards for waste disposal are set via Waste Management Regulation
2005 Three more MPAs are declared in Baa and South Ari Atoll	
2004 Maldives establishes the Atoll Ecosystem Conservation Project to promote ecosystem conservation and sustainable development on Baa Atoll	2013 Dredging or land reclamation on environmentally sensitive habitats is prohibited
2009 Three more MPAs are declared, including Hanifaru Bay, Agafaru and the South Ari Atoll	
Other	
2001 Country graduates from least developed to developing country	
2004 Earthquake near Indonesia creates a tsunami causing 82 deaths and making 15,000 homeless in the Maldives	

Source: Compiled from multiple sources.

less profitable than tourism, fisheries are essential to the Maldives for culture and tradition, employment, food security and economic diversity, especially in remote islands with few other opportunities for food security and employment. A timeline of significant conservation, tourism and fisheries events is found in Table 11.2.

Interaction between fishing and tourism and environment

Only a few direct conflicts have arisen between the tourism and fishing industries. Currently, fishermen and the Ministry of Tourism have no complaints about the relationship between the two industries (Höhne-Sparboth et al., 2013). This is a remarkable and rare example of cooperation. In many ways, the tourism market and fisheries support each other because tourism creates markets for fishermen and provides alternative employment opportunities for fishermen (Anderson, 2006). Some enterprising fishermen have developed hybrid tourist safari boats to capitalise on the intriguing and dramatic pole-and-line fishery. In this section we will explore some of interactions and conflicts.

In 1993, the first economic valuation of fisheries resources was conducted, comparing current and potential benefits from sharks, arising from the first major conflict between tourism and fisheries. Sharks had been targeted for liver oil for boat maintenance for centuries, but fishing expanded in the 1970s for highly valued shark fin and oil exports (Sinan, 2011). By the 1990s, several shark species were showing signs of overfishing (Anderson & Ahmed, 1993). The value of the shark fishery was estimated at US$1.17 million and reef sharks' value to tourism via shark-watching dives was estimated as almost double at US$2.3 million (Anderson & Ahmed, 1993), based on direct expenditure. Estimates of the value of individual sharks at top dive sites were estimated to be worth US$3,300 each for tourist value, but only US$32 for fishing value (Anderson & Ahmed, 1993), a 100-fold difference.

Although these calculations were rough and dated, and that determining how much tourism is directly caused by specific wildlife is complicated (Catlin et al., 2013), the sheer magnitude of the difference in total value was clear. This led to a shark fishing ban, creating a direct loss to the fishing industry and a benefit to the tourism industry. As fishermen lost a valuable fishery and received no direct benefits, many were reluctant to stop shark fishing (Anderson & Waheed, 2001). However, there were many indirect benefits: some shark fishermen gained employment at resorts and national services because Maldivians were paid for via revenue generated by tourism taxes (Anderson & Waheed, 2001).

Sharks and other top predators are essential for ecosystem health (e.g. Sergio et al., 2008). Interestingly, shark fishermen were also in conflict with tuna fishermen. Many pole-and-line tuna fishermen believe that oceanic sharks (especially silky sharks) help keep schools of tuna together and that shark harvest reduces tuna catches (Anderson & Ahmed, 1993). Tuna fishermen pushed for bans of shark fishing near tuna schools or fishing grounds (Sinan, 2011). The very small ray fishery (used for shark bait) also ended with the shark ban (Anderson et al., 2010).

Conflicts between resorts and tuna fishermen also arose over baitfishing locations and MPAs. Government regulation restricts fishing and other activities within resort marine areas, excluding liveaboard tourist boats as well as fishermen, causing some competition between tourism sectors. Baitfishing occurs inside the atolls, and resorts wish to protect the colourful cardinalfish, triggerfish and other species. Maldivian MPAs typically restrict all fishing other than traditional baitfishing, frequently with the focus of protecting dive sites.

Managing for reef impacts and risks

The Maldives faces a number of risks stemming from climate change (coral bleaching and sea level rise, for example) as well as the high level of dependency on tourism as its major export industry. Climate change is predicted to cause sea level rise, increased coastal erosion, increased extreme weather events (e.g. storms, droughts, heavy rainfall), ocean acidification and increased water temperature, which can lead to coral bleaching and other marine life impacts. These events will impact on both tourism and fishing. The Maldives has sought international media attention regarding their risk from climate change (especially sea level rise), including staging an underwater cabinet meeting and the movie *The Island President*. However, the country had no policies towards adapting to or mitigating climate change (Shakeela & Becken, 2015) until the ratification of the Paris Agreement in 2015, and specific plans on meeting targets have yet to be released. Natural disasters, such as the 2004 tsunami, impacted the country through immediate infrastructure and environmental effects, as well as indirect losses from reduced tourism.

In an evaluation of the threats to reefs on a global scale, the Maldives was considered to be at relatively low risk from human-caused factors, including coastal development, overexploitation and destructive fishing practices, the impact of inland pollution and erosion, and marine pollution (Bryant et al., 1999). However, the country has a very high dependency on reefs and low adaptive capacity (the ability to cope or adapt to changes) and, consequently, has very high social and economic vulnerability if reefs are damaged (Burke et al., 2011). Dredging and associated siltation can also be problematic, as well as the construction (or expansion) of islands. Although lower than many other countries, localised pollution affects many islands without waste management systems (Shakeela & Becken, 2015).

Moving forward

The primary goal for the country is clear: long-term protection of marine natural resources is required to ensure profitability from both marine ecotourism and fisheries. Healthy reef systems are essential to protecting the country from natural disasters and for resilience to environmental stressors (Hughes et al., 2007) and can mitigate some effects of climate change, such as buffering storm surges (Temmerman et al., 2013). While several policies are currently in place (described herein), further actions (evolving as new research is conducted), should be adopted. These include incentivising environmentally friendly activities such as ecotourism and sustainable fishing, and supporting conservation efforts and imposing penalties for negative actions would support these goals (Emerton et al., 2009).

The health of reef ecosystems may be stressed if tourism and fishing are not managed effectively. Reef health is also subject to other factors such as climate change, coastal erosion, invasive species, etc. However, well-developed conservation plans cost money and securing long-term funding is challenging. Education of tourists can promote willingness to pay for conservation efforts. Many tourists and Maldivian residents understand the sensitivity of marine ecosystems and are willing to pay for environmental protection and conservation in the Maldives (Domroes, 2008; Emerton et al., 2009). Tourism dollars could be used to support conservation projects. Designing policies that would secure a portion of funds from marine ecotourism, such as dive taxes, could ensure that the industries that benefit from healthy marine systems can help support the reefs' long-term health. Taxes from fisheries to improve fisheries management and conservation could also provide support, although these may be more difficult to implement.

While the sustainability of tourism and fishing are essential, social sustainability is also required. More equitable development and distribution of tourism revenue as outlined by the TMPs is crucial. Rather than restricting foreign employment, perhaps further research into understanding how to increase Maldivian employment, especially women, may be useful. Economic and social resiliency would be improved through diversification of the economy. The great dependency on tourism could have dire consequences if national, economic, political, social or environmental problems arise.

In summary, the Maldives has understood the importance of a healthy marine ecosystem for its economy and has prioritised protection of marine resources. As such, it has developed successful marine tourism and fisheries simultaneously and largely harmoniously. Although great strides have been made, further resources are required to reduce the environmental footprint of reef fisheries and tourism and ensure long-term sustainability.

Acknowledgements

The author is extremely grateful to R.C. Anderson for the insight and kindly guidance. Thanks to E. Miller, M. Nayfa, M. Hein and an anonymous reviewer for helpful and patient editing. Many thanks to the International Pole-and-Line Foundation and the Maldives Research Centre for employment and support of my work and experience in the Maldives. Thanks to all the fishermen who allowed me to accompany them on their boats and assisted my work.

References

Adam, M.S. (2006). *Country review: Maldives. Review of the state of world marine capture fisheries management: Indian Ocean*. Rome: Food and Agriculture Organization of United Nations, pp. 383–391.

Adam, M.S., Sinan, H., Jauharee, A.R., Ali, K., Ziyad, A., Shifaz, A., & Ahusan, M. (2014). *Maldives national report to the Scientific Committee of the Indian Ocean Tuna Commission, 2014*. Malé, Maldives: Ministry of Fisheries and Agriculture. Retrieved from http://www.fao.org/3/a-bf553e.pdf

Ahmed, H., Mohamed, S., & Saleem, M.R. (1997). *Exploitation of reef resources: Bêche-de-mer, reef sharks, giant clams, lobsters and others*. Paper presented at the Workshop on Integrated Reef Resources Management in the Maldives, Malé, Maldives, 16–20 March 1996.

Anderson, C., & Waheed, A. (2001). The economics of shark and ray watching in the Maldives. *Shark News, 13*(1).

Anderson, C., Huntington, T., Macfadyen, G., Powers, J., Scott, I., & Stocker, M. (2012). *Pole and line skipjack fishery in the Maldives. Public Certification Report, Version 5*. Maldives: Intertek Moody Marine.

Anderson, R.C. (2006). *Baitfish and reef fish analysis & management*. Maldives: Fisheries Outlook Study World Bank.

Anderson, R.C., & Ahmed, H. (1993). The shark fisheries of the Maldives. Retrieved from http://www.fao.org/docrep/007/ae500e/ae500e00.htm

Anderson, R.C., Adam, M.S., Kitchen-Wheeler, A.-M., & Stevens, G. (2010). Extent and economic value of manta ray watching in Maldives. *Tourism in Marine Environments, 7*(1), 15–27.

Becken, S., Hay, J., & Espiner, S. (2011). The risk of climate change for tourism in the Maldives. In J. Carlsen & R. Butler (Eds.), *Island tourism: Sustainable perspectives*. Oxford: Centre for Agriculture and Bioscience International, pp. 72–84.

Bhat, M.G., Bhatta, R., & Shumais, M. (2014). Sustainable funding policies for environmental protection: The case of Maldivian atolls. *Environmental Economics and Policy Studies, 16*(1), 45–67.

Bryant, D., Burke, L., McManus, J., & Spalding, M. (1999). *Reefs at risk: A map-based indicator of threats to the world's coral reefs* (Vol. 74). Washington, DC: World Resources Institute.

Burke, L., Reytar, K., Spalding, M., & Perry, A. (2011). *Reefs at risk revisited*. Washington, DC: World Resources Institute.

Cagua, E.F., Collins, N.M., Hancock, J., & Rees, R. (2014). Visitation and economic impact of whale shark tourism in a Maldivian marine protected area. *PeerJ, e360v1*.

Catlin, J., Hughes, M., Jones, T., Jones, R., & Campbell, R. (2013). Valuing individual animals through tourism: Science or speculation? *Biological Conservation, 157*, 93–98.

Domroes, M. (2001). Conceptualising state-controlled resort islands for an environment-friendly development of tourism: The Maldivian experience. *Singapore Journal of Tropical Geography, 22*(2), 122–137.

Domroes, M. (2008). Visitors' environment awareness in the Maldives. *Global Environmental Research, 12*, 83–92.

Emerton, L., Baig, S., & Saleem, M. (2009). *Valuing biodiversity. The economic case for biodiversity conservation in the Maldives*. Malé, Maldives: AEC Project, Ministry of Housing, Transport and Environment, Government of Maldives and UNDP Maldives.

Food and Agriculture Organization (FAO). (2014). *State of world fisheries and aquaculture 2014 opportunities and challenges*. Rome: Food and Agriculture Organization of United Nations.

Gillett, R. (2004). Aspects of fisheries management in the Maldives. *FAO/Fish Code Review,* *2,* 1–54.

Hameed, F., & Ali, M. (2001). *An overview of coastal stewardship in the Maldives.* Paper presented at the UNESCO Workshop on Furthering Coastal Stewardship in Small Islands, Canefield, Dominica, July 2001.

Henderson, J.C. (2008). The politics of tourism: A perspective from the Maldives. *Tourismos: An International Multidisciplinary Journal of Tourism, 3*(1), 99–115.

Höhne-Sparboth, T., Adam, M.S., & Ziyad, A. (2013). *A socio-economic assessment of the tuna fisheries of the Maldives.* London: International Pole & Line Foundation.

Hughes, T.P., Rodrigues, M.J., Bellwood, D.R., Ceccarelli, D., Hoegh-Guldberg, O., McCook, L., . . ., & Willis, B. (2007). Phase shifts, herbivory, and the resilience of coral reefs to climate change. *Current Biology, 17*(4), 360–365.

Jauharee, A.R., Neal, K., & Miller, K.I. (2015). *Maldives pole-and-line tuna fishery: Live bait fish review.* Wirral, UK: Centre for Marine and Coastal Studies (CMACS).

Johannes, R.E. (1978). Traditional marine conservation methods in Oceania and their demise. *Annual Review of Ecology and Systematics, 9,* 349–364.

Marine Stewardship Council (MSC). (2016). Maldives pole & line skipjack & yellowfin tuna. Retrieved from https://www.msc.org/track-a-fishery/fisheries-in-the-program/certified/indian-ocean/maldives_pole_line_tuna

Miller, K.I., Nadheeh, I., Jauharee, A.R., Anderson, R.C., & Adam, M.S. (2017). Bycatch in the Maldivian pole-and-line tuna fishery. *PLoS ONE, 12*(5), e0177391.

Ministry of Tourism (MoT). (2015a). *Maldives visitor survey.* Malé, Maldives: Ministry of Tourism.

Ministry of Tourism (MoT). (2015b). *Tourism yearbook 2015.* Malé, Maldives: Ministry of Tourism.

National Bureau of Statistics (NBS). (2015a). *Population and housing census 2014: Preliminary results – revisited.* Malé, Maldives: National Bureau of Statistics, Department of National Planning, Ministry of Finance.

National Bureau of Statistics (NBS). (2015b). *Statistical yearbook 2015.* Malé, Maldives: National Bureau of Statistics, Department of National Planning, Ministry of Finance.

Rasheed, A.A. (2014). Historical institutionalism in the Maldives: A case of governance failure. *The Maldives National Journal of Research, 2*(1), 7–28.

Rasheed, A.A. (2015). Development, development policy and governance in the Maldives: A political economy perspective. *The Maldives National Journal of Research, 3*(1), 29–51.

Sattar, S.A., Andrefouet, S., Ahsan, M., Adam, M.S., Anderson, C.R., & Scott, L. (2012). Status of the coral reef fishery in an atoll country under tourism development: The case of Central Maldives. *Atoll Research Bulletin, 590,* 163–186.

Sattar, S.A., Wood, E., Islam, F., & Najeeb, A. (2014). *Current status of the reef fisheries of Maldives and recommendations for management.* Malé, Maldives: Darwin Reef Fish Project, Marine Research Centre/Marine Conservation Society (UK).

Scheyvens, R. (2011). The challenge of sustainable tourism development in the Maldives: Understanding the social and political dimensions of sustainability. *Asia Pacific Viewpoint, 52*(2), 148–164.

Sergio, F., Caro, T., Brown, D., Clucas, B., Hunter, J., Ketchum, J., . . ., & Hiraldo, F. (2008). Top predators as conservation tools: Ecological rationale, assumptions, and efficacy. *Annual Review of Ecology, Evolution, and Systematics,* 1–19.

Shakeela, A., & Becken, S. (2015). Understanding tourism leaders' perceptions of risks from climate change: An assessment of policy-making processes in the Maldives using the social amplification of risk framework (SARF). *Journal of Sustainable Tourism, 23*(1), 65–84.

Shakeela, A., & Cooper, C. (2009). Human resource issues in a small island: Setting the case of the Maldivian tourism industry. *Tourism Recreation Research, 34*(1), 67–78.

Sinan, H. (2011). *Background report of fishery products: The Maldives*. Malé, Maldives: Ministry of Fisheries and Agriculture.

Temmerman, S., Meire, P., Bouma, T.J., Herman, P.M., Ysebaert, T., & De Vriend, H.J. (2013). Ecosystem-based coastal defence in the face of global change. *Nature, 504*(7478), 79–83.

Topelko, K.N., & Dearden, P. (2005). The shark watching industry and its potential contribution to shark conservation. *Journal of Ecotourism, 4*(2), 108–128.

UN World Tourism Organization (UNWTO). (2004). *Indicators of sustainable development for tourism destinations: A guidebook*. Madrid: UNWTO.

World Bank Group. (2016). Maldives. Retrieved from http://data.worldbank.org/country/maldives

Zubair, F.N.I., & Bouchon, F. (2014). Maldives as a backpacker's destination: Supply and demand perspectives. *Procedia – Social and Behavioral Sciences, 144*, 256–263.

Zubair, S., Bowen, D., & Elwin, J. (2011). Not quite paradise: Inadequacies of environmental impact assessment in the Maldives. *Tourism Management, 32*(2), 225–234.

12

A NARRATIVE APPROACH TO UNDERSTANDING RECREATIONAL DIVE TOURISTS' EXPERIENCES ON CORAL REEFS

Anja Pabel

Introduction

Coral reefs are well-known for their beauty and the substantial diversity of fish life they support. This makes them very attractive for marine tourism operations (Barker & Roberts, 2004). In particular, scuba diving as a marine tourism activity has increased in the past few decades, mainly due to advances in technology, equipment, education and training (Dimmock & Musa, 2015). Previous research on recreational scuba diving has focused on divers' behaviours and their impacts on coral reefs (Rouphael & Inglis, 1997; Barker & Roberts, 2004), scuba divers' motivation (Meisel-Lusby & Cottrell, 2008;) and their perceptions of attractive dive site features (Uyarra et al., 2009; Pabel & Coghlan, 2011). Research exploring the narratives of dive tourists appears to be rather limited. Such narratives, however, can help to better understand scuba divers' preferences for specific natural and trip features.

This study explores narratives, in the form of tourist blogs, to gain better insights into dive tourists' experiences at various coral reef destinations. Tourists are motivated for experiences that are remembered, and it is through storytelling that tourists are able to make sense of their experiences as memories (Moscardo, 2010) and to tell others about their tourism experiences (Bosangit et al., 2015). In accessing the recreational scuba divers' stories, it is possible to ascertain how they relate to their coral reef experiences and what they like and dislike. In turn, such information is helpful in managing diving tourism at coral reef destinations.

Defining the dive tourist

Defining what exactly scuba diving is as a tourism activity is rather challenging. Previous literature has described scuba diving as a marine ecotourism activity (Garrod & Wilson, 2003), an adventure tourism activity (Tschapka & Kern, 2013) and a special interest tourism activity (Garrod, 2008). While there is no agreed-upon international definition, the United Nations World Tourism Organization (UNWTO, 2001, p. 85) refers to dive tourists as individuals who are "travelling to destinations with the main purpose of their trip being to partake in scuba diving". In this regard, the attraction of a specific destination is uniquely focused on its dive qualities rather than any other features.

While finding a definition for dive tourism is challenging, finding an answer to exactly how many scuba divers there are worldwide is nearly impossible. The Diving Equipment and Marketing Association (2013, p. 1) estimates that "there are between 2.7 to 3.5 million active scuba divers in the US with as many as 6 million active scuba divers worldwide". However, such figures are mostly based on guesswork because there is no consensus in the dive industry as to how an active diver is actually defined (Davison, 2007). Is it the number of dives made in the past 5 years? Is it the number of diving certifications a diver holds? Is it the number of recreational overnight dive trips taken in the last 2 years? Until these and similar questions are answered, it is likely that any numbers of scuba divers on a global scale will remain guestimates. Moreover, distinctions also need to be made between scuba diving for recreational purposes and scuba diving for other purposes, i.e. for commercial or scientific reasons.

Scuba diving tourism on coral reefs

Coral reef tourism has been described by Spalding et al. (2001) as a major global industry. As the chapters in this book indicate, coral reefs are a significant pull factor for a range of activities such as wildlife viewing (see Chapter 9), coral reef restoration programmes (see Chapter 10) and fishing (see Chapter 11). Coral reefs are particularly attractive to dive tourists because they "are among the most visually impressive habitats on the planet, burgeoning with life and dazzling with colour" (Spalding et al., 2001, p. 54).

Scuba diving as a tourist activity continues to increase in popularity (MacCarthy et al., 2006; Musa & Dimmock, 2012). Much of the growth in dive tourism within the last few decades was concentrated around the world's coral reef locations. However, the dive tourism industry is very much dependent on good quality coral reefs and diverse marine life (Fitzsimmons, 2009; Lew, 2013). Previous research shows that scuba divers are willing to pay to have access to undamaged marine environments (Depondt & Green, 2006). Revenues generated through Marine Protected Areas are an invaluable income stream used for the management of numerous global coral reefs (Petrosillo et al., 2007).

From a destination perspective, various factors affect the options that dive tourists have when choosing one dive destination over another. A study by O'Neill et al. (2000) identified some of the factors that are important in ensuring a successful dive tourism experience such as clean water; healthy coastal habitats; abundant marine life and a safe, secure and enjoyable environment. Fish and coral-related attributes are also key features affecting dive enjoyment (Uyarra et al., 2009).

Research by Queiroz Neto et al. (2017) used a "hypothetical successful scuba diving destination" to investigate the importance of several destination competitiveness attributes. They identified ten underlying attributes that were valued by dive tourists, including diving operations, risk perception, diving conditions, price, destination management, big wildlife encounters, diving training, tech diving, general tourist attraction and visa policy.

Dive tourists' perceptions of coral reefs are linked to their previous diving experiences. Research by Pabel and Coghlan (2011) showed that more experienced "enthusiast" divers compared their diving experience on the Great Barrier Reef less favourably with regards to certain environmental attributes; however, they valued ease of accessibility to the dive sites. The subjectivity of such perceptions must be taken into account. Dwyer and Kim (2003) state that it is not so much the real, but the perceived environmental quality that influences the buying decisions of potential visitors. Environmental perceptions are, to a certain extent, dependent on demographic factors such as nationality, gender and age (Baysan, 2001; Uyarra et al., 2005). Moreover, Gössling et al. (2008) argues that scuba divers are not a homogenous group with identical knowledge and experiences, and therefore perceptions of marine environments tend to vary.

There has been increasing research over the last decade into service quality attributes of dive tour operators and their associated effects on the dive consumption experience (Musa et al., 2006; Pabel & Coghlan, 2011; Coghlan, 2011). Perceptions of trip quality tend to be principally linked to the actual dive experience; however, the quality of service before, during and after the dive is critical in the evaluation of a dive product (O'Neill et al., 2000). Attributes related to diving operations are therefore also highly important to scuba divers such as professional dive operations, environmental commitment of the dive operator, friendly staff/casual atmosphere, and information by the dive instructors (Queiroz Neto et al., 2017).

Furthermore, the shared experiences and the common interest in scuba diving encourage camaraderie (MacCarthy et al., 2006). This sense of camaraderie or sense of community has led to a growth of dive clubs (Kler & Tribe, 2012). Dive clubs can foster a social element that then contributes to overall diver satisfaction (Musa, 2002). The social norms displayed by the dive instructors also play an important role in validating certain environmental behaviours. A dive instructor is therefore not only responsible for the safety of divers but must also display professional role model behaviour (Andy et al., 2014).

As the previous review of literature shows, multiple aspects of dive tourism have been documented, but few studies have explored travel narratives particularly in the context of coral reefs. The aim of this study is to explore narratives, in the form of tourist blogs, to gain better insights into dive tourists' experiences at various coral reef destinations. More specifically, this study has two particular objectives:

1. To ascertain the dominant themes that emerge from the blogs.
2. To identify themes associated with negative sentiments that dive tourists mention in the blogs.

Methods

Research design

Blogs have emerged as an important data source for travel and tourism research (Xiang & Gretzel, 2010). A weblog or blog is an online diary where individuals record information, thoughts and opinions electronically to share their stories (Kay, 2003; Lawson-Borders & Kirk, 2005). On a global scale, there were approximately 181 million blogs at the end of 2011 (NMIncite, 2012). Blogs are increasingly drawn upon when naturally occurring accounts or "researcher-uncontaminated" data are needed (Volo, 2010, p. 301). This is because they allow researchers to have "a window into tourists' travel experiences" (Banyai & Glover, 2012, p. 268). An advantage of investigating blogs is that they constitute emic interpretive data collected in an unobtrusive way (Martin et al., 2006). Using blogs as a method in this study made it possible to establish what is important to dive tourists when reporting on their scuba diving experiences at coral reefs.

Procedure

A purposive sampling strategy was used to select the blogs for this study (Teddlie & Tashakkori, 2003). The blogs were selected from Mytripjournal.com and Travelblog.org. These two travel blog websites were chosen, first to allow for a broader selection of blogs than using a single travel blog site, and second, the author had direct access to the content of the blogs on these sites without having to follow any registry or membership protocols, hence all selected blogs were publicly available.

Wood's (2015) categorisation of dive destinations was used to sample five tropical coral reef areas: The Red Sea (i.e. Egypt, Sudan), the Indian Ocean (i.e. Maldives, Seychelles, Mozambique); the Indo-Pacific region (i.e. Thailand, Malaysia, Indonesia); the Pacific Ocean (i.e. the Philippines, Melanesia and Micronesia, Papua New Guinea, French Polynesia, Eastern Australia) and the Caribbean Sea (i.e. Bahamas, Mexico, Cayman Islands, West Caribbean, Tobago, Lesser Antilles, Bermuda, Netherlands Antilles). Keyword search included the words "scuba diving" followed by the name of the coral reef areas provided by Wood's categorisation. For each of the five coral reef areas, 20 blogs were selected from the two travel blog websites: Mytripjournal.com (n = 10) and Travelblog.org (n = 10).

Each blog received a number to de-identify the personal details of each blogger, but which allowed the researcher to locate a specific blog within the overall sample. Blogs written in a language other than English were excluded from analysis. Through this approach, 100 blogs were selected from May to June 2016, totalling 392 pages or 112,963 words. A sample of 100 blogs was considered sufficient for analysis after having examined the blog sampling procedures of other studies. For example, previous studies have analysed 19 blogs to gain insights on how travel contributes to self-identity and self-development (Bosangit et al., 2015); 40 blogs to gain a better understanding of tourists' experiences of Charleston (Pan et al., 2007); 114 blogs relating to positive and negative perceptions of Austria as a tourism destination (Wenger, 2008); and 120 blogs to better understand the perceived image of Hong Kong (Law & Cheung, 2010).

Analysis

The blogs selected for this study were analysed using thematic content analysis to establish common themes across the different blog episodes (Riessman, 2005). Analysis focused only on the textual material of the scuba diving blogs. While many blogs also included photos and videos, this visual material was not subjected to analysis. Many blogs included information on other features of the overall holiday; however, because these descriptions were not essential for the aim of this study, analysis focused only on the dive related episodes in each blog. Since this research is qualitative in terms of its enquiry method, statistical significance was not sought, but rather focus was given to providing depth and meaning.

While the analysis of blogs was helpful to better understand dive tourists' preferences for specific natural and trip features, blogs as a data source do not come without limitations. Bloggers represent a self-selected subgroup of tourists (Volo, 2010), and their experiences and perspectives may not be representative of all tourists. Care should therefore be taken when generalising the findings beyond the study sample. Future analysis of dive tourists' experiences could focus on dive forums, which tend to be more concise than blogs. Future studies may also investigate visual materials embedded in blogs, such as photos and videos.

TABLE 12.1 Gender and origin of bloggers

Demographic	Details	Frequency (n = 100)
Gender	Male	35
	Female	31
	Couple/group of friends	34
Origin	USA	48
	Australia and New Zealand	11
	Canada	10
	UK and Ireland	5
	France	4
	Belgium	3
	Ecuador	3
	South Africa	2
	Singapore	2
	Finland	2
	Germany	1
	Brazil	1
	Mozambique	1
	Not specified	7

Participants

Basic demographic details were collected from each blog to the extent that this information was available, including gender and tourist origin. The information is displayed in Table 12.1 and shows that 31 blogs were written by female tourists, 35 blogs by male tourists and 34 blogs where written by several people, consisting either of couples or groups of friends.

Findings

The first objective was to explore the dominant themes emerging in the blogs through thematic content analysis. Five major themes were identified including *marine life, underwater scenery, social aspects, service related attributes* and *wreck diving opportunities*. Themes are presented in no particular order below, and are substantiated with brief quotes from the dive tourists' blog episodes.

Marine life

Marine life was a major theme that emerged from the analysis of blogs. The blog entries referred to the numerous fish species encountered during dives, including groupers, blue spotted rays, barracuda, batfish, stingray, pufferfish, napoleon wrasse, snappers, stonefish, lionfish and many other species. Marine mammals were frequently mentioned as a dive highlight, including observations or interactions with dolphins and whales. Other marine fauna noted in the blogs include lobster, sea turtles, mantas, giant clams, nudibranchs and sea cucumbers. Marine fauna identification was important, for example, for this naturalist diver. "It was absolutely magical as the number and variety of fish that were out scrounging for breakfast was unreal. There are just way too many to name, I have over 50 recorded in my dive log."

Certain encounters of marine life were regarded as "once in a lifetime occurrences". One blog entry describes a meaningful encounter with manta rays: "Finally my long held dream is realized – manta rays! The sight of these graceful creatures gliding under and over us will remain one of our greatest memories". Seeing sharks was another highlight in many blog entries (see Figure 12.1). Shark species mentioned in the blogs include nurse shark, leopard shark, blacktip shark, whitetip reef shark, grey reef shark, hammerhead and tiger shark. Seeing a thresher shark was a highlight, for example, for this dive tourist: "Threshers are not sighted by divers very often, so this was a special dive for me."

Underwater scenery

Several divers were impressed with the underwater scenery of their chosen coral reef destination and gave elaborate descriptions of coral overhangs and coral pinnacles. Different types of hard and soft corals were described by numerous dive tourists. One stated: "I'm in heaven! This is a huge coral formation that

FIGURE 12.1   Shark feeding at "North Horn" at Osprey Reef on the Great Barrier
Reef

Source: Photo courtesy of Dave Moss.

has produced caves at about 125 feet." Diverse underwater scenery was also a
favourite due to the variety of marine life it attracts: "Huge coral formations
erupt from the sandy bottom and this is a great place to spot eagle rays, turtles
and often time reef sharks. The coral continues to impress as you ascend."

FIGURE 12.2   Being mesmerised by coral reefs. Diving at Gordon Reef at Ras
Mohamad National Park off the coast of El Sheikh, Egypt

Source: Photo courtesy of Terrence Cummins.

Many blogs contained information on the variety and colours of coral reefs (see
Figure 12.2). One dive tourist wrote: "The variety and intensity of colours is overwhelming. So many shades of purple, yellow, lime green, red, orange and royal
blue." Another blog entry describes: "We went down through a tunnel and end
up on the side of a huge wall of soft purple and white coral. They're just swaying
back and forth with the current on the side of the wall."

Social aspects

Another key theme that emerged from the blog entries focused on the social
aspects of recreational scuba diving. Many dive tourists mentioned they enjoyed
meeting people who shared their passion for scuba diving. This blog entry
exemplifies: "Our group of divers came from all over the blue marble. We
shared stories of experiences on our dives that day as well as diving all over the
world." The fellowship of divers was also enjoyed by this dive tourist: "We had
a fun group on board and we all got along really well, so all of my down time

was either spent hanging out and talking with other people." This dive tourist found it difficult to say farewell at the end of the dive trip: "Spending a week 24/7 with a group of 20 some people in a small space can be intense. It might sound dramatic, but we had a really fun group and they felt like good friends by the end of the week."

The social aspect also entailed the interactions between the dive tourists and dive instructors. The passion and enthusiasm of the dive instructors made a real difference to several dive tourists. This blog entry described a dive instructor as: "great, who clearly has a love for diving and is very knowledgeable about all the fish species and coral identifications". Being "passionate about the ocean and its creatures" was important when sharing the enthusiasm for scuba diving. The dive instructors' good sense of humour was also appreciated by the dive tourists. After passing a rescue diver exam, one diver "received a good natured initiation at the bar of beer and Jägermeister poured through a snorkel".

Service related attributes

Multiple blog episodes mentioned transport and service related aspects, including the comfort of the boat, quality of service and food. One dive tourist stated: "They offer the most dives in a seven day trip, excellent service, comfortable cabins, great food, and exceptionally talented dive crew." Another tourist described the room of the liveaboard diving vessel as: "small but very clean and cosy with two bunk beds, the upper having windows that look out onto the ocean". Service quality and refreshments were noted as important for day vessels: "When we got on board we were greeted with tea, Danish pastries and fresh fruit."

Wreck diving opportunities

The opportunity to participate in wreck diving was highly rated in several blog entries. Wrecks, including ship wrecks, plane wrecks and other purposely sunk structures, such as artificial reefs, clearly have appeal to dive tourists, with some going into much detail describing "underwater museums" such as the SS *Thistlegorm*, the *Numidia*, the *Umbria* and the *Aida* located at the Red Sea and several wreck diving sites at Chuuk Lagoon in Micronesia. At the SS *Thistlegorm*, tourists are able to dive "over a row of neatly lined-up, coral-crusted motorcycles which was worth the price of admission alone". The ample amount of marine life attracted to some wrecks was highlighted in this blog entry: "Going down first to the sea bed, to see one of the two locomotives, I was impressed by the abundance of sea life in the area. Plenty of bat fish and smaller reef fish, scorpion fish here and there."

Since blogs are written post experience, they also allow for an elaboration on the emotional aspects of the diving experience. The second research objective of this study was to identify themes associated with negative sentiments that dive tourists mentioned in the blogs, which is discussed in the next section.

Reasons for disappointment

Some dive tourists expressed negative sentiments in their blogs. For example, disappointment was expressed over forgetting to bring an underwater camera or not being able to see certain species that were "expected" to be present, i.e. turtles or whale sharks. Some scuba diving experiences were painful, causing a negative effect, i.e. several tourists had sinus congestions and some were "attacked" by territorial fish species. Dissatisfaction was also expressed when boat journeys took longer than expected, as well as rough conditions that caused seasickness and bad visibility. Outdated or damaged equipment caused annoyance to this dive tourist: "We picked out our equipment which I have to say is the worst equipment we have ever seen. In the end I had to wear a wetsuit that was torn in three places along the sides of my ribs; how is this supposed to keep warm? It looked like an outfit from Pirates of the Caribbean."

Overcrowding was another reason for disappointment in some blog entries: "It was too crowded with too many people kicking up sand." In the case of a dive site in Thailand, this dive tourist wrote: "This place is crazy stupid busy with dozens of tour boats at each dive site. The amount of people diving in the same spot has been a thorn in our side from the beginning, we do not enjoy this aspect of diving here." The overcrowded or "packed dive sites" were also complained about at the Red Sea, where several dive tourists observed behaviours of other divers reflecting "stunning disregard for the environment". One dive tourist wrote about her observation of another recreational diver grabbing a turtle and questioning such behaviour: "Sigh, why do so many stupid people have to ruin such awesome things? Why?"

Several blog entries raised concerns about environmental issues. For example, one dive tourist referred to the visited coral reefs as "a mess" after describing the impacts of a "recent El Niño system". On a more positive note, this tourist wrote about her observations of coral reef restoration programmes taking place in response to pollution and rising sea temperatures: "We stayed at another eco/cultural tourist place that was helping to restore the coral there. The local community has put metal structures on the ocean floor near the shore and are pumping low voltage electricity to them. The coral is growing back and fish are coming back to that area." Education was noted as one of the successful ways to minimise the impacts on coral reefs: "The local guides do a very good job educating divers on conservation. Gloves are prohibited to discourage touching and buoys are permanently placed to avoid anchor damage."

Discussion

Various aspects appeared particularly important when exploring the blogs for coral reef diving experiences, including abundance and diversity of marine life, interesting underwater scenery and the presence of wrecks. These findings concur with the existing literature on successful dive site attributes, including variety of marine flora; diversity of topography; presence of historical underwater features (wrecks); and clean water/visibility (Lew, 2013). Certain marine species were referred to as "a highlight" or "once in a lifetime experience" in the tourist blogs. Scuba diving with large pelagic fish has become an increasingly popular activity "because of the adrenaline experience they provide" (Lew, 2013, p. 9).

Since coral reefs are such vulnerable assets to the dive tourism industry (Barker & Roberts, 2004), having a sense of place attachment and stewardship may lead to minimal impact diving practices (Moskwa, 2012). While the large scale threats and the associated global decline of coral reefs have been discussed in previous chapters, there are smaller scale ways to promote environmental protection through tourism, for example by managing diver behaviours to avoid damaging coral reefs (Ong & Musa, 2011) and raising revenues to pay for the management of marine parks (Petrosillo et al., 2007). Exploring place attachment from an underwater perspective, Moskwa (2012) acknowledged that divers are bound to certain underwater landscapes that can lead to a sense of stewardship. This study found that interesting underwater scenery was also frequently mentioned in the scuba diving blogs. More experienced divers, in particular, value diverse topography in the shape of pinnacles, walls and other drop offs.

The social aspects of the scuba diving experience was another important theme that emerged from the blogs. This concurs with previous studies recognising that the shared experiences and the common interest in scuba diving lead to a sense of community (MacCarthy et al., 2006). The social element therefore has an important role to play in contributing to overall diver satisfaction (Musa, 2002). In particular the interactions between the dive tourists and dive instructors were mentioned in numerous blogs, including comments about the dive instructors' knowledge, enthusiasm and sense of humour. Presenting coral reef interpretation or safety information in an enthusiastic and entertaining manner has been associated with increased levels of satisfaction (Pabel & Pearce, 2016).

From an educational perspective, dive instructors or dive guides play an essential role in shaping the experiences of dive tourists by providing ecological knowledge about marine life and displaying environmentalist behaviours to enhance divers' understanding about the underwater world (Andy et al., 2014). Such education provided along with the diving experience can enhance people's appreciation for the marine world and change their attitudes towards contributing to the conservation of fragile coral reef ecosystems (Dearden et al., 2007).

Several blog entries indicated that satisfaction of the trip was linked to service related aspects such as transport to the dive site, customer service and the quality of rental equipment. The tangible elements of a dive experience such as hire equipment, boat and fittings, and staff play an important role in creating a sense of satisfaction (MacCarthy et al., 2006; Musa et al., 2006; Pabel & Coghlan, 2011). Therefore, taking a customer-focused approach is vital for successful dive operations, as highlighted by Queiroz Neto et al. (2017).

Opportunities for wreck diving at certain coral reef destinations were referred to as unforgettable experiences in the blog entries. Ship wrecks or plane wrecks hold high value for intrigue, as well as visual and historical value (Stolk et al., 2007; Lew, 2013). In particular, new "learner" divers and more experienced "enthusiast" divers show high interest in diving artificial reefs (Pabel & Coghlan, 2011). Wrecks in shallow, warmer waters tend to have more coral life, making them popular with divers of various skill levels. Wreck diving in greater depths tends to be more popular with more experienced divers who like the appeal of testing their skill (Lew, 2013).

In ensuring the sustainability of the dive tourism product, artificial reefs have been suggested as attractions that may offer a substitute in the event that quality dive experiences on natural reefs can no longer be guaranteed as a result of further environmental degradation due to more frequent coral bleaching episodes and pollution (van Treeck & Schumacher, 1998). Human-made underwater attractions are starting to emerge in some destinations, including the Christ of the Abyss at the Florida Keys, USA and the Underwater Sculpture Park in Grenada, Caribbean Sea. These structures tend to be placed at featureless bottoms and are useful as attractions which divert pressures from coral reefs. Novice divers have the opportunity to develop their skills by substituting natural or protected reef sites with these less fragile underwater attractions (Fitzsimmons, 2009).

Some blogs mentioned negative emotions experienced during the dives, including dissatisfaction with crowding and attacks by certain territorial fish species. Dive briefings tend to include information on preventing worst case safety scenarios, but negative emotions may also be the result due to poor visibility, waves, currents, seasickness and potentially dangerous animals (Coghlan, 2011). Previous research shows that visitors to coral reefs prefer less crowded sites or the absence of other visitors (Brander et al., 2007).

In establishing how much tourism a certain reef site can support, knowledge about the carrying capacity is pivotal. From a sustainability perspective, a reef's potential to attract dive tourists depends on its long-term health and carrying capacity. Increasing demand for coral reefs combined with commercial developments put excessive pressures on the environmental and social capacity (Doiron & Weissenberger, 2014). Concerns for crowding have important implications for managing capacity related issues because subjective negative evaluations can be the result if visitors perceive too many people in a specific area (Vaske & Donnelly, 2002). A resultant decline in reef ecosystem health caused by pressures from crowding and commercial development has the potential to reduce its long-term tourism potential.

Conclusion

This study explored dive tourists' blogs to gain better insights into their coral reef experiences. Thematic content analysis of 100 travel blogs resulted in common themes that emerged across the different blog episodes. The themes include marine life, underwater scenery, social aspects, service related attributes and wreck diving opportunities. A further theme dealt with the issues contributing to dive tourists' disappointment and concerns. Knowledge of this kind can provide important information to managers at coral reef destinations and marine tourism operators regarding what strategies could assist in improving experiences for recreational scuba divers. From a governance perspective, understanding the factors leading to concern should encourage marine tourism operators to focus on adaptive strategies, such as appropriate coral reef education to reduce any negative environmental impacts to dive sites; minimising crowding issues at frequently visited dives sites; fostering the social elements of scuba diving experiences and considering the use of artificial reefs as potential substitute experience. Such strategies are helpful in encouraging conservation outcomes and may also generate interest in scuba diving as a recreational activity.

References

Andy, L., Lee, R., & Tzeng, G. (2014). Characteristics of professional scuba dive guides. *Tourism in Marine Environments, 10*(1–2), 85–100.

Banyai, M., & Glover, T.D. (2012). Evaluating research methods on travel blogs. *Journal of Travel Research, 51*(3), 267–277.

Barker, N.H.L., & Roberts, C.M. (2004). Scuba diver behaviour and the management of diving impacts on coral reefs. *Biological Conservation, 120*, 481–489.

Baysan, S.K. (2001). Perceptions of the environmental impacts of tourism: A comparative study of the attitudes of German, Russian and Turkish tourists in Kemer, Anatalya. *Tourism Geographies, 3*(2), 218–235.

Bosangit, C., Hibbert, S., & McCabe, S. (2015). "If I was going to die I should at least be having fun": Travel blogs, meaning and tourist experience. *Annals of Tourism Research, 55*, 1–14.

Brander, L.M., van Beukering, P., & Cesar, H.S.J. (2007). The recreational value of coral reefs: A meta-analysis. *Ecological Economics, 63*, 209–218.

Coghlan, A. (2011). Facilitating reef tourism management through an innovative importance-performance analysis method. *Tourism Management, 33*(4), 767–775.

Davison, B. (2007). How many divers are there? Retrieved 23 June 2017 from https://www.undercurrent.org/UCnow/dive_magazine/2007/HowManyDivers200705.html

Dearden, P., Bennett, M., & Rollins, R. (2007). Implications for coral reef conservation of diver specialisation. *Environmental Conservation, 33*, 353–363.

Depondt, F., & Green, E. (2006). Diving user fees and the financial sustainability of marine protected areas: Opportunities and impediments. *Ocean and Coastal Management, 49*(3), 188–202.

Dimmock, K., & Musa, G. (2015). Scuba diving tourism system: A framework for collaborative management and sustainability. *Marine Policy, 54*, 52–58.

Diving Equipment and Marketing Association. (2013). Fast facts: Recreational scuba diving and snorkelling. Retrieved 23 June 2017 from http://www.dema.org/store/download.asp?id=7811B097–8882–4707–A160–F999B49614B6

Doiron, S., & Weissenberger, S. (2014). Sustainable dive tourism: Social and environmental impacts – the case of Roatan, Honduras. *Tourism Management Perspectives, 10*, 19–26.

Dwyer, L., & Kim, C. (2003). Destination competitiveness: Determinants and indicators. *Current Issues in Tourism, 6*(5), 369–414.

Fitzsimmons, C. (2009). Why dive? And why here? A study of recreational diver enjoyment at a Fijian eco-tourist resort. *Tourism in Marine Environments, 5*(2–3), 159–173.

Garrod, B. (2008). Market segments and tourist typologies for diving tourism. In: B. Garrod & S. Gössling (Eds.), *New frontiers in marine tourism: Diving experiences, sustainability, management* (pp. 31–48). Amsterdam: Elsevier.

Garrod, B., & Wilson, J.C. (2003). *Marine ecotourism: Issues and experiences.* Clevedon, UK: Channel View Publications.

Gössling, S., Linden, O., Helmersson, J., Liljenberg, J., & Quarm, S. (2008). Diving and global environmental change: A Mauritius case study. In: B. Garrod & S. Gössling (Eds.), *New frontiers in marine tourism: Diving experiences, sustainability, management* (pp. 67–92). Amsterdam: Elsevier.

Kay, R. (2003). Blogs. *Computerworld, 37*(17), 30.

Kler, B.K., & Tribe, J. (2012). Flourishing through scuba: Understanding the pursuit of dive experiences. *Tourism in Marine Environments, 8*(1/2), 19–32.

Law, R., & Cheung, S. (2010). The perceived destination image of Hong Kong as revealed in the travel blogs of mainland Chinese tourists. *International Journal of Hospitality & Tourism Administration, 11*(4), 303–327.

Lawson-Borders, G., & Kirk, R. (2005). Blogs in Campaign Communication. *American Behavioural Scientist, 49*(4), 548–559.

Lew, A.A. (2013). A world geography of recreational scuba diving. In: G. Musa & K. Dimmock (Eds.), *Scuba diving tourism* (pp. 29–51). Abingdon, UK: Routledge.

MacCarthy, M., O'Neill, M., & Williams, P. (2006). Customer satisfaction and scuba-diving: Some insights from the deep. *The Services Industries Journal, 26*(5), 537–555.

Martin, D., Woodside, A.G., & Dehuang, N. (2006). Etic interpreting of naive subjective personal introspections of tourism behaviour: Analyzing visitors' stories about experiencing Mumbai, Seoul, Singapore, and Tokyo. *Journal of Culture, Tourism and Hospitality Research, 1*(1), 14–44.

Meisel-Lusby, C., & Cottrell, S. (2008). Understanding motivations and expectations of scuba divers. *Tourism in Marine Environments, 5*(1), 1–14.

Moscardo, G. (2010). The shaping of tourist experience: The importance of stories and themes. In: M. Morgan, P. Lugosi & J.R.B. Ritchie (Eds.), *The tourism and leisure experience: Consumer and managerial perspectives. Aspects of tourism* (pp. 43–58). Buffalo, NY: Channel View Publications.

Moskwa, E.C. (2012). Exploring place attachment: An underwater perspective. *Tourism in Marine Environments, 8*(1/2), 33–46.

Musa, G. (2002). A SCUBA diving paradise: An analysis of tourism impact, diver satisfaction and tourism management. *Tourism Geographies, 4*(2), 195–209.

Musa, G., & Dimmock, K. (2012). SCUBA diving tourism – Introduction to special issue. *Tourism in Marine Environments, 8*, 1–5.

Musa, G., Kadir, S., & Lee, L. (2006). Layang Layang: An empirical study on scuba divers' satisfaction. *Tourism in Marine Environments, 2*(2), 89–102.

NMIncite. (2012). Buzz in the blogosphere: Millions more bloggers and blog readers. Retrieved 22 April 2012 from http://www.nmincite.com/?p=6531

O'Neill, M.A., Williams, P., MacCarthy, M., & Groves, R. (2000). Diving into service quality – the dive tour operator perspective. *Managing Service Quality: An International Journal, 10*(3), 131–140.

Ong, T.F., & Musa, G. (2011). An examination of recreational divers' underwater behaviour by attitude-behaviour theories. *Current Issues in Tourism, 14*(8), 779–795.

Pabel, A., & Coghlan, A. (2011). Dive market segments and destination competitiveness: A case study of the Great Barrier Reef in view of changing reef ecosystem health. *Tourism in Marine Environments, 7*(2), 55–66.

Pabel, A., & Pearce, P.L. (2016). Tourists' responses to humour. *Annals of Tourism Research*, *57*, 190–205.

Pan, B., MacLaurin, T., & Crotts, J.C. (2007). Travel blogs and the implications for destination marketing. *Journal of Travel Research*, 46(1), 35–45.

Petrosillo, I., Zurlini, G., Corlian, M.E., Zaccarelli, N., & Dadamo, M. (2007). Tourist perception of recreational environment and management in a marine protected area. *Landscape and Urban Planning*, *79*, 29–37.

Queiroz Neto, A., Lohman, G., Scott, N., & Dimmock, K. (2017). Rethinking competitiveness: Important attributes for a successful scuba diving destination. *Tourism Recreation Research*, 42(3), 356–366.

Riessman, C.K. (2005). Narrative analysis. In: N. Kelly, C. Horrocks, K. Milnes, B. Roberts, & D. Robinson (Eds.), *Narrative, memory & everyday life* (pp. 1–7). Huddersfield, UK: University of Huddersfield.

Rouphael, A.B., & Inglis, G.J. (1997). Impacts of recreational scuba diving at sites with different reef topographies. *Biological Conservation*, *82*, 329–336.

Spalding, M.D., Ravilious, C., & Green, E.P. (2001). *World atlas of coral reefs*. Prepared at the UNEP World Conservation Monitoring Centre. Berkeley, CA: University of California Press.

Stolk, P., Markwell, K., & Jenkins, J.M. (2007). Artificial reefs as recreational scuba diving resources: A critical review of research. *Journal of Sustainable Tourism*, 15(4), 331–350.

Teddlie, C., & Tashakkori, A. (2003). Major issues and controversies in the use of mixed methods in the social and behavioural sciences. In: A. Tashakkori & C. Teddlie (Eds.), *Handbook of mixed methods in social and behavioural research* (pp. 3–49). Thousand Oaks, CA: Sage.

Tschapka, M.K., & Kern, C.L. (2013). Segmenting adventure tourists: A cluster analysis of scuba divers in Eastern Australia. *Tourism in Marine Environments*, 9(3–4), 129–142.

United Nations World Tourism Organization (UNWTO). (2001). *Tourism 2020 Vision, Volume 7: Global forecasts and profiles of market segments*. Madrid: World Tourism Organization.

Uyarra, M.C., Cote, I.M., Gill, J.A., Tinch, R.R.T., Viner, D., & Watkinson, A.R. (2005). Island-specific preferences of tourists for environmental features: Implications of climate change for tourism-dependent states. *Environmental Conservation*, 32(1), 11–19.

Uyarra, M.C., Watkinson, A.R., & Cote, I.M. (2009). Managing dive tourism for the sustainable use of coral reefs: Validating diver perceptions of attractive site features. *Environmental Management*, *43*, 1–16.

van Treeck, P., & Schumacher, H. (1998). Mass diving tourism – a new dimension calls for new management approaches. *Marine Pollution Bulletin*, 37(8–12), 499–504.

Vaske, J.J., & Donnelly, M.P. (2002). Generalizing the encounter-norm-crowding relationship. *Leisure Sciences*, *24*, 255–269.

Volo, S. (2010). Bloggers' reported tourist experiences: Their utility as a tourism data source and their effect on prospective tourists. *Journal of Vacation Marketing*, 16(4), 297–311.

Wenger, A. (2008). Analysis of travel bloggers' characteristics and their communication about Austria as a tourism destination. *Journal of Vacation Marketing*, *14*, 169–176.

Wood, L. (2015). *The world's best tropical dive destinations*. Oxford: John Beaufoy Publishing.

Xiang, Z., & Gretzel, U. (2010). Role of social media in online travel information search. *Tourism Management*, *31*, 179–188.

13

TOURIST SATISFACTION AND EXPENDITURES IN A REEF-ADJOINING DOLPHIN WATCHING INDUSTRY IN LOVINA, BALI INDONESIA

Putu L. Mustika, Riccardo Welters and Naneng Setiasih

Introduction

This aim of this chapter is to examine aspects of the tourist experience and economic benefits of dolphin watching in Lovina, Bali, Indonesia in 2013. Whales and dolphins (Infraorder Cetacea, Order Cetartiodactyla) are known to inhabit deep oceans, shallow coastal waters and, some dolphin species, rivers. The sperm whales (*Physeter microcephalus*) and Cuvier's beaked whales (*Ziphius cavirostris*) are deep divers that can dive up to 2,250 m (Lee, 2014) and almost 3,000 m (Schorr et al., 2014), respectively. Estuarine and coastal species such as the Indo-Pacific humpback dolphins (*Sousa chinensis*) are usually only found in shallow waters of up to 20 m depth, and at times adjacent to mangroves (Parsons, 2004). Species such as the Irrawaddy dolphins inhabit rivers (Kreb & Budiono, 2005), coastal lagoons (Sutaria & Marsh, 2011) and estuarine waters (Yanuar et al., 2011).

Some dolphin species have been observed in association with reefs. Common bottlenose dolphins (*Tursiops truncatus*) are known to use the Turneffe Atoll (Belize) to forage (Grigg & Markowitz, 1997). Hawai'ian spinner dolphins (*Stenella longirostris longirostris*) have been observed using shallow reef lagoons to rest or socially interact (Sazima et al., 2003; Cribb et al., 2012). Some reef fish, such as the black durgons (*Melichthys niger*) and the Bermuda chubs (*Kyphosus sectatrix*), have been observed feeding on the faeces of spinner dolphins (Sazima et al., 2003), bringing more depth into the interdependence between coral reefs and dolphins.

The predictable nature of these dolphin activities has triggered wildlife tourism activities at many places, such as Hawai'i (Courbis & Timmel, 2009), Samadai Reef in Egpyt (Notarbartolo-di-Sciara et al., 2009), Moon Reef in Fiji (Cribb et al., 2012), Fernando de Noronha in Brazil and Lovina in Bali, Indonesia. Worldwide, this industry contributed USD 872 million from tickets alone in 2008 (O'Connor et al., 2009), with a good proportion trickling down to the local communities

(Mustika et al., 2012; Orams, 2013). The profit triggered by this industry is such that the number of operators in Asia in 2013 has more than doubled since 2008, and that most dolphin watching industries in seven sites in Asia have reached financial saturation (Mustika et al., 2017).

The burgeoning whale and dolphin watching industry does come at a cost to the animals. For example, vertical and horizontal avoidance (Lusseau et al., 2006), temporal habitat abandonment (Lusseau, 2004), deeper dives and compromised energy level for female dolphins (Lusseau, 2003), and lower calf survivor rate (also for general boat traffic, see Nowacek et al., 2001) have been linked to the presence of the industry. Thus, although this type of industry is traditionally considered a form of non-consumptive wildlife tourism (Roe et al., 1997), the accumulating evidence on the adverse impacts of unsustainable dolphin watching tourism has triggered some researchers to consider dolphin watching tourism as a form of consumptive (albeit still non-lethal) wildlife tourism activity (Higham et al., 2016).

The dolphin watching tourism industry in Lovina

Lovina, which is the generic name of two fishing villages in Buleleng – Kalibukbuk and Kaliasem – is about a 3-hour drive north of Denpasar (Bali, Indonesia). Although daily trips from Denpasar are possible, the mountain range in the middle of Bali is an incentive for visitors to stay overnight, because daily travel back to Denpasar through the mountain range is quite exhaustive. Its relative isolation from the hustle and bustle of south Bali makes this bucolic place renowned for its quietness. Indeed, sitting around the iconic Dolphin Statue on the beach under the starry night is a common evening activity in Lovina, although the occasional burst of rock or reggae music can also be heard from adjacent cafes.

With 83,575 visitors in 2013, Lovina is the third most visited tourist destination in Buleleng after the Banjar Hotspring and the Pulaki Temple (Disparda Bali, 2016). This number was almost double the number of visitors to the Bali Barat National Park (44,343 visitors) in the same year. Lovina is undoubtedly most renowned for its dolphin watching sector. The main target for dolphin watching is the spinner dolphins, particularly the south-east Asian subspecies (*Stenella longirostris roseiventris*), although seven other whale and dolphin species can also be found in Lovina (Mustika et al., 2013). Established in 1987, the dolphin watching sector is an important income generator in Lovina with at least 37,000 overnight dolphin watching visitors per annum in 2008/9, contributing at least USD 4.1 million per annum of tourist direct expenditures to the village (Mustika et al., 2012). A total of 179 colourfully painted dolphin watching boats were available in 2008/9; this figure increased to 192 in late 2011 when Mustika revisited the site. The boat drivers often drove the boat in a way that risked injuring the dolphins (Mustika et al., 2015), thus decreasing tourist satisfaction (Mustika et al., 2013). Since no operating licence is required for the boatmen to operate in Lovina, the government finds it difficult to apply sanctions to careless boatmen (Figure 13.1).

FIGURE 13.1 The traditional fishing boats used in Lovina (Bali, Indonesia) to take tourists to see the dolphins

Source: Photo by Mustika@James Cook University.

Unlike the association between reefs and spinner dolphins in Hawai'i, Egypt and Fiji (Notarbartolo-di-Sciara et al., 2009; Courbis & Timmel, 2009; Cribb et al., 2012), current knowledge is insufficient to understand the association between Lovina's spinner dolphins and local coral reef ecosystems. Mustika rarely saw this species in the shallow waters (less than 20 m depth) of Lovina. However, anecdotal information from a local fisher from eastern Buleleng suggests a possible resting site of this species in the eastern part of Buleleng and possibly a reef-associated site located somewhere between Lovina and Tejakula. Tejakula is an emerging, yet insufficiently studied, dolphin watching site in eastern Buleleng. The main offerings for tourists in Tejakula are snorkelling and diving, with dolphin watching as an added value.

Compared to the coral reefs in Bali Barat National Park, Lovina's coral reefs exhibit a high level of dead coral coverage (67%) and a very high mortality index (~1; Prasetia, 2013). Anecdotal reports from local dive operators mentioned that Lovina used to be one of the main diving destination in Bali before the 1997/8 mass bleaching event (Chou, 2000), which killed most of the local reefs. The vibrant development of dolphin watching in Lovina since 1989 might have compensated the income lost from the reef. Snorkelling has been identified as an added value to the dolphin watching industry, offered at the conclusion of the dolphin viewing. This snorkelling tour is usually conducted offshore from Kaliasem Village on a small fringing reef which barely accommodates 10 boats at one time. Although the snorkelling add-on is a recurring activity, its economic benefits have never been investigated.

Based on this information, this chapter aims to answer three questions as follows:

i) How does the dolphin watching industry fare in Lovina in terms of tourist experience and expenditures since 2008/9?
ii) How much spill-over does the Lovina dolphin watching industry produce to the local reef tourism industry?
iii) Is a combination of dolphin watching tourism and local reef tourism beneficial to sustain long-term local income?

Methods

We use data collected in July 2013 by the Coral Reef Alliance and Reef Check Indonesia (Mustika, 2014). This data was the first follow-up survey conducted in Lovina following Mustika's PhD research in 2008/9. Whenever possible we compared the results with the 2008/9 data that was generated from Mustika's PhD research (Mustika et al., 2012; Mustika et al., 2013).

Tourist experience

The variables included in the tourist experience analysis are: factors that attracted them to come to Lovina; the aspects they liked the best and disliked the most; the tourists' comfort level regarding the number of boats around them; tourists' opinion about boat crowding; tourists' opinion about the way the dolphin boat-men guided their trip; willingness to rejoin the trip; willingness to recommend the trip to others; and the general satisfaction level with the dolphin trip (Likert scale 1 to 10). Some qualitative questions gave us more than one answer per respondent. We thus grouped the responses into several common themes and used dummy variables to examine whether one particular theme (e.g., "the dolphins") was the answer for that question (e.g., aspects the tourists liked the most).

We treated satisfaction as an ordinal variable and used the Hanan–Karp scale (Hanan & Karp, 1989) as the benchmark. Based on Hanan and Karp, if 85–90% of the responders scored 8–10, their satisfaction level is high. If 70–80% of the responders scored 8–10, the satisfaction is medium. If 60% or fewer respondents scored 8–10, the satisfaction level is low. To facilitate categorical data analysis, we collapsed the satisfaction variable into "not so satisfied" (1–7) and "satisfied (8–10)". All statistical tests were done with $\alpha=0.05$ or a 95% confidence interval.

Expenditures

The economic benefits of the dolphin watching industry in Lovina are measured through direct tourist expenditures for ticket sales, accommodation, meals, local transportation, souvenirs and communication (telephone/internet). The relative importance of this industry to the local economy is measured through attributable expenditures, i.e., the expenditures that will be lost by the region in the absence of target resources

(Mustika et al., 2012). Specifically, this variable is informed by whether tourists would still come to the site in the absence of the main attraction, in this case the dolphins. This section will also estimate the additional revenue from the snorkelling add-on.

Calculating the number of tourists who visited the site was an important element in calculating the annual tourist expenditures for the tourist attraction. Unlike the 2008/9 data collection (Mustika et al., 2012), budget constraints prevented us from directly observing the number of tourists participating in dolphin tours in Lovina ("dolphin watchers") in 2013. However, the official visitation number in 2013 is known – 83,575 tourists, including 49,943 foreigners (Disparda Bali, 2016). We used this number and the percentage of dolphin tourists to total Lovina visitors in 2007–9 (58% from Mustika et al. (2012)) to obtain the estimated number of dolphin watchers in 2013 (48,474 watchers).

Primary direct expenditures

Primary direct expenditure was the admission fee of all tourists joining the dolphin trips (day tourists and overnight tourists) in 2013. We use the per person admission fee to calculate the primary direct expenditures (i.e., USD 6 per person in Lovina, still the same as the admission fee in 2008/9).

Annual primary direct
expenditure (admission fee)= # tourists per annum × admission fee

Auxiliary direct expenditures

The auxiliary direct expenditures included money spent on accommodation, meals, internet/communication, souvenirs and local transport over the period of stay. For each expenditure item, tourists were provided with the following expenditure range per day per person (IDR 1–150,000; IDR 151,000–300,000; IDR 301,000–450,000; IDR 451,000–600,000; and > IDR 600,000). Following Stoeckl et al. (2005) and Mustika et al. (2012), we used the mid-point of each expenditure range for the first four categories, resulting in the following expenditures: IDR 75,000; IDR 225,000; IDR 375,000; and IDR 525,000. To make the range equal, we fixed the last expenditure range at IDR 675,000. Following Mustika et al. (2016), the daily expenditures were converted to estimate the overall trip expenditures per person by multiplying these mid-points with the duration of the trip per person for overnight visitors (the trip length for day visitors was set to 1). Total auxiliary direct expenditures for Lovina was obtained by adding the total auxiliary direct expenditures of all visitors to Lovina. We separated tourists into "entirely-dedicated tourists," "partially-dedicated tourists" and "non-dedicated tourists". The first group includes tourists who would skip Lovina in the absence of the dolphins. The second group includes tourists who would reduce their length of stay in Lovina in the absence of the dolphins. The last group are tourists who would keep their length of stay as planned in Lovina in the absence of the dolphins.

Annual auxiliary
direct expenditure = # tourist per annum × length of stay (in days) ×
auxiliary expenditure per day per person

Total and attributable direct expenditures

Total tourist expenditures in Lovina was calculated as the total primary direct expenditure (i.e., ticket) and the total auxiliary direct expenditures:

Total direct expenditures = primary direct expenditure + total auxiliary
direct expenditures

Attributable direct expenditures are the potential loss in tourist expenditures in the absence of target resources – in this case, the dolphins in Lovina (Mustika et al., 2012). The attributable direct expenditures are calculated based on the expenditures of tourists who would not have come in the absence of those target resources. Because we have two dedicated groups, we provide the range of attributable expenditures from only the entirely-dedicated tourists and also from the entirely-dedicated and partially-dedicated tourists combined. The dedicated auxiliary direct expenditures are the meals, accommodation, communication, souvenirs and local transportation spent by dedicated tourists.

Total attributable
direct expenditures = primary direct expenditure + total entirely-dedicated
auxiliary direct expenditures (+ total partially-dedicated
auxiliary direct expenditures)

Results

Tourist demography and experience in 2013

In 2013, tourists from Western countries made up 65.2% of total respondents (n=116), while the remainder included Asians (34.8%) and one South American. The majority of Asian visitors were domestic visitors from Indonesia (37 out of 40 Asian visitors, or 92.5% of Asian visitors). The respondents ranged in age from 15 to 72 years, with an average of 33.9±1.16 years. The majority of respondents are highly educated, with 51% having an undergraduate education and 13.5% reporting postgraduate education (n=96). Our sample contains more female respondents (60.7% female, 39.3% male, n=107). The education level was not significantly associated with gender or nationality type.

We asked what attracted the tourists (or the motivation) to Lovina and obtained a response from all 116 respondents. The dolphins were the main reason for visiting Lovina (69.8%). The remaining reasons were, among others, the tranquillity of Lovina (6.9%) and snorkelling at the coral reefs (6%). The dolphins were also mentioned by almost 65% of respondents as an aspect they liked the most (n=112). Scenery was nominated by 23% as the main reason for their visit. Tranquillity and the friendliness of the villagers were both liked by 9.5% of respondents. Seven respondents (6.1%) mentioned the area's temples as one of the experiences they liked the most, while only four respondents (3.5%) specifically mentioned the area's coral reefs.

In terms of aspects of the tour that respondents disliked, most respondents complained about garbage on the beach and in the ocean (62.8%, n=78). Other areas of dissatisfaction were persistent vendors (19.2%). Only six respondents (7.7%) complained about the boats, such as speed or numbers.

When asked about crowding, almost 75% of respondents (n=115) thought that the number of boats was too high or far too high; on average, the tourists preferred 9.4 boats around their boats (se+0.915, n=105). A total of 63.5% of respondents (n=115) had no problem with the way the boatmen drove the boat around the dolphins, while 21.7% were neutral and 14.8% were not comfortable to extremely not comfortable with the way the boatmen drove the boat around the dolphins.

Using the Hanan–Karp scale (Hanan & Karp, 1989), less than 50% of respondents scored the dolphin watching trip between 8–10, thus placing satisfaction in the low category (Figure 13.2). Respondents who were satisfied (scales 8–10) were more likely to indicate they would rejoin the trip (Pearson Chi Square p=0.000, df=1, n=115) or recommend the trip to others (Pearson Chi Square p=0.003, df=1, n=114). Asian respondents were 3.8 times more likely to indicate that they would

FIGURE 13.2 Tourist satisfactions in Lovina for 2008/9 and 2013 (n 2008/9 = 353, n 2013 = 115)

Source: The 2008/9 data were taken from Mustika et al. (2013).

join the trip again in the future than Western respondents (Pearson Chi Square p=0.004, df=1, n=114). Asian respondents were also much more likely to indicate they would recommend the trip to others than their Western counterparts (Pearson Chi Square p=0.012, df=1, n=113, odds ratio=9.3). We found that the satisfaction level was significantly associated with the way the boatmen drove the boat and the number of dolphins they saw (ordinal regression, p=0.000, df=20).

Tourist expenditures in 2013

The length of stay varied with dolphin-oriented motivations. Entirely-dedicated tourists stayed for 1.74 days, partially-dedicated tourists stayed for 2.35 days, whereas non-dedicated tourists stayed for 4.21 days. Combined, the average length of stay for all dolphin watchers in Lovina was 2.16 days.

Our expenditure analyses revealed that the dolphins are an important factor for tourist expenditure in Lovina. In the absence of dolphins, 36.2% of respondents (n=116) would still go to Lovina despite the absence of dolphins, whereas 27.6% would reduce their length of stay and 36.2% would skip the site Lovina altogether. Tourists who claimed that the dolphins were the reason for them visiting Lovina were much more likely to reduce their length of stay or skip the site altogether in the absence of the dolphins than those who visited Lovina because of other reasons (Pearson Chi Square p=0.000, df=2, odds ratio=9.4).

The analysis of our respondents' expenditures resulted in the estimated total direct expenditures of almost USD 8.7 million for 2013. Almost 3.4% of these expenditures were spent on the dolphin tour tickets. The auxiliary direct expenditures per person per day were spent on food (21.2%) accommodation (31.1%), communication (7.1%), souvenirs (14.1%) and local transportation (26.3%) (Table 13.1). Should Lovina lose its dolphins, the region might lose 27.4–48.2% of its total direct expenditures because these are the percentages of expenditures lost from entirely-dedicated and the sum of entirely- and partially-dedicated tourists, respectively.

Spill-over to reef tourism

Almost 23% of the 2013 entirely-dedicated and partially-dedicated respondents (n=70) mentioned visiting coral reef sites as an alternative tourism activity. Thus, the expenditures shifting to adjacent reef tourism sites in the hypothetical absence of the dolphins are between USD 635,965 (entirely-dedicated tourists) and USD 956,003 (entirely- and partially-dedicated tourists), or 7.3–11% of the total direct expenditures for 2013 (also see Table 13.1).

Since the dolphin viewing in Lovina is often combined with snorkelling at an adjacent village-managed marine park, the dolphin boatmen can also obtain extra revenue from snorkelling. Some boatmen have informed us that they would take the guests once in the morning and, sometimes, once in the afternoon. Due to the small coverage of the village marine park (less than 800 m^2), we estimated a maximum of 10 boats per visit, although it is usually no more than 3 boats per visit.

TABLE 13.1 Tourist expenditures at Lovina for 2008/9 (USD, average 2008 and 2009 adjusted to 2013 inflation) and 2013 (USD, average 2013)

	2008 (in 2013 prices)	2009 (in 2013 prices)	2013
(a) Primary direct expenditure	**295,853**	**308,286**	**290,841**
Annual gross revenue per boatman (179 drivers in 2008/9, 192 drivers in 2013)	1,653	1,722	1,515
(b) Auxiliary direct expenditure†	**10,297,417**	**9,814,649**	**8,402,725**
Of which: (c) dedicated*	6,032,314	4,518,947	2,092,714–3,895,509
(d) non-dedicated	4,265,103	5,295,702	6,310,010–4,507,215
(e) Total direct expenditure (a+b)	**10,593,269**	**10,122,935**	**8,693,566**
Percentage share of primary in total direct expenditure (a/e)	2.79%	3.05%	3.35%
(f) Total attributable expenditure (a+c)	**6,328,167**	**4,827,234**	**2,383,555–4,186,350**
Percentage share of total attributable in total direct expenditure (f/e)	59.74%	47.69%	27.42–48.15%
(g) Expenditure shifted to reef if no dolphins	**n.a.**	**n.a.**	**635,965–956,003**
Percentage share expenditure shifted to reef in total attributable expenditure (g/f)	n.a.	n.a.	26.7–22.8%
Percentage share expenditure shifted to reef in total direct expenditure (g/e)	n.a.	n.a.	7.3–11.0%
† Breakdown of auxiliary direct expenditure per person per day (percentage shares in parenthesis)			
Food	18.7 (26.1%)	13.6 (24.4%)	14.9 (21.2%)
Accommodation	12.8 (17.9%)	17.7 (31.8%)	21.9 (31.1%)
Internet and telephone	7.4 (10.3%)	4.8 (8.6%)	5.0 (7.1%)
Souvenirs	14.4 (20.1%)	6.6 (11.8%)	9.9 (14.1%)
Local transportation	18.3 (25.6%)	13.1 (23.5%)	18.5 (26.3%)
Total	71.6	55.7	70.3

*Dedicated tourist expenditure presented as ranging from entirely-dedicated expenditure to the sum of entirely- and partially-dedicated.

Source: The 2008/9 expenditures were from Mustika et al. (2012).

With 3–10 boats per day visiting, the total additional tourist expenditures received from snorkelling would be between USD 21,840 to USD 72,800 per annum for the whole industry (364 working days, discounting Silent Day in March or April when no one should go out to work).

Discussion

Tourist satisfaction in 2013 was virtually the same as in 2008/9 (Figure 13.2) with no significant difference between Western and Asian respondents for both data periods. Generally, favourite themes did not vary much between the two data collection periods. In 2008 and 2009, the majority of respondents liked the dolphins and the experience with the dolphins they obtained in Lovina (Mustika et al., 2013). The tourists also liked the beautiful scenery, serenity and the authenticity of the experience and the hospitality (Mustika et al., 2013). The same results were found in 2013. The Western tourists in 2008 and 2009 had the propensity to dislike boat crowding, noisy boats and the intrusive boat behaviours around the dolphins, while their Asian counterparts were more likely to dislike the time and effort taken to find the dolphins, the low number of dolphins or the less playful animals (Mustika et al., 2013). Although the 2013 respondents also disliked these aspects, most of them complained more about the debris they found in the sea and on the beach. Mirroring the 2008/9 results (Mustika et al., 2013), satisfied tourists in 2013 were more likely to indicate that they would rejoin the trip or recommend the trip to others.

We examined tourist satisfaction against the number of dolphins seen, the number of boats, how the tourists felt about the way the boatmen drove the boat and what they thought of the numbers of boats around them. The result was in agreement with the 2008/9 results (Mustika et al., 2013) i.e., the satisfaction level of dolphin watchers was significantly associated with the way the boatmen drove the boat (i.e., they were most satisfied when the boatmen were driving considerately around the dolphins) and the number of dolphins they saw (more dolphins yielded more satisfaction).

Lovina is doing poorly compared with other marine tourism locations in Bali. The Bali Hai Cruises dolphin trips in south Bali (2008/9) garnered on average 87.4% satisfaction level (Mustika et al., 2013). In 2011, about 87% of respondents were very satisfied with the surfing experience in Uluwatu and adjacent areas in Bali (Margules et al., 2014). Lovina was also under-performing when compared to three reef-based tourism locations in Bali (Figure 13.3; Mustika, 2014).

The proportion of Asian visitors increased in 2013 compared to the 2008/9 survey (which was 27% Asians, n=387). Our data shows an increased percentage of Asian tourists participating in the dolphin watching tours compared to 2008/9. Almost all Asian visitors sampled in 2013 were domestic visitors. It is difficult to ascertain whether this is a permanent trend or because our sampling days coincided with the school holiday seasons in Indonesia, hence more domestic tourists were visiting the area.

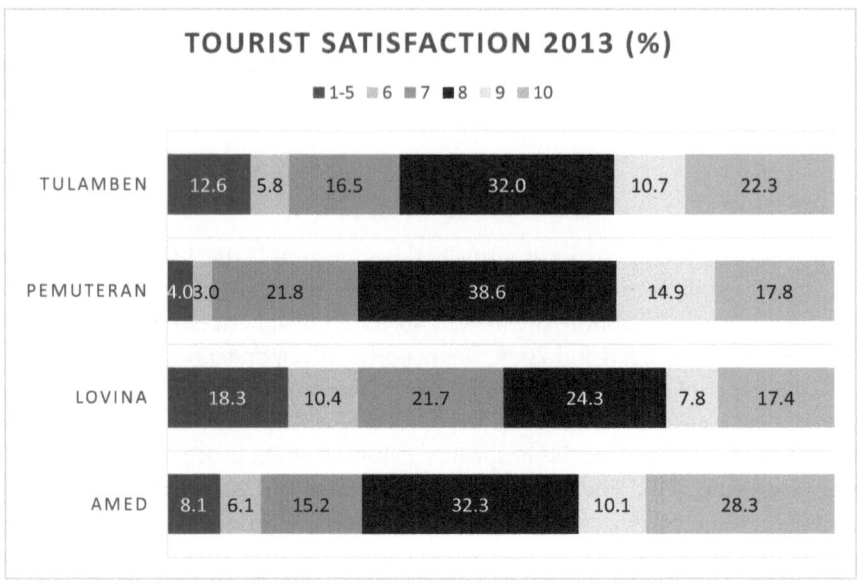

FIGURE 13.3 Lovina's tourist satisfaction when compared to three other reef-based tourism locations in Bali (2013 data)

The dolphin watching industry injected approximately USD 8.7 million into the local economy in 2013. The annual gross revenue per boatman in 2013 (USD 1,515) was lower than the 2008/9 figure (inflation-adjusted; Table 13.1). This number was lower than the annual per capita gross domestic regional product (GDRP) for the Buleleng agricultural sector in 2013 (USD 1,734; Buleleng, 2013). Since the 2008/9 data stated that the dolphin watching industry gave 1.16 times higher income than the opportunity cost (Mustika, Welters, et al., 2016), the 2013 result might support the notion that the dolphin watching industry in Lovina has, or almost, reached the saturation point as proposed by Mustika et al. (2017). Thus, options such as limiting the total number of boats in the industry should be explored to prevent further decline of per capita revenue for the boatmen and to support a more ecologically friendly dolphin watching industry (see Mustika et al., 2015).

The auxiliary direct expenditures per day in 2013 (Table 13.1) were similar to those of 2008/9, except for accommodation expenditure, which kept increasing over the previous 3 years, possibly reflecting the increasing accommodation costs in north Bali. However, the total expenditure of tourists in 2013 was at least 14% lower than in 2008/9. This trend might have been influenced by the shorter length of stay in 2013 (2.16 days) compared to 2008/9 (2.95 days; Mustika et al., 2012). However, ascertaining the factors behind shorter length of stay in Lovina is more complex and requires further research.

In the absence of the dolphins, at least 27% of total direct expenditures would flow to other tourism sites in Bali, including to adjacent reef tourism sites (7.3–11%

of total direct expenditures). Dolphin watching tourism also has considerable financial spill-over in the local reef tourism sector through the snorkelling package. Since some tourists are attracted to both the dolphins and the reefs, this result can be used by the Buleleng tourism authority to attract nature tourists who are interested in both types of activities.

Our results confirmed that the dolphins remain the drawcard for Lovina, despite the declining per capita profit for the boatmen. Despite having a dedicated Marine Protected Area established to protect the dolphins, no tangible actions have been undertaken to improve the sustainable practice of the local dolphin watching industry. Considering the presence of adjacent reef areas, the emergence of a new dolphin watching industry in east Buleleng (in the village of Tejakula) and the rapid growth of travel blogs and tourism rating sites, such as TripAdvisor and Booking.com, the Lovina dolphin watching industry needs to be managed more sustainably. Measures such as a revocable licence system and daily quota of the number of boats need to be paired with sustainable dolphin watching practices to achieve this objective.

Considering the improving condition of the reefs in Lovina in 2015 (Welly & Lazuardi, 2016), integrating dolphin watching tourism with coral reef tourism in the region as a way forward is worth exploring. This chapter shows that the combination of dolphin watching and coral reef tourism confers substantial economic benefit to the local people and the local economy. With the income from dolphin watching leaning towards saturation, the coral reefs could provide an alternative source of income, as well as provide a more resilient tourism development in Lovina.

Last, but not least, Lovina and Tejakula (in eastern Buleleng) share the same dolphin species, i.e., spinner dolphins, Fraser's dolphins (*Lagenodelphis hosei*) and short-finned pilot whales (*Globicephala macrorhynchus*; Mustika, personal observation). Considering the close proximity of the two dolphin watching sites (the distance is only about 40 km) and the thriving coral reef tourism in Tejakula, dolphin population exchange between the two sites is plausible. The two sites should therefore be managed under the same management strategies to foster sustainable dolphin watching tourism at both sites, with some spill-over effects to the local reef tourism.

Conclusion

In an area where reef tourism can be described as marginal, marine megafauna such as dolphins can provide additional income for the local community. Our study shows that dolphins remain very important to the local economy. The dolphin watching sector gives a considerable financial spill-over to the local reef tourism. The hypothetical absence of the dolphins may shift tourist expenditures to other reef tourism sites in Bali. As such, the conservation of this target species and the possible combination with local reef tourism should be made a priority to sustain local income in the future.

References

Buleleng, B. (2013). *Indikator Ekonomi Kabupaten Buleleng 2013 [Economic Indicators for Buleleng Regency 2013]*. Singaraja: Badan Pusat Statistik Kabupaten Buleleng.

Chou, L.M. (2000). Southeast Asian reefs-status update: Cambodia, Indonesia, Malaysia, Philippines, Singapore, Thailand and Viet Nam. In: C. Wilkinson (Ed.), *Status of coral reefs of the world*, pp. 117–129. Townsville, Australia: Australian Institute of Marine Science.

Courbis, S., & Timmel, G. (2009). Effects of vessels and swimmers on behavior of Hawaiian spinner dolphins (*Stenella longirostris*) in Kealake'akua, Honaunau, and Kauhako bays, Hawai'i. *Marine Mammal Science, 25*(2), 430–440. doi: 10.1111/j.1748-7692.2008.00254.x

Cribb, N., Miller, C., & Seuront, L. (2012). Site fidelity and behaviour of spinner dolphins (*Stenella longirostris*) in Moon Reef, Fiji Islands: Implications for conservation. *Journal of the Marine Biological Association of the United Kingdom, 92*(08), 1793–1798.

Disparda Bali. (2016). Jumlah kunjungan wisatawan pada obyek-obyek wisata di Bali tahun 2013 [The 2013 visitation to tourist attractions in Bali]. Retrieved 14 July 2016 from http://www.disparda.baliprov.go.id/en/statistics2

Grigg, E., & Markowitz, H. (1997). Habitat use by bottlenose dolphins (*Tursiops truncatus*) at Turneffe Atoll, Belize. *Aquatic Mammals, 23*(3), 163–170.

Hanan, M., & Karp, P. (1989). *Customer satisfaction*. New York: Amacom.

Higham, J.E., Bejder, L., Allen, S.J., Corkeron, P.J., & Lusseau, D. (2016). Managing whale-watching as a non-lethal consumptive activity. *Journal of Sustainable Tourism, 24*(1), 73–90.

Kreb, D., & Budiono. (2005). Conservation management of small core areas: Key to survival of a Critically Endangered population of Irrawaddy river dolphins *Orcaella brevirostris* in Indonesia. *Oryx, 39*(2), 178–188.

Lee, J.J. (2014). Elusive whales set new record for depth and length of dives among mammals. Retrieved 15 July 2016 from http://news.nationalgeographic.com/news/2014/03/140326-cuvier-beaked-whale-record-dive-depth-ocean-animal-science/

Lusseau, D. (2003). Male and female bottlenose dolphins *Tursiops* spp. have different strategies to avoid interactions with tour boats in Doubtful Sound, New Zealand. *Marine Ecology Progress Series, 257*, 267–274.

Lusseau, D. (2004). The hidden cost of tourism: Detecting long-term effects of tourism using behavioral information. *Ecology and Society, 9*(1). Retrieved from http://www.ecologyandsociety.org/vol9/iss1/art2

Lusseau, D., Slooten, L., & Currey, R.J.C. (2006). Unsustainable dolphin-watching tourism in Fiordland, New Zealand. *Tourism in Marine Environments, 2*(2), 173–178.

Margules, T., Ponting, J., Lovett, E., Mustika, P., & Wright, J.P. (2014). *Assessing direct expenditure associated with ecosystem services in the local economy of Uluwatu, Bali, Indonesia*. A Report for Conservation International Indonesia. Conservation International Indonesia, Save the Waves Coalition, San Diego State University. Denpasar: Conservation International Indonesia.

Mustika, P.L.K. (2014). *Survey of tourist perception and level of support for conservation in Bali*. Updated Report to the Coral Reef Alliance, December 2014. Denpasar: Coral Reef Alliance.

Mustika, P.L.K., Birtles, A., Everingham, Y., & Marsh, H. (2013). The human dimensions of wildlife tourism in a developing country: Watching spinner dolphins at Lovina, Bali, Indonesia. *Journal of Sustainable Tourism, 21*(2), 229–251. doi: 10.1080/09669582.2012.692881

Mustika, P.L.K., Birtles, A., Everingham, Y., & Marsh, H. (2015). Evaluating the potential disturbance from dolphin watching in Lovina, north Bali, Indonesia. *Marine Mammal Science, 31*(2), 808–817.

Mustika, P.L.K., Birtles, A., Welters, R., & Marsh, H. (2012). The economic influence of community-based dolphin watching on a local economy in a developing country: Implications for conservation. *Ecological Economics, 79*, 11–20. doi: 10.1016/j.ecolecon.2012.04.018

Mustika, P.L.K., Farr, M., & Stoeckl, N. (2016). The potential implications of environmental deterioration on business and non-business visitor expenditures in a natural setting: A case study of Australia's Great Barrier Reef. *Tourism Economics, 22*(3), 484–504.

Mustika, P.L.K., Welters, R., Ryan, G.E., D'Lima, C., Sorongon-Yap, P., Jutapruet, S., & Peter, C. (2017). A rapid assessment of wildlife tourism risk posed to cetaceans in Asia. *Journal of Sustainable Tourism, 25*, 1138–1158.

Notarbartolo-di-Sciara, G., Hanafy, M.H., Fouda, M.M., Afifi, A., & Costa, M. (2009). Spinner dolphin (*Stenella longirostris*) resting habitat in Samadai Reef (Egypt, Red Sea) protected through tourism management. *Journal of the Marine Biological Association of the United Kingdom, 89*(1), 211–216.

Nowacek, S.M., Wells, R.S., & Solow, A.R. (2001). Short-term effects of boat traffic on bottlenose dolphins, *Tursiops truncatus*, in Sarasota Bay, Florida. *Marine Mammal Science, 17*(4), 673–688. doi:10.1111/j.1748-7692.2001.tb01292.x

O'Connor, S., Campbell, R., Cortez, H., & Knowles, T. (2009). *Whale watching worldwide: Tourism numbers, expenditures and expanding economic benefits.* Melbourne: Economist at Large & IFAW.

Orams, M. (2013). Economic activity derived from whale-based tourism in Vava'u, Tonga. *Coastal Management, 41*(6), 481–500. doi: 10.1080/08920753.2013.841346

Parsons, E. (2004). The behavior and ecology of the Indo-Pacific humpback dolphin (*Sousa chinensis*). *Aquatic Mammals, 30*(1), 38–55.

Prasetia, I. (2013). Kajian Jenis Dan Kelimpahan Rekrutmen Karang Di Pesisir Desa Kalibukbuk, Singaraja, Bali. *Bumi Lestari, 13*(1).

Roe, D., Leader-Williams, N., & Dalal-Clayton, B. (1997). *Take only photographs, leave only footprints: The environmental impacts of wildlife tourism.* London: International Institute for Environment and Development.

Sazima, I., Sazima, C., & Silva, J.M. (2003). The cetacean offal connection: Feces and vomits of spinner dolphins as a food source for reef fishes. *Bulletin of Marine Science, 72*(1), 151–160.

Schorr, G.S., Falcone, E.A., Moretti, D.J., & Andrews, R.D. (2014). First long-term behavioral records from Cuvier's beaked whales (*Ziphius cavirostris*) reveal record-breaking dives. *PLoS One, 9*(3), e92633.

Stoeckl, N., Smith, A., Newsome, D., & Lee, D. (2005). Regional economic dependence on iconic wildlife tourism: Case studies of Monkey Mia and Hervey Bay. *The Journal of Tourism Studies, 16*(1), 69–81.

Sutaria, D., & Marsh, H. (2011). Abundance estimates of Irrawaddy dolphins in Chilika Lagoon, India, using photo-identification based mark-recapture methods. *Marine Mammal Science, 27*(4), E338–E348. doi: 10.1111/j.1748-7692.2011.00471.x

Welly, M., & Lazuardi, M.E. (2016). *Kondisi Biofisik dan Sosial Ekonomi Pesisir Bali 2015* (p. 169). Denpasar: Coral Triangle Center.

Yanuar, A., Tjiu, A., Suprapti, D., Syahirsyah, Saniswan, Y., & Widjaya, I. (2011). *Discovery of Irrawaddy dolphin Orcaella brevirostris population and habitat in Kubu Raya waters, West Kalimantan: A preliminary survey of Irrawaddy dolphins in salt and brackish waters.* Jakarta: WWF Indonesia.

14

TOURISM DEVELOPMENT AND IMPACTS ON REEF CONSERVATION IN BRAZIL

Fernanda de Vasconcellos Pegas, Guy Castley and Ambrozio Queiroz Neto

Introduction

Extending across 7,400 km of coastline, Brazil's beaches are the nation's primary tourist attraction (Pegas et al., 2015). The north-east region, in particular, is home to 90% of all Brazil's coral reefs and the only reef formations in the south Atlantic region. Despite its long coastline and natural resource potential, and recently being ranked number one in natural resources by the World Economic Forum (WEF) in the *Travel and Tourism Competitiveness Report* (WEF, 2015), Brazil is still not recognised as an international scuba diving destination. Relevant publications in the dive industry, such as the *Encyclopaedia of recreational diving* (Professional Association of Diving Instructors [PADI], 2008), do not cover diving sites in Brazil (or South America). Therefore, Brazilian diving destinations, including reefs, remain largely unknown outside of Brazil. To highlight coral reef destinations and the environmental challenges these destinations are facing, this chapter provides an overview of the current situation of Brazil's coral reef tourism.

Brazilian coastline overview

Brazilian coral reefs extend along 3,000 km of coastline from the Parcel de Manuel Luís Maranhão (0° 53' S, 44° 16' W) to the reefs of Viçosa (18° 01' S, 39° 17' W). Reefs also occur around oceanic islands such as those in Atol das Rocas Biological Reserve and Fernando de Noronha National Park. Within this region there are a number of popular diving destinations (Figure 14.1), as well as important refuge areas (Box 14.1). Rio de Janeiro, with 32 diving spots, is the state with the greatest number of marine diving areas (WannaDive, 2016). Most of Brazil's diving spots are located in coastal tourist areas. Pegas et al. (2015) identified 195 coastal tourist areas along Brazil's Littoral Pleasure Periphery (LPP), with 59% (116) of these localities in the northern region.

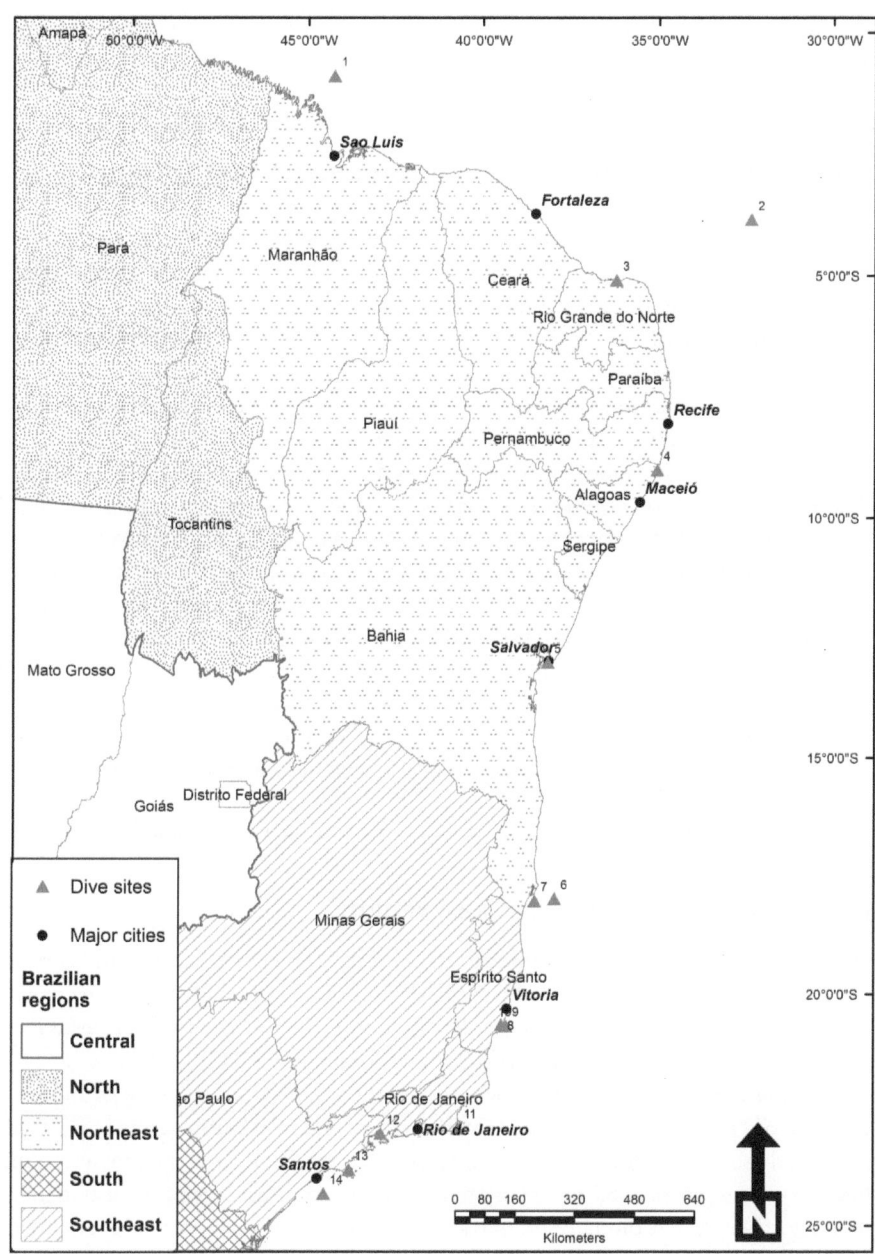

FIGURE 14.1 Popular diving destinations in Brazil

Numbers on the map from north to south are 1) Parcel de Manuel Luís Maranhão, 2) Fernando de Noronha National Park, 3) Porto de Galinhas, 4) Maragogi, 5) Banco de Panela, 6) Abrolhos Marine National Park, 7) Viçosa,

8) Guarapari, 9) Ilha Rasa, 10) Victory 8B wreck, 11) Ilha do Cabo Frio, 12) Angra do Reis, 13) Ilha Bela, 14) Laje de Santos.

Box 14.1 Atol das Rocas: an important reef refuge

The Atol das Rocas Biological Reserve was designated as Brazil's first marine protected area in 1979. The atoll is volcanic in origin and largely comprised of coralline algae as opposed to corals and has raised questions about its true status as an "atoll" or "annular algal reef". Together with the Fernando de Noronha National Park, it forms part of an important World Heritage Area and diving is commonplace. The atoll is home to a great variety of endemic and migratory wildlife species and is an important rookery site for threatened marine turtle populations, such as the green turtle (*Chelonia mydas*). The site was also recently recognised as a Ramsar wetland of international significance because it regularly supports more than 150,000 individuals of approximately 30 bird species. Although access to the atoll is restricted, it does lie within historical fishing grounds and tourism to the region requires monitoring. Research teams such as the Tamar Project provide ongoing monitoring for turtle populations on the atoll.

Sources: WWF, 2017 (https://www.worldwildlife.org/ecoregions/nt0123); Ramsar, 2016 (http://www.ramsar.org/news/brazil-designates-atol-das-rocas-biological-reserve) and Bellini et al., 1996.

Coastal tourism in the north-east region

At the national level, the total contribution of the travel and tourism industry to the Brazilian gross domestic product (GDP) is about 9.6%, with 94.7% generated by the domestic tourism sector. Over 3.1 million jobs (3.1% of total) are provided by this industry (World Tourism and Travel Council [WTTC], 2015).

The north-east region, which encompasses the region from the state of Bahia to the state of Maranhão, has experienced the fastest rate of tourism growth in Brazil, particularly since the mid-1990s. This intensive growth has led to concerns about the health of the region's coral reef ecosystems. Recent tourism development initiatives in the north-east are a direct result of the Action Program for Tourism Development (PRODETUR, Programa de Desenvolvimento do Turismo). PRODETUR is a collaboration between state and local governments, domestic (i.e., Banco do Nordeste do Brasil BNB) and international (i.e., the Inter-American Development Bank) financial agencies and tourism agencies (e.g., Ministry of Tourism) to promote development in the north-east region (Pegas et al., 2015). Public and private investment was responsible for the development of 30% of all coastal hotels and resorts built in Brazil between 2006 and 2014 (BSH International, 2008, 2011).

Bahia is the top ranking state in terms of overall financial investment and tourism infrastructure growth within the PRODETUR. Under the PRODETUR, 78.5% of all investment allocated for the north-east region (equivalent to US$5.6 million) was allocated to Bahia. One outcome was the generation of 49,000 jobs within the state (Pegas et al., 2015). Significant impacts have been identified at the municipal level. For instance, in 2010 over 50% of jobs in the municipalities of Mata de São João (Bahia) and Maragogi (Alagoas) were generated by the tourism industry (Brazilian Institute of Geography and Statistics [IBGE], 2011a). Other socioeconomic benefits linked with the tourism industry via PRODETUR include the development of sewage and water treatment facilities and improvements in transportation services (Pegas & Stronza, 2010; IBGE, 2011b).

Coral reefs and the diving industry in Brazil

Scuba diving tourism is a multi-billion dollar industry and a very important niche market for coastal destinations, especially for those located in tropical zones (Tapsuwan & Asafu-Adjaye, 2008; Lew, 2013; Hein et al., 2015). Dive tourism has received considerable attention from researchers, with studies exploring sustainability (Dimmock & Musa, 2015; Doiron & Weissenberger, 2014), diver attitudes, preferences and motivations (Edney, 2012; Giglio et al., 2015), as well as potential impacts from recreational dive tourism (Zakai & Chadwick-Furman, 2002; Hasler & Ott 2008; Guzner et al., 2010; Lamb et al., 2014; Bravo et al., 2015). These clearly demonstrate the importance of this niche tourism market in recent years. Divers are typically well educated and high income earners (Dearden et al., 2006; Dimmock & Musa, 2015) and most are environmental friendly and value operators who undertake socioenvironmental practices (Queiroz Neto et al., 2017). Because of the growing sense of environmental consciousness displayed by divers, scuba diving tourism organisations appear to be placing an increasing focus on the sustainable use of these natural resources (Chapter 15 looks at similar issues in relation to snorkelling).

Information about the diving industry in Brazil is inconsistent and remains limited. In contrast to other countries, dive-training organisations treat their statistical reports as confidential. In one of the few reports available, an estimated 65,000 divers undertook an average of 12 dives per year (Ministério do Turismo, 2005). According to this report, approximately 15,000 new divers are added to this number every year, generating US$5 million in equipment sales and divers spending about US$13 million on travel. Due to a lack of data, the current statistics of the industry in Brazil remain unknown.

Diving locations are spread throughout coastal regions but the states with the greatest number of marine diving spots are Rio de Janeiro (32), Santa Catarina (24), São Paulo (23) and Pernambuco (16). All other states (n=6) together provide access to a further 35 sites. Some of Brazil's most popular diving destinations are Recife, Maragogi, Arraial do Cabo, Angra dos Reis, Ilha Bela, Victory 8B shipwreck, Porto de Galinhas, Fernando de Noronha Marine National Park, Banco da Panela, Angra dos Reis and Abrolhos Marine National Park (PADI, 2016; see also Box 14.2).

Box 14.2 Diving in Brazil's marine national parks

Abrolhos Marine National Park is a UNESCO Biosphere Reserve and is part of the largest and most biodiverse coral reef complex in the South Atlantic (Leão et al., 2003). It has also been listed as a Ramsar site since 2010. Its conservation helps sustain the populations of various IUCN Red Listed species, such as leatherback (*Dermochelys coriacea*), hawksbill (*Eretmochelys imbricata*), loggerhead (*Caretta caretta*) and green turtles (*Chelonia mydas*), coral species (e.g., *Millepora nitida*) and humpback whales (*Megaptera novaeangliae*) (de Vasconcellos Pegas et al., 2015). In addition, the islands of the archipelago support at least 32 species of seabird and provide an important breeding ground for migratory species (Ramsar, 2012).

The park is also popular among recreational divers (Giglio et al., 2015; Giglio et al., 2016), and tourism impacts are managed by limiting visitor numbers to the area. All visitors must be accompanied by trained guides who have been accredited by the management authority, the Instituto Chico Mendes de Biodiversidade (ICMBio). Despite these regulations, damage to the reef continues to occur (Giglio et al., 2016). As in other marine protected areas, Abrolhos' reefs remain threatened by illegal fishing, litter and wild-life stress caused by tourist activities. North of Bahia and 86 km offshore, the Parcel Manoel Luiz State Marine Park is the largest bank of corals in South America. As in Abrolhos, the park is dotted with various shipwrecks that contribute to the diving experience. Unlike most of the best dive sites in Brazil, the park has limited diving tourism activities due to local currents and distance from the coast (Fundação de Amparo à Pesquisa do Estado de São Paulo [FAPESP], 2016).

Tourism development and reef conservation in Brazil

Although a widespread phenomenon, the impacts caused by coastal tourism development and coastal development in general (e.g., beachfront houses, marinas and cargo ports) vary in scale, intensity and type (Pegas et al., 2015). One of these impacts is the growth of both permanent and temporary populations in tourist areas. Driven by employment and income opportunities and white sand beaches, 32 million of the north-east's total population of 53 million live along the coast (Brasil Escola, 2016). Along Brazil's reef belt (i.e., coastline between the states of Bahia and Rio Grande do Norte), there has been substantial urban growth, with over 15 million people now living along this section of the LPP. Seasonal population growth is common in some of the popular coastal communities as people travel to these areas for recreational purposes. In these localities, the social and environmental impacts of coastal tourism can be severe due to the sporadic increase in population numbers during the peak tourism season. In the beachfront village of

Praia do Forte, Mata de São João Municipality, the permanent population of 2,500 doubles during the summer months (Pegas & Stronza, 2010). This sporadic, and often poorly managed, urban growth places pressure on local infrastructure (e.g., water resources, waste management facilities) and natural resources (e.g., water pollution) and can affect local reef ecosystems. In addition, tourism development in the north-east region has exacerbated the long history of poverty and social displacement in the region, generating both direct and indirect impacts on onshore and reef ecosystems.

In Bahia, tourism development is associated with deforestation, water and air pollution, and sewage release in mangroves (Andrade, 2008; Limonad, 2007). In Pernambuco, illegal settlements and environmental degradation takes place in areas of great conservation importance, such as coastal dunes (Costa et al., 2008). In Ceará, tourism has been linked with illegal fishing practices, sedimentation, illegal development in protected areas, and mangrove and dune deforestation (Souza, 2005). In Paraíba, poor tourism management over the past 15 years, including a lack of observance of carrying capacity limits, is causing reef degradation (de Sousa Melo et al., 2006; Amorim & Sassi, 2009). Direct tourism impacts on reefs observed in the region are caused by divers and bathers walking on the reefs, which impacts coral (Júnior & Lima, 2014; Torres, 2016) and reef fish communities (Da Silva, 2015). Furthermore, impacts on coral reefs through dive tourism can also be driven indirectly by the desire of divers to view cryptic species (e.g., seahorses) associated with reefs (Uyarra & Côte, 2007).

On a local scale, one of the reef areas threatened by unsustainable tourism practices is the Abrolhos Marine National Park. The park is visited by 1,300 divers on an annual basis, with about 9,100 dives per year distributed across 15 dive spots (Giglio et al., 2016). Based on these visitation numbers, Giglio et al. (2016) estimate that, on an annual basis, divers touch corals over 74,000 times, cause coral damage about 9,000 times, and raise sediment onto these corals almost 11,000 times. Initiatives aimed at reducing the frequency and magnitude of impacts take place via a required course for diving guides who operate within the park's boundaries. Established in 2011, the course is conducted on an annual basis as part of the park's management plan (ICMBio, 2011). Ongoing monitoring of dive tourism impacts on coral reefs is important because these niche activities are known to have impacts of coral reef health (Torres, 2016).

The beach of Porto de Galinhas in the state of Pernambuco is also a popular reef tourism destination. In early 2000, the beach was visited by approximately 60,000 people on a monthly basis, with 69% visiting the nearby reefs to dive, walk and snorkel (Ministério do Meio Ambiente [MMA], 2003). A more recent study suggests the presence of 1,200 divers on an annual basis (Barradas et al., 2012).

In spite of these observed negative impacts, tourism development, including dive tourism, is also a catalyst for environmental conservation initiatives. At the regional level, the establishment of the PRODETUR is associated with the conservation of 16,524 ha of coastal habitat (Saab, 1999), including the establishment of coastal and inland Areas of Environmental Protection (APAs); six protected areas in Bahia and one in Rio Grande do Norte; conservation of nine lagoons; dune restoration programmes in Ceará; and restoration of the Reginaldo Salgadinho Valley in Maceió. Table 14.1 illustrates the conservation areas that protect coral reef ecosystems in Brazil.

In theory, environmental regulations and marine protected areas (MPAs) form the cornerstone of marine environmental conservation efforts because they contribute to increased size, density, diversity and biomass of various functional

TABLE 14.1 Marine protected areas by state, year of establishment and policy level

Conservation Area	State	Year	Classification
Parque Estadual Marinho do Parcel do Manuel Luís	Maranhão	1991	Ramsar Site; State Park
Reserva Biológica do Atol das Rocas	Rio Grande do Norte	1979	Biological Reserve; UNESCO site
Parque Nacional Marinho de Abrolhos	Bahia	1983	National Park
APA de Fernando de Noronha: Rocas, São Pedro and São Paulo	Pernambuco	1986	Environmental Protection Area
Parque Nacional Marinho de Fernando de Noronha	Pernambuco	1988	National Park; UNESCO site
APA do Litoral Norte	Bahia	1992	Area of Environmental Protection
APATinharé-Boipeba	Bahia	1992	Area of Environmental Protection
APA Ponta da Baleia/ Abrolhos	Bahia	1993	Area of Environmental Protection
APA Costa dos Corais	Pernambuco and Alagoas	1997	Area of Environmental Protection
APA Municipal Recifes de Pinaúnas	Bahia	1997	Municipal Area of Environmental Protection
Parque Municipal Marinho do Recife de Fora	Bahia	1997	Municipal Park
Parque Municipal Marinho da Coroa Alta	Bahia	1998	State Park
APA da Baía de Todos os Santos	Bahia	1999	Area of Environmental Protection
Parque Municipal Marinho do Recife de Areia	Bahia	1999	Municipal Park
Parque Estadual Marinho da Areia Vermelha	Paraíba	2000	State Park
Reserva Extrativista Marinha de Corumbau	Bahia	2000	Extractive Reserve
APA da Baía de Camamu	Bahia	2002	Area of Environmental Protection
APA Estadual dos Recifes de Corais	Rio Grande do Norte	2011	State Environmental Protection Area

Source: ICMBio (2009).

groups of marine species (Halpern, 2003). However, the benefits of these areas are affected by the nature and intensity of any activity (legal or illegal) that may take place including tourism. The effectiveness of MPAs has received widespread attention, and some papers address specific benefits to coral reefs and their communities (Magdaong et al., 2014; Selig & Bruno, 2010; van der Meer et al., 2015).

In a Brazilian context, the success of implementing MPAs to protect biodiversity and fisheries resources is regarded as poor (Gerhardinger et al., 2011) and zonation within MPAs may be critical to managing current and future impacts. For example, although the APA Costa dos Corais, established in 1997, has the longest coastal coverage among MPAs and is the first protected area to protect coral reefs

TABLE 14.2 Primary legislation that targets and/or includes conservation of coral reef ecosystems in its specifications

Conservation initiative (established year; level)	
Federal Law 9.605 (1998; Federal)	Jail time and fine for an individual who kills, harasses, hunts, captures and uses wildlife that is native or in migration, without permission, licence or authorisation of the respective environmental authority or in violation of such if acquired.
Normative Ruling 13 (2005; Federal)	The use of chemical substances, anaesthetics, toxins or substances that cause irritation to marine wildlife is prohibited; the removal and/or any actions that cause physical damage to corals and reef species is prohibited.
Federal Decree 6.514 (2008; Federal)	The exploration of coral reefs without authorisation of the environmental agency in charge or in violation of such if granted is prohibited. In this event, the violator is fined.
Normative Ruling 204 (2008; Federal)	Harvesting of reef species can only take place upon licence granted by the Instituto Brasileiro do Meio Ambiente (IBAMA). As of 2016, no company has yet been granted this authorisation or authorisation to sell Brazilian reefs. This Ruling also prohibits the collection, sale, transport and holding of pieces of corals and rocks.
Conduct Adjustment Term (2010; Federal)	Establishes codes of conduct and visitation rules at Picãozinho and Ponta do Seixas reefs in the state of Paraíba. Changes include a maximum number of visitors at one time and tourist guides on board each boat.
Licence to operate (2012; Local)	Licence to operate granted to Vanco Brasil Exploração e Produção de Petróleo e Gás Natural Ltda. Exploration areas are Canario, Jandaia and Sabia (25° 59' 32.56" S / 46° 7' 49.6" W). Licence prohibits drilling on corals and algae banks.
National Action Plan for the Conservation of Coralíneo Environments (2016; Federal)	Plan protects 52 endangered marine species. Among the 10 objectives is the implementation of sustainable practices on Extractive Reserves and establishment of codes of conduct on marine tourism areas.

in Brazil, it is also currently threatened by tourism (Araújo & Bernard, 2016). The APA do Litoral Norte in northern Bahia is stratified into zones that determine the level of urban development and management practices within the APA. In addition to the protection offered by these MPAs, coral ecosystems are also protected by environmental legislation (Table 14.2).

In practice, the conservation effectiveness of MPAs and environmental legislation to the long-term conservation of marine resources in these tourist destinations is questionable (Barbosa et al., 2010; Gerhardinger et al., 2011; Hein et al., 2015; Pegas et al., 2015). As Araújo & Bernard (2016) have highlighted, tourism is likely to be the single greatest threat facing Brazil's largest MPA (APA Costa dos Corais). Enforcement of conservation legislation in the APAs and other protected areas (e.g., poaching control) varies among and within regions due to insufficient government funding to hire and train personnel and to purchase and maintain monitoring equipment (de Marques & Peres, 2015; Pegas & Castley, 2016). As is the case with many other countries, Brazilian conservation management agencies continue to grapple with the challenges of effective use of limited revenues to achieve their conservation objectives (Gerhardinger et al., 2011). However, financial restrictions may vary from one region to the next. As Araújo & Bernard (2016) noted, the APA Costa dos Corais is currently on a sound financial footing and able to support management activities.

Another conservation challenge lies with regulations about tourist use of the reefs, which often lack accurate identification and integration of tourism carrying capacity of reefs in reef management strategies (Lopes et al., 2014; Ríos-Jara et al., 2013). Visitor carrying capacity of a diving spot is linked with visitor behaviour. Understanding diver behaviour is therefore important when determining the carrying capacity of the reef and when implementing management strategies aimed at reducing impacts of recreational diving (Giglio et al., 2016). Behavioural changes can be achieved through pre-dive educational briefings (Camp & Fraser, 2012) that encourage diver behaviour and compliance based on the norms of diver profiles, dive objectives and characteristics of the dive site (Smith et al., 2010; Giglio et al., 2015; ICMBio, 2011).

Weak enforcement and poorly drafted management strategies are strong indicators that the establishment of protected areas alone does not guarantee successful achievement of conservation goals over time. Marine management strategies should therefore carefully identify direct and indirect threats to reef ecosystems as part of the management plan. Mitigation initiatives should be implemented across government levels depending on the type of conservation area, as well as its users (e.g., divers, and dive and tourist guides). This includes, but is not limited to, providing pre-dive briefings encompassing ecological aspects of corals for divers, bathers and boat operators to reduce reef degradation through their behaviours (e.g., educational activities at Abrolhos Marine National Park); establishing diving zones for beginner divers over sand bottoms or places with low coral abundance; and restricting visitation numbers based on carrying capacity at each reef formation (e.g., as at Picãozinho and Ponta do Seixas reefs).

As a wildlife watching activity, divers tend to prefer areas that are larger with limited numbers of fellow divers and bathers. As such, maintaining a sustainable carrying capacity for reefs is not only important for the reefs but also for diver satisfaction and maintaining a viable diving industry. Conversely, restricting the number of dive sites and increasing the number of visitors can jeopardise the diving industry at that location (Brander et al., 2007). Other conservation strategies that can mitigate tourist impacts on reefs are the establishment and enforcement of environmental laws (e.g., anti-poaching laws, collection of coral and marine animals), delimiting areas for boat circulation and anchoring, and reducing litter (e.g., providing garbage disposal areas, establishing garbage collection systems). Implementing these interventions to minimise tourism impacts on coral reefs will also require research and monitoring to assess the effectiveness of the strategy that is development.

Conclusion

Brazil's LPP is a popular tourist destination with both the domestic and international tourist markets. The north-east region is the nation's beach tourism node as well as the location of most reef formations and the most visited dive spots. Despite limited information on the number of divers and diving revenue at the state and national levels, this industry is here to stay. Reef formations are experiencing increasing visitation linked to the thriving beach-going tradition. Although economic benefits are important outcomes of this new industry, tourism also generates detrimental impacts on the same reefs it is dependent on. Reef impacts caused by the diving industry range in scale, type and location. Finding a common ground between economic development and sustainable use of reef resources requires more than legislative regulation. It needs to include effective law enforcement, development of management plans that address the needs of all stakeholders (e.g., local communities, recreational divers, conservation agencies and tourism operators), local support and involvement, diver education and visitor behaviour that is not detrimental to marine resources. In Praia do Forte, Bahia, community support for sea turtle conservation practices has been paramount for the recovery of three endangered species and conservation of reef species and coastal ecosystems (Pegas & Stronza, 2010). Although an important conservation component, community participation in decision-making is often lacking (Pegas et al., 2015; Oliveira & Silva, 2016). Changes in tourist behaviour take time and, in some locations, may not occur without strict and efficient environmental regulation and education. Current political instability and shortages in government funding for conservation purposes highlight the need for the diving industry to contribute to reef conservation in Brazil. After all, positive diving experiences are directly linked with marine biodiversity encountered during the dive. A healthy and protected marine area not only contributes to a positive diving experience but also to the long-term conservation of the social fabric of the locations visited and the natural marine resources on which they depend.

References

Amorim, T.P.L. & Sassi, R. (2009). Microssimbiontes associados à Millepora alcicornis (Linnaeus, 1758) (Cnidaria, Hydrozoa) dos recifes costeiros de Picãozinho, João Pessoa-PB, 2009. [Microsymbionts associated with Millepora alcicornis (Lennaeus, 1758) (Cnidaria Hdrozoa) of the coastal reefs of Picãozinho, João Pessoa-PB, 2009]. Masters Thesis from the Universidade Federal da Paraiba. Retrieved from http://www.revista.ufpe.br/tropical oceanography/artigos_completos_resumos_t_d/42_2_2014_11_cavalcante.pdf

Andrade, J. (2008). Programa Berimbau: Iniciativa político-institucional de regulacao de conflitos socioambientais na aarea de influencia de Costa do Sauıpe-Bahia [Berimbau program: Political institutional initiatives for the regulation of socioenvironmental conflicts in the Costa do Sauıpe-Bahia area]. *RAC-Eletronica*, *2*, 426–448.

Araújo, J.L., & Bernard, E. (2016). Management effectiveness of a large marine protected area in Northeastern Brazil. *Ocean & Coastal Management*, *130*, 43–49.

Barbosa, S., Formagio, C., & Barbosa, R. (2010). Areas protegidas, uso e ocupaçãao do solo, qualidade de vida e turismo no litoral norte paulista: Algumas reflexçoes sobre o municıpio de Ubatuba [Protected areas, use and land settlement, quality of life and tourism in the paulista north coast: Some reflexions about the municipality of Ubatuba]. *Caderno Virtual de Turismo*, *10*(2), 121–137.

Barradas, J., Amaral, F., Hernandez, M., Flores-Montes, M., & Steiner, A. (2012). Tourism impacts on reef flats in Porto de Galinhas Beach, Pernambuco, Brazil. *Arquivos de Ciencias do Mar*, *45*(2), 81–88.

Bellini, C., Marcovaldi, M.A., Sanches, T.M., Grossman, A., & Sales, G. (1996). Atoll das Rocas Biological Reserve: second largest *Chelonia* rookery in Brazil. *Marine Turtle Newsletter*, *72*, 1–2.

Brander, L.M., Van Beukering, P., & Cesar, H.S. (2007). The recreational value of coral reefs: A meta-analysis. *Ecological Economics*, *63*(1), 209–218.

Brasil Escola. (2016). População do Nordeste [Population of the Northeast]. Retrieved from http://brasilescola.uol.com.br/brasil/populacao-nordeste.htm

Bravo, G., Marquez, F., Marzinelli, E.M., Mendez, M.M., & Bigatti, G. (2015). Effect of recreational diving on Patagonian rocky reefs. *Marine Environmental Research*, *104*, 31–36.

Brazilian Institute of Geography and Statistics (IBGE). (2011a). Pesquisa de servi cos de hospedagem 2011: Municıpios das capitais, regioes metropolitantas das capitais e regioes integradas de desenvolvimento [Research about hospitality services 2011: Counties, metropolitan regions and development integrated regions]. Retrieved from https://ww2.ibge.gov.br/home/estatistica/economia/comercioeservico/psh/2011_todas_regioes/default.shtm

Brazilian Institute of Geography and Statistics (IBGE). (2011b). Atlas geografico das zonas costeiras e oceanicas do Brasil [Geographical atlas of Brazil's coastal and oceanic zones]. Brasília: IBGE. Retrieved from http://biblioteca.ibge.gov.br/visualizacao/livros/liv55263.pdf

BSH International. (2008). Investimentos no Brasil: Hotéis & resorts 2008 [Investments in Brazil: Hotels and resorts 2008]. Retrieved from http://www.bsh.com.br/sys/down load/investimentos_2008.pdf

BSH International. (2011). Investimentos no Brasil: Hotéis & resorts 2011 [Investments in Brazil: Hotels and resorts 2011]. Retrieved from http://www.bsh.com.br/sys/down load/relatorio_investimentos_no_brasil_2011.pdf

Camp, E., & Fraser, D. (2012). Influence of conservation education dive briefings as a management tool on the timing and nature of recreational SCUBA diving impacts on coral reefs. *Ocean & Coastal Management*, *61*, 30–37.

Costa, M., Araújo, M., Cavalcanti, J., & Souza, S. (2008). Verticalizacao da praia da Boa Viagem (Recife, Pernambuco) e suas consequencias socioambientais [Verticalization of Boa Viagem beach (Recife, Pernambuco) and socioenvironmental consequences]. *Revista da Gestao Costeira Integrada, 8*(2), 2008–2233.

Da Silva, I.G.L. (2015). Impactos do turismo na ictiofauna de recifes do nordeste do Brasil. [Tourism influence in the ichthyofauna of corals in Northeast Brazil]. Undergraduate Thesis. Centro de Biociências, Universidade Federal do Rio Grande do Norte, Natal, RN. Retrieved from https://monografias.ufrn.br/jspui/bitstream/123456789/2001/1/TCC%20Isabela%20Guimarães%202015.pdf

de Marques, A.A.B., & Peres, C.A. (2015). Pervasive legal threats to protected areas in Brazil. *Oryx, 49*(1), 25–29.

de Sousa Melo, R., da Silva, M., de Lima, E., & Nishida, A.K. (2006). Estimativa da capacidade de carga recreativa dos ambientes recifais da Praia do Seixas (Paraíba-Brasil). [Recreation carrying-capacity of reef ecosystems at the Praia do Seixas (Paraiba-Brasil)]. *Turismo-Visão e Ação, 8*(3), 411–422.

de Vasconcellos Pegas, F., Grignon, J., & Morrison, C. (2015). Interdependencies among traditional resource use practices, sustainable tourism, and biodiversity conservation: A global assessment. *Human Dimensions of Wildlife, 20*(5), 454–469.

Dearden, P., Bennett, M., & Rollins, R. (2006). Implications for coral reef conservation of diver specialization. *Environmental Conservation, 33*(4), 353–363.

Dimmock, K., & Musa, G. (2015). Scuba diving tourism system: A framework for collaborative management and sustainability. *Marine Policy, 54*, 52–58.

Doiron, S., & Weissenberger, S. (2014). Sustainable dive tourism: Social and environmental impacts: The case of Roatan, Honduras. *Tourism Management Perspectives, 10*, 19–26.

Edney, J. (2012). Diver characteristics, motivations, and attitudes: Chuuk Lagoon. *Tourism in Marine Environments, 8*(1/2), 7–18.

Fundação de Amparo à Pesquisa do Estado de São Paulo (FAPESP). (2016). Recifes na Foz do Rio Amazonas. *[Reefs at the Amazon River Delta]*. Retrieved from http://revistapesquisa.fapesp.br/wp-content/uploads/2016/01/064-067_Corais_239.pdf?b6926f

Gerhardinger, L.C., Godoy, E.A.S., Jones, P.J.S., Sales, G., & Ferreira, B.P. (2011). Marine protected dramas: The flaws of the Brazilian national system of marine protected areas. *Environmental Management, 47*, 630–643.

Giglio, V.J., Luiz, O.J., & Schiavetti, A. (2015). Marine life preferences and perceptions among recreational divers in Brazilian coral reefs. *Tourism Management, 51*, 49–57.

Giglio, V.J., Luiz, O.J., & Schiavetti, A. (2016). Recreational diver behavior and contacts with benthic organisms in the Abrolhos National Marine Park, Brazil. *Environmental Management, 57*(3), 637–648.

Guzner, B., Novplansky, A., Shalit, O., & Chadwick, N.E. (2010). Indirect impacts of recreational scuba diving; patterns of growth and predation in branching stony corals. *Bulletin of Marine Science, 86*(3), 727–742.

Halpern, B.S. (2003). The impact of marine reserves: Do reserves work and does reserve size matter? *Ecological Applications, 13*(1), S117–S137.

Hasler, H., & Ott, J.A. (2008). Diving down the reefs? Intensive diving tourism threatens the reefs of the northern Red Sea. *Marine Pollution Bulletin, 56*, 1788–1794.

Hein, M.Y., Lamb, J.B., Scott, C. & Willis, B.L. (2015). Assessing baseline levels of coral health in a newly established marine protected area in a global scuba diving hotspot. *Marine Environmental Research, 103*, 56–65.

Instituto Chico Mendes de Biodiversidade (ICMBio). (2009). Recifes de coral [Coral reefs]. Retrieved from http://www.mma.gov.br/estruturas/205/_publicacao/205_publicacao 29112010034950.pdf

Instituto Chico Mendes de Biodiversidade (ICMBio). (2011). Abrolhos capacita condutores de ecoturismo subaquático [Abrolhos provides capacity training to subaquatic ecotourism guides]. Retrieved from http://www.icmbio.gov.br/portal/ultimas-noticias/20-geral/2377-abrolhos-capacita-condutores-de-ecoturismo-subaquatico

Júnior, G., & Lima, E. (2014). Resposta da comunidade de tardígrados ao impacto do pisoteio associado ao turismo nos recifes de Porto de Galinhas (Ipojuca, PE). [Response of the community of Tardigrada in response to the trampling impact caused by tourist activities in the reefs of Porto de Galinhas (Ipojuca, PE)]. *Masters degree from the Universidade Federal de Pernambuco.* Retrieved from http://repositorio.ufpe.br/bitstream/handle/123456789/16240/Disserta%C3%A7%C3%A3o%20Edivaldo%20Lima%20PPGBA_2015.pdf?sequence=1&isAllowed=y

Lamb, J.B., True, J.D., Piromvaragorn, S., & Willis, B.L. (2014). Scuba diving damage and intensity of tourist activities increases coral disease prevalence. *Biological Conservation, 178*, 88–96.

Leão, Z.M., Kikuchi, R.K., & Testa, V. (2003). Corals and coral reefs of Brazil. In J. Cortes (Ed.), *Latin America coral reefs* (pp. 9–52). Oxford: Elsevier.

Lew, A.A. (2013). A world geography of recreational scuba diving. In G. Musa & K. Dimmock (Eds.), *Scuba diving tourism* (pp. 29–51). London: Routledge.

Limonad, E. (2007). O fio da meada: Desafios ao planejamento e a preserva͵c~ao ambiental na Costa dos Coqueiros (Bahia) [Challenges in the development and environmental preservation at the Costa dos Coqueiros (Bahia)]. *Revisa Electronica de Geografia y Ciencias Sociales, 11*(245), 1–16.

Lopes, R., Soares, I., & de Araújo, J. (2014). Área de Proteção Ambiental dos Recifes de Corais – Area dos Parrachos de Maracaju/RN: Desafios para o uso sustentável. [Environmental Protection Area of the Recifes de Corais – Area of the Parrachos de Maracaju/RN: Challenges for sustainable use]. *Caminhos de Geografia, 15*(51), 215–236.

Magdaong, E.T., Fujii, M., Yamano, H., Licuanan, W.Y., Maypa, A., Campos, W.L., Alcala, A.C., White, A.T., Apistar, D., & Martinez, R., (2014). Long-term change in coral cover and the effectiveness of marine protected areas in the Philippines: A meta-analysis. *Hydrobiologia, 733*(1), 5–17.

Ministério do Meio Ambiente (MMA). (2003). Diagnóstico do turismo nos municípios do Cabo de Santo Agostinho, Ipojuca e São José da Coroa Grande – Relatório Final. [Tourism assessment in the Cabo de Santo Agostinho, Ipojuca and São José da Coroa Grande Municipalities – Final Report]. Retrieved from http://www.cprh.pe.gov.br/downloads/Sistematizacao%20das%20Informacoes.pdf

Ministério do Turismo (MTur). (2005). Excelência em Turismo – aprendendo com as melhores experiências internacionais: México mergulho – Relatório técnico. [Excellence in tourism – learning from the best international experiences: Mexico diving – Technical report] Brasilia: MTur.

Oliveira, W.A. & Silva, C.B. (2016). A percepção da comunidade de Perobas (RN) sobre o desenvolvimento do Turismo. [Perception of the community of Perobas (RN) about tourism development]. *Revista Brasileira de Ecoturismo, 9*(1), 112–132.

Pegas, F., & Castley, J.G. (2016). Private reserves in Brazil: Distribution patterns, logistical challenges, and conservation contributions. *Journal for Nature Conservation, 29*, 14–24.

Pegas, F., & Stronza A. (2010). Ecotourism and sea turtle harvesting in a fishing village of Bahia, Brazil. *Conservation & Society, 8*, 15–25.

Pegas, F., Weaver, D., & Castley, G. (2015). Domestic tourism and sustainability in an emerging economy: Brazil's littoral pleasure periphery. *Journal of Sustainable Tourism, 23*(5), 748–769.

Professional Association of Diving Instructors (PADI). (2008). *Encyclopedia of recreational diving* (3rd ed.). Rancho Santa Margarita, CA: PADI.

Professional Association of Diving Instructors (PADI). (2016). Brazil. Retrieved from http://www.padi.com/scuba-vacations/brazil

Queiroz Neto, A., Lohmann, G., & Scott, N. (2017). Re-thinking destination competitiveness: The investigation of customer value in scuba diving tourism. Presented at CAUTHE Conference: Council for Australasian Tourism and Hospitality Education, Dunedin, New Zealand, 8–10 February.

Ramsar. (2012). Wetland tourism: Brazil – Abrolhos Marine National Park. Retrieved from https://www.ramsar.org/sites/default/files/documents/pdf/case_studies_tourism/Brazil/Brazil_Abrolhos_EN.pdf

Ríos-Jara, E., Galván-Villa, C.M., Rodríguez-Zaragoza, F.A., López-Uriarte, E., & Muñoz-Fernández, V.T. (2013). The tourism carrying capacity of underwater trails in Isabel Island National Park, Mexico. *Environmental Management, 52*, 335–347.

Saab, W. (1999). Considerações sobre o desenvolvimento do setor de turismo no Brasil [Considerations about tourism development in Brazil]. *BNDES Setorial, Rio de Janeiro, 10*, 285–312.

Selig, E., & Bruno, J. (2010). A global analysis of the effectiveness of marine protected areas in preventing coral loss. *PLoS ONE, 5*(2), E9278.

Smith, K., Scarr, M., & Scarpaci, C. (2010). Grey nurse shark (*Carcharias taurus*) diving tourism: Tourist compliance and shark behaviour at Fish Rock, Australia. *Environmental Management, 46*(5), 699–710.

Souza, M. (2005). Análise do turismo em Aquiraz-Ceará: Política, desenvolvimento e sustentabilidade [An analysis of the tourism in Aquiraz-Ceará: Policies, development and sustainability] (Unpublished master's thesis). Universidade Federal do Ceará, Fortaleza. Retrieved from http://www.repositorio.ufc.br/handle/riufc/16011

Tapsuwan, S., & Asafu-Adjaye, J. (2008). Estimating the economic benefit of SCUBA diving in the Similan Islands, Thailand. *Coastal Management, 36*(5), 431–442.

Torres, D. (2016). Influência do turismo na comunidade de corais em recifes do nordeste do Brasil. [Tourism influence in the community of corals in Northeast Brazil]. Undergraduate Thesis. *Centro de Biociências, Universidade Federal do Rio Grande do Norte, Natal, RN*. Retrieved from https://monografias.ufrn.br/jspui/bitstream/123456789/2585/1/TCC-Daniel.pdf

Uyarra, M.C., & Côte, I. (2007). The quest for cryptic creatures: Impacts of species-focused recreational diving on corals. *Biological Conservation, 136*(1), 77–84.

van der Meer, M.H., Berumen, M.L., Hobbs, J.P., & van Herwerden, L. (2015). Population connectivity and the effectiveness of marine protected areas to protect vulnerable, exploited and endemic coral reef fishes at an endemic hotspot. *Coral Reefs, 34*(2), 393–402.

WannaDive. (2016). Brazil: South America. Retrieved from http://www.wannadive.net/spot/South_America/Brazil/index.html

World Economic Forum (WEF). (2015). *The travel & tourism competitiveness report 2015.* World Economic Forum, Geneva. Retrieved from http://www3.weforum.org/docs/TT15/WEF_Global_Travel&Tourism_Report_2015.pdf

World Tourism and Travel Council (WTTC). (2015). Travel and tourism: Economic impact 2015 Brazil. Retrieved from https://www.wttc.org/-/media/files/reports/economic%20impact%20research/countries%202015/brazil2015.pdf

Zakai, D., & Chadwick-Furman, N.E. (2002). Impacts of intensive recreational diving on reef corals at Eilat, northern Red Sea. *Biological Conservation, 105*, 179–187.

15

IMPACTS OF VESSEL-BASED DAY TOURS ON CORAL REEFS

Observations from snorkel tours in Maui, Hawai'i

Brooke A. Porter

Introduction

This chapter suggests a set of best practices for promoting reduced-impact snorkel tourism on coral reefs. Coastal tourism draws direct benefits from coral reef ecosystems both aesthetically and physically. Coral reefs create numerous tourism opportunities, the most obvious being dive and snorkel tourism. There are other tourism activities that also depend on viewing coral reef ecosystems, such as from glass-bottom boats and submarine tours. Coral reefs can also support recreational activities such as surfing in the form of reef breaks. Marine aquariums, hotels and tourist-frequented shopping areas may also depend on [the extraction of] coral reef ecosystem resources for tourism products in the form of aquaria. Reefs also influence beach tourism, forming protective barriers and creating sought-after lagoons and tide pools. The role of coral reef ecosystems in tourism is diverse. As a result of the consumptive nature of marine resource use, much attention has been given in the literature to the biological vulnerabilities of coral reefs, the associated fisheries and to the social vulnerabilities of those dependent upon them (e.g., Klint et al., 2010; Moberg & Folke, 1999; Pomeroy et al., 2006).

With the exception of sport fishing, most forms of marine tourism have the potential to add a secondary and non-extractive value to marine resources. For this reason marine tourism and, perhaps more specifically, marine ecotourism has been suggested as an agent for conservation (Garrod & Wilson, 2004; Hooker & Gerber, 2004; Orams, 2004; Orams & Forestell, 1995; Quiros, 2007). Although tourism may act as an agent for positive environmental change, the issues associated with an increase in tourism to an area have the potential to ruin that area through misuse or overuse of the resources (Brown & Hall, 1999; Cesar et al., 2003). For example, Cater and Cater (2007) argue that attaching a value to nature may create more unsustainable activities as others try to exploit the resource. Others argue that coastal and marine tourism may increase extractive marine resource use through an increased demand for seafood in adjoining destinations. However, as a study in Zanzibar revealed, apart from seafood, tourism had a substantial influence on extractive uses of fisheries for the souvenir trade and had created an increased demand for shark teeth and jaws (Gössling et al., 2004). This is supported by Dias et al. (2011) who found a strong link between tourism and the shell curio trade as visitors look to acquire a "piece" of their vacation experience; a tangible memory to bring home.

One of the major goals of tourism development is to increase revenues by encouraging increased tourism. However, increased tourism also requires additional investment in infrastructure. In a marine tourism context this may mean dredging, the construction of sea walls, run-off and non-point source pollution. Unless regulated, the consequences of development may affect coastal and near shore environments. Coastal engineering, for example, often prioritises short-term benefits over long-term impacts and outcomes (Sedrati & Anthony, 2014). In addition to infrastructure, an increase in visitor numbers requires an increase in food and other inputs. The majority of peripheral areas characteristic of coastal and island destinations are dependent on external markets (Brown & Hall, 1999), and providing the necessary goods for subsistence therefore increases the carbon footprint vis-à-vis transportation and packaging. More people also equates to more effluent. Beyond general waste management, human effluent often carries pharmaceuticals, including those used in contraception, as well as other endocrine disrupting chemicals (Kusk et al., 2011). More visitors in coastal destinations also means more visitors wanting to directly view reef ecosystems. For example, in Hawai'i nearly 44% of visitors participate in snorkelling and/or diving activities (Hawai'i Tourism Authority [HTA], 2015). Although there are many potential environmental impacts associated with marine tourism, this chapter focuses on the direct impacts of vessel-based day snorkel tours to coral reef ecosystems in Hawai'i. Various operator efforts towards marine conservation practices are also discussed and operator size is considered as an important variable.

Tourism scale

Mass marine tourism has been associated with overcrowding and other negative outcomes (Inglis et al., 1999; Roman et al., 2007; Szuster et al., 2011), although it is worth noting that crowding was a perception-based measure in many of these studies rather than a measure of sustainability. Cater and Cater (2007) devoted a subsection of their text on marine tourism to the scale of tourism activity (e.g., mass vs. niche). In doing so, they presented the findings of a number of researchers who outlined examples of large or mass tourism operators that have challenged this perceived negative link between sustainability and scale. In particular, they defined the ability of mass tour operators to introduce "sophisticated environmental management strategies to ecotourism that are beyond the capability of most traditional small-scale operations" (Cater & Cater, 2007, p. 12). Similarly, Lück (2002) found large package tour operators are able to make significant contributions towards conservation and community-based operations at the destination level. Preferably, the contributions to conservation from mass tourism must be proportional to its scale and, thus, greater than those of a small-scale or niche tourism industry (Buckley, 2003). While the risks associated with mass scale tourism are greater, with respect to direct impacts to the marine environment, so is the potential to introduce beneficial technologies and systems as well as positive environmental behaviours. The next section considers environmental and conservation inputs resulting from the operations of a large marine tour operator in Maui, Hawai'i.

Snorkel tourism in Maui

Tourism is a major business in Hawai'i. The HTA estimated tourism sales at US$25.2 billion in 2015 (HTA, 2016) with over 170,000 people employed in the industry. The blue waters of Hawai'i make the marine realm enticing and coral reefs are often visible. Nearly 25% of the marine fauna in Hawai'i is endemic (Eldredge & Evenhuis, 2003). In addition, the mild climate and leeward areas make Maui an ideal location for coral reef tourism. Day-trip snorkel tours in Maui are popular, with multiple tours running daily out of the island's two main harbours: Lahaina Harbour and Ma'alaea Harbour. Most Maui vessel-based tours, if not all, take visitors to snorkel sites where coral reefs account for the most benthic cover (personal observation, 2001–2010).

Methods

For the following section of the chapter, I will draw from nearly a decade of personal experience working in and alongside the marine tourism industry in Maui, Hawai'i. I worked with the Pacific Whale Foundation (PWF), a non-government organisation (NGO) dedicated to marine conservation, education and research, for a total of 7 years between 2001 and 2010. PWF also operates ocean eco-cruises currently advertised under the following categories: dolphin watch; snorkel; sunset & cocktail; whale watch (seasonal); specialty & holiday (PWF, 2016). Within the organisation, I served in various capacities, including marine naturalist guide, marine educator, community outreach coordinator and director of conservation. In addition to a reflective analysis, secondary data taken from tour operator/organisation websites is used to support the ideas and concepts presented in this chapter. Debating the roles of vessels sizes and capacities, I will explore various components of the vessel-based coral reef snorkel tour, drawing largely from my previous professional affiliations with PWF. It is important to note that although the PWF's mission remains focused on marine education, conservation and research, company practices are likely to have evolved and adapted since my professional affiliation with the organisation.

Findings

Marine vessels as day-tour platforms

In Maui the most common, and popular, coral reef snorkel tours are vessel-based. There are a number of reefs accessible from shore; however, shore accessibility may be challenging, especially for novice snorkellers due to rocks and/or sea conditions. Private charters are readily available; however, the majority of Maui visitors participate in commercial group tours. Commercial snorkel tour operators on Maui differ widely based on factors that include the length of the excursion, snorkel site(s) visited, price, on-board activities (e.g., meals, waterslides), interpretive programmes and carrying capacity (which is operator-dependent and vessel-dependent). In general, most commercial tours range in passenger capacity from about 20 to 149. The definition of mass tourism varies in the literature and may also be considered subjective. For the purposes of this discussion, a vessel-based tour with a capacity of 25 passengers or more is considered as a mass tourism vessel.

Coral reefs are vulnerable to sudden environmental changes, including rapid temperature changes, which in some cases may lead to coral bleaching events (Becken, 2005). One of the factors causing bleaching is climate change. Actions by marine operators to reduce climate changing emissions is a positive step towards mitigating climate change, and thus a positive contribution towards coral reef health. Most Maui tour vessels are powered by marine diesel engines, with some vessels also using wind power via sails. Although a larger passenger capacity requires a larger vessel, the amount of carbon emissions produced is dependent on a number of factors in addition to vessel size, including vessel design and age as well as operator practices. For example, some Maui operators (e.g., PWF) maintain voluntary reduced maximum speeds of 15 knots or less during months that correspond with the migration and presence of the North Pacific humpback whales. To reduce emissions, some operators have experimented with alternative fuels. Biodiesel is available on the island and some marine operators have used (e.g., PWF) or are currently using (e.g., Aqua Adventures Maui) biodiesel or biodiesel blends. PWF, despite their well-publicised environmental commitment, found using biodiesel resulted in significantly more oil changes and extra oil filters, and was also linked to other mechanical failures. Although PWF continues to promote the general use of and uses biodiesel in their land vehicles, using biodiesel in a marine capacity was found to create more, rather than less, environmental impact (Kaufman, 2012).

While alternative fuels constitute a positive step towards reducing carbon dioxide emissions, operating entirely on wind or solar power, regardless of carrying capacity, is the most environmentally ideal scenario. However, navigating tight spaces such as harbours and mooring at dive sights is challenging under sail, especially with larger vessels. In addition, the unpredictability of the weather and seas in Hawai'i makes a wind-only scenario unlikely for operators. Further, during whale season (November– May), regulations require operators to cease movement within

100 metres of a whale. For vessels under sail, this means dropping sails every time a whale is sighted. Another alternative is solar technology; however, the current cost of upgrades to solar power may be prohibitive for many operators. Therefore, much like public transportation, a greater capacity per snorkel tour could be seen as a strategy to reduce per-head emissions. This assumption is based on the finite number of marine tour operator permits that are available and currently in use in Maui (and the state). Given the limit on the number of commercial tour vessels in Maui, consolidating tours to reduce the number of vessels could also be a useful strategy to reduce emissions. It is worth noting that doing so could raise issues associated with perceived crowding (Inglis et al., 1999; Roman et al., 2007; Szuster et al., 2011); however, this issue is outside the scope of this chapter. In summary, the ideal vessel platform in the context of environmental impact would: a) be powered by supplemental wind or solar power or, as an alternative, an "environmentally efficient" alternative fuel blend; b) require a vessel be filled to near or maximum capacity.

Irrespective of vessel efficiencies, all vessels must be cleaned and equipment (hopefully) disinfected between uses. The majority of the cleaning processes take place in harbours with run-off flowing directly into the sea. With the general shift towards environmentalism, more cleaning products are becoming available. The "eco-friendliness" of a product can be difficult to discern; however, there are agencies dedicated to reporting on this issue (e.g., Environmental Working Group, Consumer Reports). In Maui, locally produced, environmentally friendly cleaning products are available (e.g., Maui Pure Island Products). Operators such as PWF purchased these items in bulk and also refill cleaning containers to reduce overall plastic waste. To determine the best cleaning products and practices, an operator will need to devote time and resources to researching the best available choice. This may not be feasible for smaller operators. While there are specialised products available for disinfecting of snorkel equipment, consultations with a local water quality expert indicated that using bleach (sodium hypochlorite) as a disinfectant, and allowing the mixture to off-gas outdoors for at least 24 hours before dumping, was an environmentally appropriate practice (personal experience, 2007).

Site traffic

Lamb et al. (2013) demonstrated a correlation between intensive site use resulting from coral reef dive tourism and increased disease in coral reefs in Thailand. They noted that direct injury to the reef from diver contact was correlated to an increase in coral disease. In addition to trampling or direct contact with the reef, snorkellers can introduce a number of pharmaceuticals and chemicals to reef ecosystems via topical personal care products (e.g., lotions or sunscreen) and urine (which may contain contraceptives and antibiotics) (Kusk et al., 2011). The nanoparticles found in most commercial sunscreens also have a negative impact on benthic organisms (Botta et al., 2011). Danovaro et al. (2008) found that common sunscreen ingredients cause rapid and complete bleaching of hard coral species. Likewise, Downs et al. (2014) found that the introduction of benzophenone-2 (BP-2), a common additive to UV blocking personal care products, caused deformations in coral polyps, negatively affecting their future reproductive capabilities, and further increased rates of coral bleaching and coral death. Noting that 90% of reef tourism is concentrated in about 10% of the global reef areas, these high-use areas are under threat from sunscreen contamination resulting from tourism activities (Danovaro et al., 2008). In addition to harmful sunscreen compounds, there are other chemicals introduced by tourists. Bhandari et al. (2015) studied the effects of commonly excreted oestrogenic chemicals on fish life cycle. In particular, the research found that exposure to bisphenol A (BPA), a compound found in plastics and food packaging and 17-α ethinylestradiol (EE2) (an ingredient in oral contraceptives), both led to negative transgenerational consequences, including reduced fertility rate in fish. Intersex, or the presence of both male and female characteristics, is another effect of oestrogenic compounds on aquatic animals (Iwanowicz et al., 2016). Combined with trampling, the introduction of chemical compounds has significant and detrimental impacts to coral reef ecosystems. Consolidating visitors into a smaller number of vessels has the potential of reducing emissions per head; however, a greater concentration of visitors to a snorkel site is less than ideal and the problems caused by sunscreens will not be alleviated.

There are a number of potential strategies for managing site impact. In Maui, some popular snorkel sites such as Molokini (a crescent-shaped island) are regulated by a first-come, first-serve day-use mooring system (Department of Land and Natural Resources [DLNR], 2016). Moorings are an excellent alterative to anchors which, if set incorrectly, can drag across reefs damaging years of growth in minutes (Dinsdale & Harriot, 2004). In protected sites such as Molokini, anchoring is prohibited, thus the number of visitors is regulated by the number of moorings. In other unregulated areas, operators tend to avoid crowding, either anchoring far from other vessels or choosing an alternative site to increase visitors' satisfaction (personal observation, 2001–2010). In terms of managing specific anthropogenic factors associated with site use, trampling and contact with the reef are easily minimised aboard Maui-based snorkel tours, where the majority of tours visit reefs that are mostly at least 2.5 metres below the surface. In addition, many tour operators frequently remind guests to avoid deliberate and accidental contact with the reef. Mitigating the impacts of sunscreen is possible. On Maui, some operators promote and sell reef-safe sunscreens and legislation has been introduced to ban the sale of oxybenzone (House Bill 600 introduced in 2017). Some operators, including PWF, are encouraging the use of alternative sun protection such as rashies (an athletic shirt made of synthetic materials that protects the wearer against rashes caused by abrasion, stings or sunburn) and wetsuits. An acceptable mitigation strategy for human excrement will require the collection of waste on board boats and then discharge into land-based sewerage systems. Previously, a potential, albeit slightly bizarre strategy, was developed through a brainstorming session during my time at PWF in regards to launching a "Don't Pee in the Sea" campaign in attempts to reduce localised impacts to coral reefs (A. Rillero, personal communication), although it was not implemented.

Educational programmes aboard marine tour vessels

Recent research confirms that many tourists are interested in learning about the places they visit (Hrycik & Forestell, 2013; Lück, 2015). In relation to snorkel tours, there is potential for a spill-over education effect that creates positive transferable behaviours (van Beukering & Cesar, 2004). In relation to the scale of operations, two outcomes can be identified in the area of audience education. Large operators have the potential to educate, inform and inspire large numbers of visitors and through this contribute to a greater global good. Smaller operators are able to intimately connect with visitors, thus encouraging lasting changes in environmentally positive behaviours.

Being at sea is a novel experience for many. Therefore, a marine-based snorkel tour has the ability to create dynamic disequilibrium by introducing visitors to stimulating and ever-changing coral reef ecosystems (Forestell, 1993). In terms of a tour platform, another notable feature of these vessels is that they hold the audience captive. This "temporary captivity" is unique in that it reduces variables in the tour experience. For example, a low visitor to staff ratio offers staff or crew the ability to more easily monitor visitor behaviour and to discourage behaviours (e.g., coral trampling) that may be detrimental to the marine environment. Again, factoring in an effective staff to visitor ratio, staff have the ability to model appropriate environmental behaviours (e.g., recycling). Research has found the role of interpreter/naturalist to have significant influence on the overall audience perception of a tour (Borges de Lima, 2016; Christie & Mason, 2003). Because the abilities of a guide are varied and dependent not only on training and knowledge, but also on personality, speaking ability and likeability (Forestell, 1993), it is more helpful to shift the discussion to a concept of *meaningful guide contact*. Meaningful contact is subjective in nature; therefore, it is more effective to identify an effective guide to visitor ratio. The demographics of snorkel tours are dynamic and will change throughout the tour depending on the phase of the tour (Forestell, 1993), comfort of passengers (e.g., seasickness, exhaustion, foul weather) and passengers' general interests. Many tour vessels broadcast information by means of an on-board audio system. The audibility of the information largely depends on the vicinity of the passenger to speakers. If guests are too close to a speaker they may complain of noise pollution; too far, and the wind and waves make hearing difficult (personal experience, 2001–2010). In a perfect scenario, each passenger would have a personal interaction with a guide. We will assume that over the course of an average 6-hour tour (excluding time in the water and other work responsibilities), a guide/crew would have the ability to foster personal interactions with at least four passengers per hour for at least 4 hours of the trip, or roughly 15 guests per trip. Thus, to meet this theoretical interactive minimum, the guide to visitor ratio would translate to 1:15. This ratio is given based on the assumption that the reduction of ecological impact is a result of consequential contact with a tour guide (Borges de Lima, 2016; Littlefair & Buckley, 2008).

Power in numbers

This section will shift the discussion from specific tour features and components to a broader discussion of the environmental impacts of coral reef tourism from the vantage of education and conservation awareness. Larger tour operators are able to significantly influence the market through the development of innovative tourism products and technologies or volume purchasing power (Lück, 2002). In Maui, similar to most islands, the majority of supplies are imported making many "green" choices cost-prohibitive. Larger operators have been able to purchase products (e.g., bio-compostable flatware and rubbish bags) in bulk thus improving their environmental practices.

In the case of snorkel tours on Maui, the influence of larger operators has resulted in a number of environmental success stories that directly and indirectly affect coral reef ecosystems. For example, PWF was instrumental in petitioning for effluent pumping stations at the main Maui harbours and was one of the first operators to voluntarily pump their vessel sewage via a company-owned pump truck (Schorr, 2008). PWF advertises additional "Conservation Victories" that include: (1) banning tobacco products and smoking (and thus, cigarette butt litter) on all Maui County beaches and parks; (2) stopping Hawai'i Superferry (a controversial high-speed, interisland ferry that began operations without conducting an Environmental Impact Statement (EIS)); (3) Maui County captive cetacean ban; (4) plastic bag ban; and (5) providing public and scientific testimony that contributed to the endangered species designation for Hawai'i insular false killer whales (PWF, 2015). It is apparent that large-scale operators have great monetary and temporal resources to devote to activism and the advancement of marine conservation legislation, a benefit that may not be available to smaller operators with fewer employees and financial resources.

Discussion and conclusions

This chapter focused on the major components of marine snorkel tours as a method to contextualise the scale of tourism and effects on coral reefs. Ultimately, in terms of its effect on coral reef ecosystems, operator size matters just as much as is does not. Given the diversity of the industry and the variability of snorkel tours, it may be more helpful to explore a series of best practices. Using the points of discussion from this chapter, I have suggested a best practices recommendation for snorkel day tours. Table 15.1 is presented as a future guide for consumer use and as a reference tool for operators.

The components presented in Table 15.1 are not necessarily straight forward. Further, their focus is on environmental sustainability and they do not consider visitor satisfaction, which may be affected by issues such as crowding (Inglis et al., 1999; Roman et al., 2007; Szuster et al., 2011). Factors including weather and the presence of large marine mammals, such as the humpback whale, will significantly influence the choices in vessel power in Hawai'i. However, only two of the

TABLE 15.1 Best practices: snorkel tour guidelines for coral reef conservation

Component	Undesirable	Better	Best
Staff/crew	No standards for guides	Trained staff as guides	Experienced, educated, trained staff as guides
*Vessel power	*Diesel only	Alternative fuel blend, wind/solar power	Primarily wind/solar power
Carrying capacity	*Private charter	*Small–medium capacity vessel (<60 pax)	*High capacity vessel (+60 pax)
Dive site traffic	High site traffic	Medium site traffic	Low site traffic
Educational programmes	No programmes	Organised programmes	Organised programmes, staff training
Sunscreen policy	No policy	Rashies available for rent, reef-safe sunscreen encouraged or available for purchase	Only reef-safe sunscreen or alternative sun protection (e.g., rashies) allowed
Guide to visitor ratio	Less than 1:15	1:15	Greater than 1:15
Effluent waste	Always dumps at sea	Sometimes dumps at sea	Pumps at land-based pump stations
Products on vessels	Single-use plastics, non-recyclable flatware	Recyclable flatware	Bio-compostable flatware or reusable flatware
Cleaning products	No standards	Eco-friendly products	Eco-friendly, locally sourced products

Note. The asterisks indicate carrying capacity as directly related to the use of regular diesel for power. For example, carrying capacity would become less significant or insignificant for a vessel travelling with alternative fuels and sail power, respectively.

ten components presented in the best practices model are directly related to scale. Operator size becomes relevant in terms of emissions vis-à-vis carrying capacity and in-site traffic. As discussed, biofuel, wind and solar provide promising alternatives to diesel and will hopefully render the argument regarding emissions and scale to be a moot point in the near future. Regarding site traffic, while a larger operator may temporarily increase site traffic, site traffic must be viewed as a broader, island-wide issue because there are no regulations to prohibit different operators from visiting the same site during a single day. Previous research on dive tourism has suggested dive sites do have a sustainable carrying capacity that can minimise reef degradation, but yet these numbers are likely as low as 3,000–6,000 dives per site per year (Hawkins et al., 2005; Zakai & Chadwick-Furman, 2002). Identifying acute carrying capacity is not only site specific, but requires longer term studies. On Maui, these numbers are likely to be larger given the level of contact that snorkel tours have with the reef, but lower than that from dive tours due to the depth of most reefs. Future research on reef carrying capacity for snorkellers is necessary, as is research on the role of specific anthropogenic factors (e.g., non-reef-safe sunscreen and urine). In the interim, it is recommended that operators a) implement education and outreach programmes that include information on marine conservation and the general marine environment; b) enforce a reef-safe sunscreen policy; c) refrain from dumping effluent at sea; d) utilise the most environmentally friendly cleaning products and e) explore options for reducing general waste aboard vessels (e.g., cutlery). Looking ahead, it is imperative that operators realise their ability to impact an audience, as well as individual visitors realise their purchasing power as consumers.

Acknowledgement

A special thanks to Pacific Whale Foundation for nearly a decade of amazing experiences that formed the basis for this chapter. To all the staff who came before me, to those with whom I worked and those that have and will come after me, I doubt there is a more dedicated group of ocean warriors. Also a very special thanks to Linda Porter, whose comments and inputs significantly improved the quality of this chapter.

References

Becken, S. (2005). Harmonising climate change adaptation and mitigation: The case of tourist resorts in Fiji. *Global Environmental Change, 15,* 381–393.

Bhandari, R.K., vom Saal, F.S., & Tillitt, D.E. (2015). Transgenerational effects from early developmental exposures to bisphenol A or 17α-ethinylestradiol in medaka, *Oryzias latipes. Scientific Reports, 5,* 1–5.

Borges de Lima, I. (2016). Pivotal role of tour guides for visitors' connection with nature: Conceptual and practical issues. *International Journal of Humanities and Applied Sciences, 5,* 18–22.

Botta, C., Labille, J., Auffan, M., Borschneck, D., Miche, H., Cabie, M., . . ., & Bottero, J. (2011). TiO$_2$-baesd nanoparticles released in water from commercialized sunscreens in a life-cycle perspective: Structures and quantities. *Environmental Pollution, 159,* 1534–1550.

Brown, F., & Hall, D. (1999). *Case studies of tourism in peripheral areas.* Bornholm: Research Center of Bornholm.

Buckley, R. (2003). *Case studies in ecotourism.* Wallingford, UK: Centre for Agriculture and Bioscience International.

Cater, C., & Cater, E. (2007). *Marine ecotourism: Between the devil and the deep blue sea.* Wallingford, UK: Centre for Agriculture and Bioscience International.

Cesar, H., Burke, L., & Pet-Soede, L. (2003). The economics of worldwide coral reef degradation. Retrieved from http://eprints.uberibz.org/48/1/Rappor03.pdf

Christie, M.F., & Mason, P.A. (2003). Transformative tour guiding: Training tour guides to be critically reflective practitioners. *Journal of Ecotourism, 2,* 1–16.

Danovaro, R., Bongiorni, L., Corinaldesi, C., Giovannelli, D., Damiani, E., Astolfi, P., . . ., & Pusceddu, A. (2008). Sunscreens cause coral bleaching by promoting viral infections. *Environmental Health Perspectives, 116,* 441–447.

Department of Land and Natural Resources (DLNR). (2016). *Day-use moorings.* Retrieved from http://dlnr.hawaii.gov/dobor/day-use-moorings/

Dias, T.L., Neto, N.A.L., & Alves, R.R. (2011). Molluscs in the marine curio and souvenir trade in NE Brazil: Species composition and implications for their conservation and management. *Biodiversity and Conservation, 20,* 2393–2405.

Dinsdale, E.A., & Harriot, V.J. (2004). Assessing anchor damage on coral reefs: A case study in selection of environmental indicators. *Environmental Management, 33,* 126–139.

Downs, C.A., Kramarsky-Winter, E., Fauth, J.E., Segal, R., Bronstein, O., Jeger, R., . . ., & Loya, Y. (2014). Toxicological effects of the sunscreen UV filter, benzophenone-2, on planulae and in vitro cells of the coral, *Stylophora pistillata. Ecotoxicology, 23,* 175–191.

Eldredge, L.G., & Evenhuis, N.L. (2003). Hawaii's biodiversity: A detailed assessment of the numbers of species in the Hawaiian Islands. Retrieved from http://hbs.bishopmuseum.org/pdf/op76.pdf

Forestell, P. (1993). If Leviathan has a face, does Gaia have a soul? Incorporating environmental education in marine eco-tourism programs. *Ocean & Coastal Management, 20,* 267–282.

Garrod, B., & Wilson, J.C. (2004). Nature on the edge? Marine ecotourism in peripheral coastal areas. *Journal of Sustainable Tourism, 12,* 95–120.

Gössling, S., Kunkel, T., Schumacher, K., & Zilger, M. (2004). Use of molluscs, fish, and other marine taxa by tourism in Zanzibar, Tanzania. *Biodiversity and Conservation, 13,* 2623–2639.

Hawai'i Tourism Authority (HTA). (2015). *2015 Annual Visitor Research Report.* Retrieved from http://files.hawaii.gov/dbedt/visitor/visitor-research/2015-annual-visitor.pdf

Hawai'i Tourism Authority (HTA). (2016). Hawaii tourism fact sheet. Retrieved from http://files.hawaii.gov/dbedt/visitor/visitor-research/2016-annual-visitor.pdf

Hawkins, J.P., Roberts, C.M., Kooistra, D., Buchan, K., & White, S. (2005). Sustainability of scuba diving tourism on coral reef of Saba. *Coastal Management, 33,* 373–387.

Hooker, S.K., & Gerber, L.R. (2004). Marine reserves as a tool for ecosystem-based management: The potential importance of megafauna. *BioScience, 54,* 27–39.

Hrycik, J.M., & Forestell, P.H. (2013). Change in focus of attention among whale-watch passengers as a function of temporal phase of the tour. *Tourism in Marine Environments, 8,* 189–198.

Inglis, G., Johnson, V., & Ponte, F. (1999) Crowding norms in marine settings: A case study of snorkeling on the Great Barrier Reef. *Environmental Management, 24,* 369–381.

Iwanowicz, L.R., Blazer, V.S., Pinkey, A.E., Guy, C.P., Major, A.M., Munney, K., . . ., & Sperry, A. (2016). Evidence of estrogenic endocrine disruption in smallmouth and largemouth bass inhabiting Northeast U.S. national wildlife refuge waters: A reconnaissance study. *Ecotoxicology and Environmental Safety, 124,* 50–59.

Kaufman, G. (2012, 1 March). Response to: Disappointing raft trip . . . dishonest [Response to Trip Advisor Review]. Retrieved from https://www.tripadvisor.com/ShowUserReviews-g815379-d519580-r125421760-Pacific_Whale_Foundation-Maalaea_Maui_Hawaii.html

Klint, L.M., Jiam, M., Law, A., Delacy, T., Filep, S., Calgaro, E., . . ., & Harrison, D. (2010). Dive tourism in Luganville, Vanuatu: Shocks, stressors, and vulnerability to climate change. *Tourism in Marine Environments, 8,* 91–109.

Kusk, K.O., Krüger, T., Long, M., Taxvig, C., Lykkesfeldt, A.E., Frederiksen, H., . . ., & Bonefeld-Jørgensen, E.C. (2011). Endocrine potency of wastewater: Contents of endocrine disrupting chemicals and effects measured by in vivo and in vitro assays. *Environmental Toxicology, 30,* 413–426.

Lamb, J.B., True, J.D., Piromvaragorn, S., & Willis, B.L. (2013). Scuba diving damage and intensity of tourist activities increases coral disease prevalence. *Biological Conservation, 178,* 88–96.

Littlefair, C., & Buckley, R. (2008). Interpretation reduces ecological impacts of visitors to world heritage site. *AMBIO: A Journal of the Human Environment, 37*(5), 338–341.

Lück, M. (2002). Large-scale ecotourism – A contradiction in itself? *Current Issues in Tourism*, 5, 361–370.

Lück, M. (2015). Education on marine mammal tours – But what do tourists want to learn? *Ocean & Coastal Management*, 103, 25–33.

Moberg, F., & Folke, C. (1999). Ecological goods and services of coral reef ecosystems. *Ecological Economics*, 29(2), 215–233.

Orams, M.B. (2004). The use of the seas for recreation and tourism. In H.D. Smith (ed.), *The oceans: Key issues in marine affairs* (pp. 161–173). Dordrecht, The Netherlands: Springer.

Orams, M.B., & Forestell, P.H. (1995). From whale harvesting to whale watching: Tangalooma 30 years on. In O. Bellwood, H. Choat, & N. Saxena (eds.), *Recent advances in marine science and technology '94* (pp. 667–673). Townsville, Australia: James Cook University of North Queensland.

Pacific Whale Foundation (PWF) (2015). Our conservation victories. Retrieved from https://www.pacificwhale.org/conservation/

Pomeroy, R.S., Ratner, B.D., Hall, S.J., Pimioljinda, J., & Vivekananadan, V. (2006). Coping with disaster: Rehabilitating coastal livelihoods and communities. *Marine Policy*, 30, 786–793.

Quiros, A. (2007). Tourist compliance to a Code of Conduct and the resulting effects on whale shark (*Rhicodon typus*) behaviour in Donsol, Philippines. *Fisheries Research*, 84, 102–108.

Roman, G.S., Dearden, P., & Rollins, R. (2007). Application of zoning and "limits of acceptable change" to manage snorkelling tourism. *Environmental Management*, 39, 819–830.

Schorr, S.J. (2008, 29 May). Pump Don't Dump makes progress. *Kihei Community Association*. Retrieved from http://www.pumpdontdump.com/info/print6870_pumpdont dumpmakesprogress.pdf

Sedrati, M., & Anthony, E.J. (2014). Confronting coastal morphodynamics with counter-erosion engineering: The emblematic case of Wissant Bay, Dover Strait. *Journal of Coastal Conservation*, 18, 483–494.

Szuster, B.W., Needham, M.D., & McClure, B.P. (2011). Scuba diver perceptions and evaluations of crowding underwater. *Tourism in Marine Environments*, 7, 153–165.

van Beukering, P.J.H., & Cesar, H.S.J. (2004). Ecological economic modelling of coral reefs: Evaluating tourist overuse at Hanauma Bay and algae blooms at the Kihei Coast, Hawai'i. *Pacific Science*, 58, 243–260.

Zakai, D., & Chadwick-Furman, N.E. (2002). Impacts of intensive recreational diving on reef corals at Eilat, northern Red Sea. *Biological Conservation*, 105, 179–187.

PART IV

Indigenous use, the media and the way forward

16

INDIGENOUS CORAL REEF TOURISM

Henrietta L. Marrie

Introduction

An estimated 850 million people live within 100 kilometres of a coral reef and many directly benefit from the economic, social and cultural services reefs provide (Burke et al., 2011). Coral reefs support many economically important fish species and provide food for hundreds of millions of people. Coral reefs also provide protection for coastal areas from storms and erosion, and generate jobs and income from fishing, tourism and recreation (World Wildlife Fund for Nature [WWF], 2015). In the so-called Coral Triangle, more than 120 million people – about one-third of the inhabitants – depend directly on local marine and coastal resources for their income, livelihoods and food security (Asian Development Bank, 2014). The Coral Triangle refers to an area in the South East Asia/Western Pacific region bounded by Indonesia, the Philippines, Papua New Guinea and the Solomon Islands. This region has around 83% of the range of Indo-Pacific corals and reef fish and is regarded as the world's greatest coral diversity hotspot. The region also attracts tens of millions of visitors every year, creating an industry estimated to be worth US$12 billion annually, with earnings shared between travel operators, tour guides, hotels, diving operations and countless other businesses (Pet-Soede et al., 2011). The aim of this chapter is to explore the social, cultural, environmental and economic impacts of the global tourism industry on local, coral reef-dependent Indigenous peoples and their communities, as well as the nature of their participation within this industry, taking into account the diversity of historical and local factors experienced by Indigenous peoples within the major coral reef regions of the world.

Also in this chapter, the language of Article 8(j) of the Convention on Biological Diversity (CBD) is used to identify the peoples who are traditionally and historically associated with the world's coral reefs, namely Indigenous and local communities embodying traditional lifestyles and whose knowledge, innovations and practices are relevant to the conservation and sustainable use of biological diversity – in this case, coral reef biodiversity. The local communities referred to in the Convention, in most instances, are Indigenous to their region, for example the traditional fishers of the coastal and island communities of the nations that share the Red Sea (including Egypt, Saudi Arabia, Eritrea, Sudan and Yemen), and the inhabitants of the many islands of the Indonesian Archipelago or the small island developing States of Oceania (Secretariat of the Convention on Biological Diversity [SCBD], 1998). However, in recognising the indigeneity of most local communities, and in order to simplify the terminology for the purposes of this chapter, the term *Indigenous coral reef communities* (ICRC) will be used, except where it is necessary to maintain a distinction between Indigenous peoples and local communities. Indigenous peoples and local communities, sometimes referred to as tribal peoples are also the subjects of two other international instruments, namely the UN Declaration on the Rights of Indigenous Peoples (UNDRIP) and the International Labour Organisation Convention (C169) concerning Indigenous and Tribal Peoples in Independent Countries (United Nations Centre for Human Rights, 1994). For a brief discussion on the definition of "Indigenous", see Cohen (1999, pp. 4–6).

Coral reefs form the mainstay of many local traditional economies, particularly of those countries generally referred to as "small island developing states". Many of these island communities, particularly in Oceania and the Caribbean, were once subjected to European colonial occupation and governance, becoming dependent protectorates or trust territories. While many have regained their independence, such as the Marshall Islands (from the US in 1986), Kiribati (from the UK in 1979), Barbados (from the UK in 1966) and Mauritius (from Portugal in 1968), others remain trust territories, for example French Polynesia, American Samoa and the Cook Islands. The uniqueness and extreme fragility of marine and coastal biological diversity of the small island developing states is also recognised under the CBD, as is their disproportionate responsibility and their limited capacity to conserve and manage their biological resources (SCBD, 1998).

Coastal and marine tourism is an important economic sector and in some regions a major component of the livelihoods of coastal communities. On a global basis, tourism and related economic activities generate 9.8% of gross domestic product (GDP) and employ 277 million people, or 1 in every 11 jobs (World Travel and Tourism Council [WTTC], 2015). With 80% of all tourism based near the sea (Honey & Krantz, 2007), this growth can result in major environmental, cultural, social and economic implications (WWF, 2015). With growth of 3.5% per annum forecast through to 2030, tourism will continue to be a major industry, especially in the Tropics (United Nations World Tourism Organization [UNWTO], 2012).

Who are the ICRC?

The ICRC, in terms of their cultures and languages, geo-political and national status, and histories, are an extraordinarily diverse group. This diversity provides challenges and opportunities for the tourism industry and ICRC. Indigenous peoples and local communities are responsible for most of the linguistic and cultural diversity of peoples associated with coral reefs. While not all the languages are associated with coastal and island communities, the countries in the Coral Triangle indicate something of the linguistic and cultural diversity associated with coral reefs. Papua New Guinea, the most linguistically diverse country in the Coral Triangle, has 847 languages, Indonesia 655, Philippines 153, Vanuatu 105 and the Solomon Islands 69 languages (Maffi, 1999). Much of this linguistic diversity is masked by the adoption of the language of the colonisers or the dominant society as the official language, for example in French Polynesia and the Federated States of Micronesia.

The varying political status of Indigenous peoples and local communities associated with the world's coral reefs also needs to be considered because it impacts their ability to control access to, manage and enjoy their marine estates – factors that can have an impact on or influence tourism visitation and experience and, ultimately, the nature of the benefits that flow to Indigenous peoples and local communities. The Indigenous peoples subject to the *UN Declaration on the Rights of Indigenous Peoples (2007)* are those who have lost sovereignty over their traditional lands and waters as a result of invasion and colonisation, mostly by European nations, generally as the result of their expansion into the New World starting in the fifteenth century. Such peoples include the Indigenous peoples of Australia, the Philippines, Hawai'i and the Americas. While they have lost control over their territories, for the most part they have been able to retain, or recover some rights by virtue of constitutional and/or statutory recognition, treaties, agreements and territorial settlements with their colonising powers. The Panamanian government in 1938 and 1953 passed legislation giving political recognition and autonomy to the native Kuna and boundary definition for their homeland of Kuna Yala on the north-east coast of Panama. This ensured the Kuna's tenure and sovereignty, guaranteeing direct management of their terrestrial and marine resources as an integral part of their cultural patrimony (Muller et al., 2003). The local communities include the "tribal peoples" referred to in *ILO Convention (169) concerning Indigenous and Tribal Peoples in Independent Countries* (United Nations Centre for Human Rights, 1994) in countries such as Indonesia, Malaysia, India, the countries of the Red Sea and the many small island states of Oceania. Thus the nature of the interactions with tourists and tourism operators and the involvement and benefits that flow to ICRC will depend to a large extent on whether ICRC either:

- hold, retain or have had restored traditional (marine) title/tenure over their traditional marine estates (i.e., effectively the extent to which they are the "owners" of the coral reefs in which tourism activities take place); or
- retain traditional connections/affiliations with coral reefs over which they do not exercise legal title, while, being involved with or impacted by tourism activities (over which, in some circumstances, they exercise little or no control).

The impacts of colonisation on ICRC should be not ignored, both in terms of the peoples themselves and their coral reef estates. These impacts included disease, warfare, slavery, forced religious conversion, dispersal and alienation of traditional land and marine estates and, for the surviving populations, ethnic and social exclusion often motivated by racism embedded in discriminatory laws and practices (Steiner, 1999; Hickey, 2007). In many cases traditional knowledge, cultures and practices that should be a part of the coral reef tourism experience have been severely modified, transformed or simply lost or neglected. Many communities were decimated by disease and warfare. For example, in 1552 Bartolome de las Casas, in his publication *The Devastation of the Indies*, estimated that some 15 million Indigenous people were slaughtered in the Americas during the half century following Columbus's arrival in the New World (Cohen, 1999). In the South Pacific, the population of Vanuatu declined from an estimated half million in the pre-contact period of the early 1800s to 45,000 by the 1940s (Hickey, 2007). Under a system of indentured labour (but which frequently involved kidnapping – colloquially referred to as "blackbirding"), an estimated 62,000 Pacific Islanders were brought from New Caledonia, the Solomon Islands, Vanuatu (then known as New Hebrides) and Fiji between 1863 and 1904 to work on the Queensland sugar plantations in Australia (Docker, 1970; Andrew & Cook, 2000). In 1862 the estimated population of Rapa Nui (Easter Island) was 6,000, but declined to 111 people by 1877 having been ravaged by small-pox and tuberculosis (Teave & Cloud, 2014). In the 150 years following the arrival of Captain James Cook in 1778, Native Hawai'ians were subject to a loss of more than an estimated 90% of their population as a result of disease, social disintegration and spiritual alienation (Cohen, 1999). The European and American trade in slaves from Africa transformed the demographics of the coastal and island communities of the Americas and the Caribbean, again severely impacting the Indigenous peoples. The Tainos, the Indigenous people of many of the Caribbean islands were mostly wiped out by the Spanish and now survive as Maroons, communities of mixed ancestries of Amerindian peoples and Africans who escaped slavery (Agorsah, 1994). Traditional belief systems were disrupted by the often forced introduction of new religions. One example is the native Rapa Nui religion, which was eliminated by the Catholic missions established in 1866 (Teave & Cloud, 2014). The legacy of colonisation lives on in the present poor physical and mental health status of Indigenous peoples around the world (Cohen, 1999).

Colonial impacts also extended to the marine/coral reef environments through coastal development, agriculture, commercial fishing and pollution. Thus, colonial impacts have led to modified traditional coral reef behaviours, including the introduction of new technologies that enable more efficient hunting and fishing, sometimes dramatically altering the way ICRC manage and exploit coral reef resources – and even think about their reefs as the old ways disappear (Poepoe et al., 2007).

While the contribution of ICRC to, and the necessity of their involvement in the conservation and sustainable use of marine resources is belatedly becoming more widely recognised (Haggan et al., 2007), there appears to be little analysis of the impacts of tourism on their cultures and economies, or indeed of their participation in the tourism industry itself. An attempt will therefore be made to explain the environmental, economic, social and cultural impacts, both positive and negative, that tourism has on ICRC, recognising the geo-political and cultural diversity of those communities and their different abilities and capacities to manage, participate in and derive benefit from tourism.

Traditional values associated with coral reefs

Many ICRC have associations with their reef and marine estates that go back for many millennia. The Yidinji people, whose traditional estates include the section of Australia's Great Barrier Reef east of Cairns, have oral histories that pre-date the formation of the Great Barrier Reef, and perform ceremonies associated with the rising sea levels that led to its formation (Attenborough, 2015). Chief Roi Mata's domain is a World Heritage listed cultural site in Vanuatu that celebrates a thirteenth century paramount chief whose social reforms continue to be relevant to contemporary local society.

ICRC, particularly small island communities whose islands are often surrounded by coral reefs, are particularly dependent upon their reefs not only for their everyday subsistence, but also everyday material goods. Reefs and coral cays provide a wide range of foods in addition to fish, crustaceans and shellfish, which includes turtle meat and eggs, dugong, and sea birds and their eggs. Various types of shell (pearls and pearl shell), including turtle shell, are used for trade, gifts, utensils (containers, cutting tools), weapons (stingray barbs for spear tips) and for personal adornment (shell necklaces, bracelets), which can also be indicators of personal status. Both marine foods and artefacts are important components of coral reef tourism.

Over generations, Indigenous reef communities have also developed sets of beliefs and customs associated with their marine estates in which particular species and places have special meaning, evolving lifestyles founded on a "conservation ethic" based on centuries (even millennia) old wisdom of holistic, integrated management of marine and terrestrial ecosystems (Nga Kai O Te Moana, Ministry of Maori Development, 1993). Implicit in this is the notion that the marine world can be as important as the terrestrial world. In traditional Hawai'ian societies, the "principle of dualism" – that things of the cosmos are presented as pairs of opposites – allowed Hawai'ians to develop an organised conceptual world view with paired opposites that depended on each other to complete the whole. One example of this pairing process is that many marine organisms have a terrestrial counterpart, the opposite inherent on land versus that found in the marine (Steiner, 1999).

ICRC in many parts of the world have strong traditions of designation areas that are restricted to ceremonial, religious and conservation purposes. In effect this system created a system of protected areas that facilitated the preservation of nature as well as culture (Verschuuren et al., 2010). These protected areas predated the Yellowstone model which dominated national and international legislation and policies on protected areas through a large part of the twentieth century. The strategy behind establishing protected areas using the Yellowstone model involves regulating human access and use of specific areas that have been set aside to protect nature. The natural areas set aside by Indigenous and local communities included mountains, lakes and rivers as well as groves and were regarded as sacred or areas subject to seasonal closures because of their importance as nesting, breeding or spawning grounds. Many of these were important for ecological, landscape, biological, ecological or fragility reasons. There are strong parallels between the areas set aside by Indigenous communities and those set aside by the declaration of contemporary protected area. Areas that had been set aside by Indigenous communities were usually under the control of spiritual leaders or traditional institutions that were vested with a body of customary regulations, and norms were usually defined and enforced to ensure compliance with rules for governing traditional access and use of these areas (Oviedo, 2003).

The central feature of traditional marine resource management and tenure systems was the ability to control, restrict and regulate access to marine resources and habitats, and implement the rotational use of marine resources for management purposes (Mulrennan, 2007). In Vanuatu traditional beliefs such as the spatial and temporal separation of agricultural and fishing practices, as well as totemic affiliations, have been instrumental in managing marine resources. Areas of symbolic significance and spatial-temporal "refugia" create extensive networks of protected freshwater, terrestrial and marine areas (Hickey, 2007). In Hawai'i, as Poepoe et al. (2007, p. 128) point out:

> Many natural processes that affect fish distribution are monitored by the community, but the most important of these are seasons and the moon. The moon was as essential in scheduling the activities of the ancient Hawai'ians as clocks are to modern man. The moon calendar is a predictive tool based on awareness of natural cycles and their relationship to fishing and farming success. Its wisdom reflects lifetimes of observations and experiences by many generations of Hawai'ians in their quest for survival. Modern-day people of Hawai'i still refer to the calendar to plan fishing and planting activities, and a popular form of the calendar is published annually.

The relationship of ICRC to their reefs are also expressed in the particular cultural and spiritual values that have evolved over time, and as expressed in art and ceremony. Ceremonies frequently focused on particular totemic species (turtle, dugong, whale shark, dolphin, tuna), important natural phenomena such as the winds, moon and the stars (as navigational aids), and celebrations of their seafaring prowess. Many coastal and island communities, such as the people of the island communities of Polynesia, Melanesia and Micronesia developed considerable seafaring skills, which also enabled them to be tourists. Using regionally distinctive single and double-hulled canoes capable of voyages over hundreds of kilometres, islanders navigated using their knowledge of the stars, prevailing winds, tides, ocean currents and wave patterns (Haddon & Hornell, 1975; Lewis, 1978).

In contemporary times these values are also often portrayed in their visual arts, ceremonies and cultural performances, which bring international focus to island

and coastal communities. The tradition-focused prints, head-dresses and masks of the Torres Strait Islanders have generated exhibitions such as *Ilan Pasin (this is our way): Torres Strait Art* (Mosby, n.d.). For the Yolngu of north-east Arnhem Land, traditional bark paintings of *sea country* also serve a political purpose to assert their rights over their marine estates, as in the exhibition *Saltwater: Yirrkala Bark Paintings of Sea Country* (Buku-Larrnggay Mulka Centre, 1999), and other bodies of Oceanic artwork also provide social, cultural and political commentaries (Chiu, 2004).

Coral reef issues for Indigenous and local communities

Although ICRC share their concerns with the wider community about global impacts on coral reefs such as climate change, coral bleaching, ocean acidification and rising sea levels, as well as more local impacts due to marine pollution, over-fishing, increased coastal development, terrestrial sedimentation (particularly as a result of agricultural development and mining) and, in some localities, tourism, their more immediate concerns focus on the:

- status of their traditional marine tenure in terms of their legal rights to control, manage access to and determine the kinds of activities that take place in regard to their coral reefs, including access by tourism operators;
- lack of recognition, respect for and exercise of their traditional knowledge, that leads to, *inter alia*, over-exploitation of their marine resources;
- restrictions on their traditional enjoyment of their reefs for hunting, fishing and other cultural purposes;
- environmental concerns; and
- inclusion in the design and management of marine protected areas.

These concerns are borne out by what might be termed the Great Barrier Reef Marine Park paradox, which will conclude this section.

Ownership of traditional marine tenure

For ICRC, the recognition and exercise of their rights to own, control access to, enjoy and manage their coral reef estates are crucial. Many ICRC are still recovering from the colonial legacy of dispossession of their lands and waters, which saw traditional marine tenure systems and usage rights completely disregarded. As Johannes points out with regard to Oceania:

> the value of [traditional] marine tenure was not generally appreciated by Western colonizers. It not only ran counter to the [European] tradition of freedom of the seas which they assumed to have universal validity, but it interfered with their desire to exploit the islands' marine resources – a right they took for granted as soon as they planted their flags.
>
> *(Johannes, 1978, as cited in Mulrennan, 2007, p. 187)*

Traditional systems of marine tenure are slowly being recognised through enforcement of treaty rights, constitutional recognition, agreements and legislated recognition of sea rights. In Australia, under separate state land rights laws and the *Native Title Act 1993* (Cth), this recognition is now extending to Indigenous involvement in marine and coastal protected areas through the establishment of traditional use marine resource areas or TUMRAs. These rights also extend to the protection and enjoyment of cultural sites located within traditional marine estates, as well as the observance of cultural practices associated with those sites.

Lack of respect for traditional knowledge

The Conference of the Parties to the CBD, in decision IV/9, has recognised "that traditional knowledge should be given the same respect as any other form of knowledge in the implementation of the Convention" (SCBD, 1998, p. 69). Popular myths about traditional knowledge should also be dispelled. As Poepoe et al. (2007) caution in relation to Hawai'ian traditional knowledge, but is generally true of bodies of traditional knowledge around the world:

> Traditional knowledge and practice should not be interpreted as static, rigid or non-changing. "The culture lives on through its practitioners" and cultural activities have a strong sense of "place". Tradition as it exists in the world of contemporary Ho'olehua homesteaders, is an accumulation of knowledge and behavioural norms that have strong roots in culture, local history and experience, and which is being constantly verified and augmented. It is legitimate in its own right and does not need to be recast in the idiom of Western industrial society or verified through the methods of contemporary government resource managers.
>
> *(Edith Kanaka'ole Foundation, 1995, as cited*
> *in Poepoe et al., 2007, p. 122)*

Local traditional knowledge has an important role to play in enhancing the tourist experience, for example identifying the best times and places to fish for certain species, reefs that should be avoided during spawning seasons, optimising diving experiences and adding a cultural dimension to the tourism experience. It can also extend to personal safety, alerting tourists to certain dangers associated with particular locations – tides, rips, venomous and inedible species.

Restrictions on traditional enjoyment of coral reefs for hunting, fishing and other cultural purposes

In regard to the sustainable use of components of biological diversity, Article 10(c) of the CBD protects and encourages "customary use of biological resources in accordance with traditional cultural practices that are compatible with conservation or sustainable use requirements". However, Poepoe et al.'s (2007, p. 120) observations in relation to Hawai'i can be applied more generally: "Overfishing by a growing population that no longer recognizes traditional conservation practices has greatly contributed to the decline in inshore fisheries."

Conservationists are also often at loggerheads with traditional hunters, particularly with respect to iconic species such as dugong and turtle, overlooking the fact that species decline is also a result of habitat destruction from agricultural pollutants and sedimentation, commercial fishing bycatch, marine pollution (particularly plastic bags and bottles), entanglement in ghost-nets and commercial shipping (including cruise ships) – which are all introduced threats to local marine environments (Kershaw, 2015). Traditional hunting methods often attract adverse media attention and debate rages about modern-day hunters forgoing traditional technologies and methods in favour of modern technologies – guns, motor boats, sonar fish finders, scuba equipment, etc. In Australia, restrictions with regard to particular species are placed on traditional marine resource use rights to the extent that members of coastal reserve communities and native title holders, under a permit system, can take dugong and turtle, even in protected areas, but only for non-commercial local community consumption. Such permit systems are usually managed by the communities themselves as part of community ranger responsibilities in conjunction with the relevant state authorities. However, critics are quick to point out the inherent contradiction of hunting vulnerable and endangered species in protected areas.

Environmental concerns

Of great concern to both ICRC and the tourism industry is the state of the reefs. Currently the world's reefs are classed as being in a high risk category because of local threats, including overfishing and destructive fishing, marine pollution, increased coastal development, terrestrial sedimentation and tourism in some areas. The impacts of global scale threats, such as global warming, ocean acidification and rising sea levels, will further compound local threats. Burke et al. (2011) have estimated that as much as 75% of all coral reefs will be at high risk by 2050. In the Oceania region, Australia has the largest area of coral reefs, with 67% considered at low risk from local threats. In South Asia, an estimated 66% of the Maldives coral reefs have been assessed as being at low risk from local threats. In Indonesia, only 4% of coral reefs are considered to be at low risk from local threats while in the Philippines only 1% of coral reefs are considered to be at low risk from local threats (State of the Tropics, 2014).

The greatest local threats to coral reefs in South East Asia are overfishing and destructive fishing, with almost 85% of reefs threatened by these hazards (Burke et al., 2012). Destructive fishing includes the use of explosives or poisons that often result in significant collateral damage and slow recovery rates for both reef animals and corals (Saila et al., 1993). Given that this region is also located in the Coral Triangle, biodiversity has also been placed at risk by these widespread threats (State of the Tropics, 2014).

Of particular concern to the tourism industry is coral bleaching. This issue also caused extensive concern in Australia with regard to the coral bleaching that occurred in the northern section of the Great Barrier Reef Marine Park in 2016 and 2017, which became something of an issue in the context of the then forthcoming federal election (Chandler, 2016). Also of particular concern to low lying coral atoll countries, such as Tuvalu, Kiribati and the Maldives, are rising sea levels. Rising sea levels also increase their vulnerability to the destructive impacts of intense tropical weather systems, which further threaten the viability of such communities, creating a new class of refugees – climate refugees.

ICRC involvement in Marine and Coastal Protected Areas

For many ICRC, the resolution of these issues can be found in the establishment of marine parks and protected areas. One of the principal solutions to address coral reef degradation is to create a national and international network of marine and coastal protected areas (MCPAs). While many MCPAs are small and fragmented, their effective protection is able to provide local communities with benefits that extend well beyond the area that is protected (Spalding et al., 2001). However, the extent of MCPAs is much lower in tropical regions (State of the Tropics, 2014). Of the eight nations in Central America, five had more than 10% of their territorial waters under an MCPA by 2010, while the three largest MCPAs (Australia's Great Barrier Reef Marine Park, North Western Hawai'ian Island Coral Reef Reserve and the Phoenix Islands Protected Area, Kiribati) are in Oceania. All other tropical regions have less than 5% of marine territorial waters protected (State of the Tropics, 2014).

Many studies have emphasised the importance of stakeholder involvement in both the design and management of MCPAs. It has also been shown that participation assists with compliance and helps in the development of regulations that take note of cultural sensitivities. The involvement of community stakeholders also ensures that valuable information about socioeconomic and cultural properties, as well as biological aspects of ocean use, are included in management practices (Vierros, 2004). The Tun Mustapha Park, a vast MCPA encompassing almost 1 million hectares and 50 islands off Sabah, Malaysia, contains a globally significant mix of coral reefs, mangroves and seagrasses and is an important passage for fish, turtles and other marine mammals. The region also supports the livelihoods of 80,000 people, with approximately 80,000 tonnes of fish caught each day. The park is managed in collaboration with local communities. Community-based natural resources management in the park is a key tool for conserving and sustainably using marine resources while improving livelihoods in the Coral Triangle (WWF, 2015). However, the benefits of ICRC involvement in MCPA management are not just restricted to ecological benefits – ICRC can also be involved in activities such as surveillance, reporting of illegal activities (poaching, drug trade, illegal visitors) and maintaining border security.

Protected areas have also been found to be an important element in strategies to eradicate hunger, disease, poverty and degradation of the environment. However, positive outcomes are only able to be achieved if changes are made to traditional economic and subsistence activities by the local and Indigenous communities that are involved. This may require the development of new types of culturally appropriate employment that capitalise on traditional skills, particularly where this involves the replacement of unsustainable traditional resource use practices. As Peteru (1999) observed, a number of traditional practices may become unsustainable because of indirect pressures on Indigenous and local communities and their resource bases. The introduction of protected area status is able to offer new forms of employment opportunities, including tour guiding, employment at interpretative centres, park rangers, tourist-oriented production of artefacts and the provision and maintenance of infrastructure. Most of these new employment opportunities will require training which should include the acquisition of science-based management skills to complement traditional knowledge. This approach to training and education will enable participants to more effectively undertake environmental/social/cultural impact assessments, strategic planning, biodiversity monitoring, parataxonomy and habitat rehabilitation (Marrie, 2004). Phelan (2007) reports on traditional fishers in the Injinoo Aboriginal Community (northern tip of Cape York Peninsula, Australia) integrating catch data and biological information to highlight trends in the harvest of *Protonibea diacanthus*, resulting in a self-imposed community ban in consultation with other communities through a regional agreement to restore the sustainability of the fishery.

Protected area status enhances tourism potential, such as the Tubbataha Reefs Natural Park in the Philippines (Marris, 2016), although it is by no means a prerequisite. The Maldives has only some reefs under formal protection; however, there are a number of informal protections in place that ensures this key tourism drawcard remains viable (State of the Tropics, 2014). MCPAs with World Heritage status enhance their drawing power as tourism destinations. Perhaps the best known of these is Australia's Great Barrier Reef Marine Park, but others include the lagoons of New Caledonia (Engelhardt, 2016) and the Palawan World Heritage Area in the Philippines (Sherwell, 2016).

Great Barrier Reef Marine Park paradox

In Australia, the coral reef coastline of the Great Barrier Reef extending 2,300 kilometres along the Queensland coast, includes the traditional marine estates of about 70 Aboriginal Traditional Owner clan groups, with connections to country dating back over 60,000 years. Despite having a regulatory environment that recognises Traditional Owner connections to *sea country*, including the input of Aboriginal people in decision-making, management and the establishment of TUMRAs and marine park Indigenous land use agreements (ILUAs), participation by Aboriginal people in the tourism industry associated with the Great Barrier Reef is almost non-existent, despite abundant opportunities. This situation is very different from the level of Aboriginal involvement in tourism in the adjacent Wet Tropics World Heritage Area (Marrie & Marrie, 2014). Furthermore, and herein lies the paradox, there appears to be little or no Aboriginal interest in the industry itself other than to ensure cultural heritage sites, hunting rights and that practices are respected and regulated.

Low Aboriginal participation in coral reef tourism also appears to be the result of a lack of reef transport infrastructure and equipment. Currently, Aboriginal participation in coral reef tourism is generally restricted to onshore cultural activities (cultural performances that celebrate stories of important reef-associated species, such as turtle, dugong, sharks and crocodiles) and participation in local festivals. One example of a festival of this type is the Gimuy Fish Festival held annually in Cairns.

Coral reef tourism: motivations and attractions

Tourists, and the kinds of tourism products and services they support, are extremely diverse. At the local level, much depends on whether they are domestic or international – from low-budget backpackers and caravanning retirees to those participating in the high-end tourism market, taking packaged tours, arriving on cruise liners or staying at 5-star resorts. The quality of the experience and engagement with local ICRC will likewise vary enormously and depend much on their motivation.

With regard to coral reef tourism, many tourists are attracted by particular species, which can also be culturally significant to the local Indigenous people. For the islanders of Donsol in the Philippines, the whale shark is their "patron animal" and is celebrated with an annual festival. Whale sharks attract tourists and lead to spending in the local economy. The benefits accrue to local communities as well as whale sharks and other marine species only when local people recognise that they need to conserve their biggest asset (WWF, 2015). The recognition of the contribution of scuba divers coming to see sharks in a number of Fijian communities has led to a ban on shark fishing in a 30-mile "shark corridor". Shark tourism contributed US$42.2 million to the country's economy in 2010 (Vianna et al., 2011).

Many tourists also succumb to the lure and mystique of the exotic, coupled with such terms as "paradise" and "mystery" – terms that are used extensively to market particular tourism destinations. Examples include the Fijian island of Taveuni (Belcher, 2016) and Upolu in Western Samoa (Marshall, 2016). Early European explorer accounts of the exotic lands and cultures "discovered", particularly in the South Pacific, excited the imagination of their scientific and scholarly colleagues, as well as that of their wider communities through the tales of sailors and the exhibition of exotica for popular entertainment. Great scientific expeditions were undertaken to collect specimens (human, biological and ethnological) to add to the exotic collections of natural history museums throughout Europe, Russia and the Americas, adding vastly to the collective knowledge of human kind. Great theories about race, evolution, anthropology, art and aesthetics, and human behaviour were propounded on the basis of these discoveries and resulting collections, further fuelling the popular imagination, which still remains the motivation behind much of the marketing of such destinations.

Much also depends on how certain types of coral reef tourism experiences are promoted by travel writers and the media, for example the marketing of one of Fiji's great scuba diving attractions, Taveuni Island, where the Great White Wall dive is promoted as being one of the world's best dive sites. A recent article in *Club Marine* promotes Taveuni as the "Mystery Island of Fiji", but, geography aside, acknowledges that Taveuni's greatest asset is its people. Taveuni hospitality is second to none – the locals are warm, friendly and hospitable – and it praises local Fijian foods (cooked in the *lovo* – Fijian earth oven) and local Fijian culture where local female staff members perform the *Meke*, a traditional Fijian dance portraying local stories and legends (Belcher, 2016). This contrasts with New Caledonia, where an emphasis is placed on sophisticated French flair and a French Riviera-style in-harbour stay (Engelhardt, 2016) – no reference at all to the cultures of the local Indigenous peoples. In the Caribbean, the remnants and ruins of the colonial past may be as much a drawcard as the local cultures.

Impacts of tourism on Indigenous and local communities and their coral reef habitats

Tourism can have undoubted economic flow-on effects in terms of jobs (particularly in the provision of services), businesses (bait and tackle shops, vehicle hire, etc.) and infrastructure development and maintenance (airports, marinas, resorts, pontoons, jetties, wharves, camping grounds, etc.) for local people. However, tourism can also be a fickle industry impacted by seasonal conditions – many tourists tend to avoid tropical destinations during the monsoon season, natural disasters (floods, tsunamis and volcanic eruptions), human generated disasters (chemical and oil spills from coastal shipping), currency fluctuations, political instability and global economic downturns (SCBD, 2004). Unsustainable tourism activities may precipitate social tensions and degradation by encouraging local prostitution, drug abuse and harassment of tourists themselves. When tourism flows cease or are interrupted by either natural or human crises, communities that have a heavy reliance on tourism become vulnerable, particularly when faced with an unanticipated loss of income and jobs.

In circumstances where tourism infrastructure, such as hotels, is owned by offshore corporations, there has been criticism that the bulk of profits are remitted to overseas entities (Deda, 2004). This has to be balanced by the positive benefits of offshore investment, including jobs and the construction of infrastructure that could not be funded by local capital. The exclusive nature of some tourism developments has also been criticised as generating negative social impacts, particularly when local communities are excluded or disenfranchised (Loper et al., 2008). Fortunately, increasing recognition of the importance of sustainable ecotourism has led to increased efforts by businesses and governments to engage communities and to protect coral reefs, while continuing to provide opportunities for jobs and economic development (State of the Tropics, 2014).

Tourism can also have a highly complex impact on cultural values. Calkin (2016) notes the corrosive nature of Caribbean tourism, which creates an unsustainable industry with no connection to local culture. A number of authors have observed that tourism activities may lead to intergenerational conflicts caused by the aspirations of younger members of communities who have had greater contact with and been affected by the behaviour of tourists. Furthermore, they may affect gender relationships through offering different employment opportunities to men and women. This may lead to an erosion of traditional practices, including cultural erosion and disruption of traditional lifestyles. Traditional practices and events may also be influenced by tourists' preferences. The Indigenous art market is sometimes categorised into "fine" and "tourist" art: the latter focused more on the souvenir trade with affordability and portability guiding production. Similarly, resort-style cultural performances might be "packaged" specifically for tourist consumption and may bear little or no relation to contemporary cultural practices and realities. Moreover, tourism development can lead to Indigenous and local communities losing access to their traditional land, resources and sacred sites. Loss of access to resources which are integral to the maintenance of traditional knowledge systems and traditional lifestyles can lead to loss of traditional culture (Deda, 2004).

While coral reef tourism carries the danger of leading to reefs being "loved to death", it can also provide the economic incentive for more effective management. Intense tourism use has parallels with other unsustainable activities, including overfishing (State of the Tropics, 2014). Other physical impacts on coral reefs that may stem from tourism development include marine pollution from tourism activities, infrastructure construction that removes coastal vegetation and leads to increased nutrient and sediment loads, overfishing to supply tourist restaurants and high intensity use of dive sites (Tratalos & Austin, 2001).

Indigenous and local community involvement in coral reef tourism

As the State of the Tropics report (2014) has highlighted, many nations, including small island nations, depend on coral reefs for a significant percentage of their economic activity. In some areas reef-related tourism is expanding rapidly. In the Maldives, for example, 38% of national GDP is from tourism, with over 60% of tourists visiting for scuba diving. It is apparent that healthy coral reef systems are essential if economic activity of this type is to continue (Burke et al., 2011; Cesar et al., 2003; Loper et al., 2008).

ICRC can generally benefit from tourism because of the economic development opportunities it brings, as a vehicle for cultural education and maintenance and as a means to gain recognition of and respect for their cultures. Economic opportunities in the tourism industry include:

- employment;
- training and skills development for young people in the tourism and hospitality industries (as guides and interpreters);

- stand-alone community-based cultural tourism enterprises; and
- joint ventures between ICRC and mainstream tourism enterprises enabling them to surmount such difficulties as access to the capital required to invest in tourism infrastructure, boats, training, etc.

Many ICRC stress the importance of owning and operating their own tourism ventures. Tourism is also seen as a means to achieve cultural and social aspirations. In Australia, tourism can facilitate Traditional Owners getting back into and maintaining contact with their traditional estates, including marine estates. It is also seen as a vehicle for increasing the cultural awareness of wider society and overseas visitors about local Indigenous cultures (Marrie & Marrie, 2014). In terms of ICRC involvement in tourism, the nature of the tourism activities can be seen in those that take place at coral reef sites (*in situ* activities) and those that take place away from coral reef location (*ex situ* activities). *In situ* activities typically include fishing; scuba diving; whale-watching; underwater photography; swimming with iconic species such as whale sharks, dolphins and manta-rays; and collecting (corals, sea-shells, aquarium species). They may also include cultural experiences via an Indigenous guide or interpreter (for example, on tourist boats). *Ex situ* activities that typically take place away from reefs include cultural performances at the local resort or cultural centre; enjoyment of local cuisine at restaurants and local markets; and annual festivals.

With regard to *in situ* tourism activities, local traditional ecological knowledge can have a valuable role in enhancing the tourism experience. Reliant on marine/coral reef resources for their livelihoods and ultimately survival, local fishers and marine foragers have developed detailed knowledge regarding how the distribution and abundance of marine animals vary from year to year with types of habitat, season, weather, time of day, stage of tidal cycles, lunar phase and other factors. These groups are also able to provide observations about migratory behaviours of marine animals that contribute to these changing distributions and abundance. This is particularly important in nearshore tropical waters in understanding reef fish spawning aggregations. Groupers, snappers, jacks, emperors, mullets, bonefish, rabbitfish, surgeonfish and other species of coral reef fish use the same location and season and moon phase to aggregate when spawning each year (Johannes & Neis, 2007; Poepoe et al., 2007). Such *in situ* knowledge can be important for tourists eager to dive the reefs, photograph marine activity, or fish for prized food species – knowing when certain areas should be closed/avoided to protect breeding species or open for visitation, and optimal times of the year to observe migratory species, such as whales, whale sharks and nesting turtles. It is this kind of knowledge that can considerably enhance, even contribute to the success of a visitor's stay.

The international regulatory environment: Indigenous and local communities and tourism

The CBD has created the most effective platform for the regulation of Indigenous coral reef tourism. In 2004 the SCBD issued two sets of guidelines relevant to Indigenous coral reef tourism: the *Guidelines on Biodiversity and Tourism Development* (SCBD, 2004) and the *Akwe: Kon Voluntary Guidelines for the Conduct of Cultural, Environmental and Social Impact Assessment Regarding Developments Proposed to Take Place on, or which are Likely to Impact on, Sacred Sites and on Lands and Waters Traditionally Occupied or Used by Indigenous and Local Communities* (SCBD, 2004). The tourism guidelines were followed in 2015 with a manual for their application, *Tourism Supporting Biodiversity: A Manual on Applying the CBD Guidelines on Biodiversity and Tourism Development* (SCBD, 2015). These sets of guidelines serve to emphasise the status of the CBD as the most important international treaty for addressing many of the concerns of the world's Indigenous peoples (Bruggemann et al., 2002), remembering that declarations, such as the UNDRIP, have no binding legal effect in international law (Marrie, 2009).

The *CBD Guidelines on Biodiversity and Tourism Development* (SCBD, 2004) is comprehensive in its scope and contains multiple references to Indigenous and local communities, addressing such matters as:

- poverty reduction;
- protection of Indigenous livelihoods;
- supporting effective participation and involvement in the development, operation and monitoring of tourism activities taking place on lands and waters traditionally occupied by them;
- access to infrastructure;
- sharing of benefits arising from tourism;
- involvement in impact assessments (with reference to the *Akwe: Kon Guidelines*);
- respect for customs and application of traditional knowledge, innovations and practices;
- prior informed consent requirements; and
- inclusion in monitoring and evaluation of tourism impacts, including in the development of indicators and early warning systems (SCBD, 2004).

The *Guidelines* encourage parties to build the capacity of Indigenous and local communities not only for participation in the tourism industry, but also for monitoring the environmental, social and cultural impacts on communities. Although these guidelines are voluntary, they nevertheless establish aspirational standards and a basic reference point, or "checklist", for those Indigenous and local communities, including ICRC, engaging with the tourism industry.

Conclusion

ICRC, as interested parties and stakeholders in coral reef tourism, are extremely diverse in their cultures, histories and present-day political status. Nevertheless they share the common global concerns around those factors (global warming, ocean acidification, coral bleaching, marine pollution) that impact the coral reefs on which they depend, wholly or partly, for their livelihoods. As an identifiable group of communities, they also share many common characteristics and issues: insecure tenure of their marine estates and inability to control access to and use of traditionally used marine resources; disregard for cultural practices such as their own sustainable use practices regarding coral reef marine resources; marginalisation by the industry itself; and inequitable sharing of the financial, economic and social benefits that the industry might bring to the community. However, each local ICRC also has its own specific concerns and issues that they must negotiate with governments, tourism operators and developers and, until recently, without the aid of a set of internationally agreed standards to which they can refer. It is also emphasised that the relationship between ICRC and coral reef tourism in general is little researched despite the potential mutual benefits that ICRC, tourists and tourism operators can enjoy, and it is hoped that this chapter may stimulate such research.

References

Agorsah, E.K. (1994). *Maroon heritage: Archaeological, ethnographic and historical perspectives.* University of the West Indies, Kingston, Jamaica: Canoe Press.

Andrew, C., & Cook, P. (Eds.). (2000). *Fields of sorrow: An oral history of descendants of the South Sea Islanders (Kanakas).* Mackay, Queensland: MacPrint Australia Pty.

Asian Development Bank (ADB). (2014). *Regional state of the Coral Triangle – Coral Triangle marine resources: Their status, economies and management.* The Philippines: Asian Development Bank.

Attenborough, D. (Presenter). (2015). *The Great Barrier Reef* [television documentary miniseries]. United Kingdom: Atlantic Productions. Available from http://www.abc.net.au/tv/programs/david-attenboroughs-great-barrier-reef/

Belcher, A. (2016). Tempting Taveuni: Fiji's Taveuni Island offers divers and sportfishers a host of colourful and lively options. *Club Marine, 31*(1), 110–119.

Bruggemann, J., Hernandez, M., Rodriguez, E., Soler, J., & Tapper, R. (Eds.). (2002). *Biodiversity and tourism in the framework of the convention on biological diversity: The case of the Tayrona National Park, Colombia.* Germany: German Federal Agency for Nature Conservation.

Buku-Larrnggay Mulka Centre. (1999). *Saltwater: Yirrkala bark paintings of sea country: Recognising Indigenous sea rights.* Yirrkala, Northern Territory, Australia: Buku-Larrnggay Mulka Centre, in association with Jennifer Isaacs Publishing, Neutral Bay, New South Wales, Australia.

Burke, L., Reytar, K., Spalding, M., & Perry, A. (2011). *Reefs at risk revisited.* Washington, DC: World Resources Institute.

Burke, L., Reytar, K., Spalding, M., & Perry, A. (2012). *Reefs at risk revisited in the Coral Triangle.* Washington, DC: World Resources Institute.

Calkin, J. (2016). The caring Caribbean: A luxury resort on St Kitts has an innovative green philosophy. *The Weekend Australian, Travel & Indulgence,* June 25–26, pp. 8–9.

Cesar, H., Burke, L., & Pet-Soede, L. (2003). *The economics of worldwide coral reef degradation.* Arnhem, The Netherlands: Cesar Environmental Economics Consulting.

Chandler, J. (2016). Grave barrier reef: The coral bleaching signals a defining environmental shift. *The Monthly,* June, 40–45.

Chiu, M. (Curator). 2004. *Paradise now? Contemporary art from the Pacific.* New York: Asia Society.

Cohen, A. (1999). *The mental health of Indigenous peoples: An international overview*. Geneva: Department of Mental Health, World Health Organization.

Deda, P. (2004). Tourism in protected areas: Reducing conflicts and enhancing synergies. In *Biodiversity issues for consideration in the planning, establishment and management of protected area sites and networks* (CBD Technical Series No. 15, pp. 143–147). Montreal: Secretariat of the Convention on Biological Diversity.

Docker, E.W. (1970). *The Blackbirders: The recruiting of South Seas labour for Queensland 1863–1907*. Sydney and London: Angus & Robertson.

Edith Kanaka'ole Foundation (EKF). (1995). Draft Ke Kalai Maoli Ola No Kanaloa, Kaho'olawe Cultural Use Plan. Hawai'i: Consultants to the Kaho'olawe Island Reserve Commission, Hawai'i.

Engelhardt, L. (2016). Bon voyage: New Caledonia, where sophisticated French flair rubs noses with laid-back Pacific island charm. *Club Marine, 30*(6), 56–64.

Haddon, A.C., & Hornell, J. (1975). *Canoes of Oceania*. Honolulu: BP Bishop Museum.

Haggan, N., Neis, B., & Baird, I.G. (Eds.). (2007). *Fishers' knowledge in fisheries science and management* [Coastal Management Source Books 4]. Geneva: UNESCO Publishing.

Hickey, F.R. (2007). Traditional marine resource management in Vanuatu: Worldviews in transformation. In N. Haggan, B. Neis, & I.G. Baird (Eds.), *Fishers' knowledge in fisheries science and management* [Coastal Management Source Books 4], (pp. 147–168). Geneva: UNESCO Publishing.

Honey, M., & Krantz, D. (2007). *Global trends in coastal tourism*. Washington, DC: Centre on Ecotourism and Sustainable Development (CESD) and World Wildlife Fund for Nature (WWF).

Johannes, R.E., & Neis, B. (2007). The value of anecdote. In N. Haggan, B. Neis, & I.G. Baird (Eds.), *Fishers' knowledge in fisheries science and management* [Coastal Management Source Books 4], (pp. 41–58). Geneva: UNESCO Publishing.

Kershaw, P.J. (Ed.). (2015). *Sources, fate and effects of microplastics in the marine environment: A global assessment*. Report by the Joint Group of Experts on the Scientific Aspects of Marine Environmental Protection (GESAMP), No. 90. London: International Maritime Organization.

Lewis, D. (1978). *We, the navigators: The ancient art of landfinding in the Pacific*. Canberra: Australian National University Press.

Loper, C., Pomeroy, R., Hoon, V., McConney, P., Pena, M., Sanders, A., . . . , & Wanyonyi, I. (2008). *Socioeconomic conditions along the world's tropical coasts: 2008*. Silver Spring, MD: National Oceanic and Atmospheric Administration (NOAA), Global Coral Reef Monitoring Network (GCRMN) and Conservation International (CI).

Maffi, L. (1999). Linguistic diversity: Introduction. In D.A. Posey (Ed.), *Cultural and spiritual values of biodiversity: A complementary contribution to the global biodiversity assessment* (pp. 21–35). London: Intermediate Technology Publications, and Nairobi, Kenya: United Nations Environment Program.

Marrie, H.L. (2004). Protected areas and Indigenous and local communities. In *Biodiversity issues for consideration in the planning, establishment and management of protected area sites and networks* (CBD Technical Series No. 15, pp. 106–110). Montreal: Secretariat of the Convention on Biological Diversity.

Marrie, H.L. (2009). The UNESCO Convention for the Safeguarding of the Intangible Cultural Heritage and the protection and maintenance of the intangible cultural heritage of Indigenous peoples. In L. Smith & N. Akagawa (Eds.), *Intangible heritage* (pp. 169–192). London: Routledge.

Marrie, H.L., & Marrie, A.P.H. (2014). Rainforest Aboriginal peoples and the wet tropics of Queensland World Heritage Area: The role of Indigenous activism in achieving effective involvement in management and recognition of cultural values. In S. Disko & H. Tugenhadt (Eds.), *World Heritage Sites and Indigenous peoples' rights* (pp. 340–375). Copenhagen: International Work Group for Indigenous Affairs.

Marris, S. (2016). Diving with a conscience: At Tubbataha Reefs Natural Park, in the Philippines, environmentally conscious travellers dive one of the world's most spectacular sites while supporting local coral reef conservation efforts. *Club Marine, 30*(6), 114–122.

Marshall, C. (2016). Soft landings in paradise. *The Weekend Australian*, June 25–26. *Travel & Indulgence*, p. 12.

Mosby, T. (Exhibition Curator). (n.d.). *Ilan Pasin (this is our way): Torres Strait Art*. Cairns, Queensland: Cairns Regional Gallery.

Muller, S.A., Castillo, G., Castillo, B.D., Solis, R.G., Andreve, J., & Castillo, A. (2003). *Biodiversity and tourism: The case for the sustainable use of the marine resources of Kuna Yala, Panama*. Bonn, Germany: Federal Ministry for the Environment, Nature Conservation and Nuclear Safety.

Mulrennan, M.E. (2007). Sustaining a small-boat fishery: Recent developments and future prospects for Torres Strait Islanders, Northern Australia. In N. Haggan, B. Neis, & I.G. Baird (Eds.), *Fishers' knowledge in fisheries science and management* [Coastal Management Source Books 4], (pp. 183–198). Geneva: UNESCO Publishing.

Nga Kai O Te Moana, Ministry of Maori Development. (1993). Customary Maori fisheries. In D.A. Posey. (Ed.), (1999), *Cultural and spiritual values of biodiversity: A complementary contribution to the global biodiversity assessment* (pp. 422–423). London: Intermediate Technology Publications, and Nairobi, Kenya: United Nations Environment Program.

Oviedo, G.T. (2003). *Lessons learned in the establishment and management of protected areas by Indigenous and local communities in South America*. WCPA Ecosystems, Protected Areas and People (EPP) Project. Gland, Switzerland: International Union for Conservation of Nature.

Peteru, C. (1999). Western Samoan views of the environment. In D.A. Posey. (Ed.), *Cultural and spiritual values of biodiversity: A complementary contribution to the global biodiversity assessment* (pp. 421–422). London: Intermediate Technology Publications, and Nairobi, Kenya: United Nations Environment Program.

Pet-Soede, L., Tabunakawai, K., & Dunnais, M.A. (2011). *The Coral Triangle* [photobook]. Manilla, The Philippines and Gland, Switzerland: Asian Development Bank and World Wildlife Fund for Nature.

Phelan, M.J. (2007). Tropical fish aggregations in an Indigenous environment in northern Australia: Successful outcomes through collaborative research. In N. Haggan, B. Neis, & I.G. Baird (Eds.), *Fishers' knowledge in fisheries science and management* [Coastal Management Source Books 4], (pp. 169–181). Geneva: UNESCO Publishing.

Poepoe, K.K., Bartram, P.K., & Friedlander, A.M. (2007). The use of traditional knowledge in the contemporary management of a Hawai'ian community's marine resources. In N. Haggan, B. Neis, & I.G. Baird (Eds.), *Fishers' knowledge in fisheries science and management* [Coastal Management Source Books 4], (pp. 119–143)]. Geneva: UNESCO Publishing.

Saila, S.B., Kocic, V.L., & McManus, J.W. (1993). Modelling the effects of destructive fishing practices on tropical coral reefs. *Marine Ecology Progress Series, 94*(1), 51–60.

Secretariat of the Convention on Biological Diversity (SCBD). (1998). *A programme for change: Decisions from the Fourth Meeting of the Conference of the Parties to the Convention on Biological Diversity, Bratislava, Slovakia 4–5 May 1998*. Montreal: SCBD.

Secretariat of the Convention on Biological Diversity (SCBD). (2004). *Guidelines on biodiversity and tourism development*. Retrieved from https://www.cbd.int/doc/publications/tou-gdl-en.pdf

Secretariat of the Convention on Biological Diversity (SCBD). (2015). *Tourism supporting biodiversity: A manual on applying the CBD guidelines on biodiversity and tourism development.* Retrieved from https://www.cbd.int/tourism/doc/tourism-manual-2015-en.pdf

Sherwell, P. (2016). A world of water: Head to Palawan in The Philippines for stunning seascapes. *The Weekend Australian*, June 18–19. *Travel & Indulgence*, pp. 12–13.

Spalding, M.D., Ravilious, C., & Green, E.P. (2001). *World Atlas of Coral Reefs.* Prepared at the United Nations Environment Program World Conservation Monitoring Centre. Berkeley, CA: University of California Press.

State of the Tropics. (2014). *State of the Tropics 2014* [Report]. Cairns, Australia: James Cook University.

Steiner, W.W.M. (1999). The loss of cultural diversity and marine resource sustainability: The impact in Hawai'i. In D.A. Posey. (Ed.), *Cultural and spiritual values of biodiversity: A complementary contribution to the global biodiversity assessment* (pp. 417–421). London: Intermediate Technology Publications, and Nairobi, Kenya: United Nations Environment Program.

Teave, E., & Cloud, L. (2014). Rapa Nui National Park, Cultural World Heritage: The struggle of the Rapa Nui People for their ancestral territory and heritage, for environmental protection, and for cultural integrity. In S. Disko & H. Tugenhadt (Eds.), *World Heritage Sites and Indigenous peoples' rights* (pp. 402–421). Copenhagen: International Work Group for Indigenous Affairs.

Tratalos, J.A., & Austin, T.J. (2001). Impacts of recreational SCUBA diving on coral communities of the Caribbean island of Grand Cayman. *Biological Conservation, 102*(1), 67–75.

United Nations Centre for Human Rights. (1994). *A compilation of international instruments* [Volume 1, Second Part, Universal Instruments]. Geneva: United Nations Centre for Human Rights, pp. 475–489.

United Nations World Tourism Organization (UNWTO). (2012). *UNWTO tourism highlights, 2012 edition.* Madrid: United Nations World Tourism Organization.

Verschuuren, B., Wild, R., McNeely, J.A., & Oviedo, G. (Eds.). (2010). *Sacred natural sites: Conserving nature & culture.* London: Earthscan.

Vianna, G.M.S., Meeuwig, J.J., Pannell, D., Sykes, H., & Meekan, M.G. (2011). *The socioeconomic value of the shark-diving industry in Fiji.* University of Western Australia, Perth: Australian Institute of Marine Science.

Vierros, M. (2004). Some considerations on marine and coastal protected areas network design. In *Biodiversity issues for consideration in the planning, establishment and management of protected area sites and networks* (CBD Technical Series No. 15, pp. 52–57). Montreal: Secretariat of the Convention on Biological Diversity.

World Travel and Tourism Council (WTTC). (2015). *Travel and Tourism: Economic Impact 2014.* World Travel and Tourism Council. Retrieved from www.wttc.org/-/media/files/reports/economic%20impact%20research/economic%20impact%202015%20summary_web.pdf

World Wildlife Fund for Nature (WWF). (2015). *Living Blue Planet Report: Species, habitats and human well-being.* Gland, Switzerland: WWF International, and London: Zoological Society of London.

17

MEDIA IN CORAL REEF TOURISM MANAGEMENT

Indications from online travel magazines

Anja Pabel and Glen Croy

Introduction

Coral reefs are some of the most fragile environments on the planet, and also the most susceptible to change. Unfortunately, they have been in decline for some time (Hughes et al., 2003), and change is occurring faster than reefs can cope with on two fronts. First, climate change-induced coral bleaching presents a massive threat to coral reefs (Hoegh-Guldberg et al., 2007). Climate change also has an effect on weather systems, which can lead to an increased prevalence and intensity of wind storms such as hurricanes and tropical cyclones (Webster et al., 2005), which can damage coral reef ecosystems (De'ath et al., 2012). Second, reefs attract tourism development and tourists. As other chapters in this book have highlighted, tourist activities on coral reefs can cause significant damage through poor boating management, poor quality coastal developments, inappropriate fishing and pollution (Saphier & Hoffmann, 2005; Danovaro et al., 2008; Sarmento & Santos, 2012). Although sometimes part of the problem, humans can also be part of the solution.

This chapter focuses on the impact of tourism in changing natural environments, and particularly how these tourism impacts could be better managed through understanding the media's role in generating expectations of experiences and behaviours. This understanding might also assist in modifying tourist behaviours that are considered to negatively impact coral reefs.

A typical tourism experience can be divided into five phases (Fridgen, 1984). The first phase is anticipation, in which travel and experience motivations are developed, destinations are shortlisted and a decision is made. Second, the tourist travels to the destination. Third, the tourist has the destination experience, followed by travelling home (fourth), before, fifth, recollecting the experience. In the third phase, when the tourist is on-site, they have already committed significant resources of time and money to satisfy previously developed motivations and experience expectations. At this point, management efforts to eliminate or mitigate tourist behaviours are likely to have a marginal effect. Consequently, the phase to best manage the experience is during the anticipation phase when expectations are being created (Moyle & Croy, 2009). However, if anticipation-phase management interventions do not align with tourists' desires, the destination may be eliminated from the tourists' shortlist of possible destinations. This could be problematic for destinations that have developed a relationship dependent on coral reefs (Pabel & Coghlan, 2011; Ramis & Prideaux, 2013).

Additionally, due to the broad media exposure reefs have gained as tourist attractions, and the recreational experiences they offer, the role of the destination to influence destination experiences might well be overstated (Cater, 2009). In this context, the media can be very influential. Conversely, Destination Management Organisations (DMOs) may have limited influence and may be constrained by their stakeholders in implementing strategies to initiate changes to the destination's profile. In this context, the authors' aim is to explore how reefs are represented in the media, and through this representation examine how DMOs may attempt to influence the destination and tourists for better reef outcomes.

Tourism and media

The media has many significant tourism roles in protected areas (Moyle & Croy, 2009; Cini & Saayman, 2013; Saarinen, 2016). Most prominent has been the media's role in raising awareness and constructing images of destinations and the potential tourist experiences available in protected areas. The destination images constructed may not accurately reflect the actual destinations, although they are the ones used by tourists to inform their anticipation-phase motivations, expectations and decisions. Specifically, studies have shown that it is the audience perceived destination image, often informed by multiple sources, that is important (Jin et al., 2015). Tourists' perceived destination image can generate demand for experiences that the area cannot provide, especially in conservation environments already negatively affected by tourism (Kihima, 2014). This is exemplified by reef sites that have suffered significant damage through tourism activity (Danovaro et al., 2008; Sarmento & Santos, 2012), and as a consequence are unable to give tourists the level of satisfaction they were expecting (Prideaux et al., 2017).

At the same time, some more pristine reefs may not have a developed image and have a much less developed tourism activity (Maingi et al., 2016). Protected areas may benefit from not having a prominent image, thus limiting the probable tourism impact. Many protected areas lack a prominent image despite considerable attempts to develop one (Žugić et al., 2015; Maingi et al., 2016). Indeed, tourist demand for a protected area is often used as justification for the area's protection and funding and also as a direct funding source for protection efforts (Crompton, 2008, 2009). However, if tourist demand does materialise, it is difficult to then change the perceived destination image and reverse or lessen the demand (Crompton, 2009; Kihima, 2014).

Destination image

The disparity between the perceived versus the portrayed image is important. Gartner (2001), in his definition of destination image, explicitly drew the distinction between the perceived image a person holds and the image attempted to be portrayed by the destination. The emphasis on the perceived image is due to its influence in tourist behaviour and decision making (Lin et al., 2007; Moyle & Croy, 2009; Kruger & Saayman, 2010). The perceived image is comprised of two components, the cognitive which is developed first, and the affective, which is added to the cognitive (Lin et al., 2007; Cini & Saayman, 2013; Jin et al., 2015). The cognitive component reflects the known "facts" of the destination, whereas the affective is the developed "feelings" and emotional associations with the place. In constructing an overall image, tourists use a range of sources. One of the main sources are friends and family, whilst information directly from the destination is considered a less popular source (Moyle & Croy, 2009; Žugić et al., 2015; Maingi et al., 2016). As such, the image portrayed by the destination may have limited potential to influence tourists' perceived image.

Sources also have different levels of credibility, and with higher levels of credibility comes increased influence on the image created (Gartner 1993; Moyle & Croy, 2009). The media is generally seen to have higher levels of credibility and consequent image influence than destination efforts. Saarinen (2016), for example, reported that perceptions of wilderness areas are created through the media, and these direct tourist behaviours. Unfortunately for destinations, their contributions are often perceived as the least credible, additionally limiting the potential influence on image creation (Gartner, 1993). Furthermore, once an image has been created, it is a difficult and long process to change it (Croy, 2010; Kihima, 2014). For example, Kihima (2014, p. 56) stated that "the idea of what is beautiful and worthy of admiration has not radically changed over the past four decades". Overall, the media has high perceived credibility levels and greater reach and engagement, which collectively enables a more complex destination image to be created. Once the image is created, it can have a lasting effect on tourist behaviour.

The framing effect

The framing effect is the way that the media draws the public's eye to specific events and topics. Research on framing effects focuses on interpretations, attitudes and opinions that emerge as a result of the influence that different presentations have about certain issues (Lecheler & De Vreese, 2012). The way in which certain topics are presented creates a frame for that information, and how that information is perceived by individuals can cause changes to how destinations appear. In this regard, the media is sometimes criticised for misinterpreting particular events or for producing distorted images of some destinations, especially in the context of risks and disasters (Chew & Jahari, 2014; Walters et al., 2016). However, due to the media's perceived credibility, what is presented is influential in the creation and change to the audience's perceptions. As a consequence, when the media presents a distorted image that is adopted by the audiences (due to the media's perceived credibility), it is critiqued by those that see the disconnect between the projected image and the "actual" image.

In his book on framing analysis, Goffman (1974, p. 24) describes how individuals make sense of the world around them using schemes of interpretation called "primary frameworks." Depending on how the media portrays certain topics and events, the type of frame that is employed "provides a way of describing the event to which it is applied" (Goffman, 1974, p. 24). In the tourism context, research on framing effects typically sets out to investigate positive and negative aspects of a topic or event, which is referred to as an equivalence or valence framing effect (Kapuscinski & Richards, 2016). For example, previous tourism studies applied framing to determine how message frames influence booking intention and trust (Sparks & Browning, 2011); to evaluate tourists' attitudes to environmental protection (Kim & Kim, 2014); and to investigate hotel guests' intention to reuse hotel linen (Blose et al., 2015). These studies showed that variations in message presentations can lead to changes in the audiences' perceptions and attitudes.

Thus, the media's presentation of coral reefs is a potential driver of the audience's attention and interpretation, forming a basis to their perceived image and consequent behaviours. If the media presents a distorted image of coral reefs, the audience is also likely to have a distorted image, creating management issues for DMOs in coral reef destinations.

Managing tourism through media

Given the influence of media on potential tourists' experiences, there is a need to manage these images. Previous studies in natural and protected areas have demonstrated opportunities to manage on-site behaviours through pre-site media exposure (Moyle & Croy, 2009).

Media created images of places and experiences help frame people's perceptions and expectations, which are then acted out as tourist behaviours and achieve satisfaction at the destination. Tourists' communications to others then generate further awareness, perceptions and expectations. Through this image-driven process, sometimes referred to as the hermeneutic circle (Jenkins, 2003; Ponting, 2009), tourists wish to recapture the images and experiences they were exposed to pre-site when on-site, to demonstrate post-site. As indicated, this presents two potential intervention points. The first is in the creation of pre-site imagery and the second occurs when the tourists are on-site.

Nonetheless, within the coral reef context, changing tourist behaviour could create collateral damage for the tourism-reliant surrounding areas (Fitzsimmons, 2009; Pabel & Coghlan, 2011). In these contexts, in addition to protecting and enhancing the reef system, consideration of local communities is also needed. This requires another image management process, which is referred to as positioning, to garner community support for a changing emphasis and possibly further funding

(Crompton, 2000, 2008, 2009). Managing tourist image and positioning through the media is complex. However, given the media's reach and credibility, it is an important component of reef tourism management.

For tourist image management, a four-step model is presented (Croy, 2010). This starts with assessing the destination image (actual and ideal), periodically measuring the image, comparing the actual (measured) image with the ideal image and, finally, developing strategies to address the image differences. What is evident in some coral reef destinations is that an enticing perceived image has generated tourist demand for a range of activities. Unfortunately, increased demand and the impacts this has caused, has led to a shift in the actual image moving away from the perceived image. This indicates the emergence of an image management problem. Furthermore, there are challenges in identifying the ideal image due to the complexity of stakeholders' demands. For example, it appears in some cases that DMOs are attempting to boost further tourist demand using coral reefs as primary attractors, whilst reef protection agencies are attempting to limit demand and impacts (Coghlan et al., 2017). Combined with a mismatch between the perceived and real image, this presents a further problem and level of complexity.

The media is a suggested means to align the images. However, the more obvious the destination's role, the less credible and influential any media strategies will be (Gartner, 1993; Croy, 2010). Croy (2010) recommended that the media be used as part of the solution to influence the image through the use of hosted media familiarisation programmes. In these programmes the media is introduced to stories and experiences at coral reefs that are aligned with the real image and desired portrayal. In addition, when media portrayals are not aligned with the real image, destination sources should connect with the distorted perceived image in an effort to align it with the desired one. As an example, a hosted travel writer is introduced to the scientific monitoring and protection efforts of a particular reef, as well as its eco-accredited tourism businesses, with the aim of demonstrating that in spite of the effects of climate change tourist are still able to visit the reef subject to specific behavioural constraints. The travel writer then publishes the story which is written in a style that is aligned with the desired image. Through an image management plan, the destination can then align with and further promote that media image. Continuing the example, another story emerges that presents a distorted image of the reef's condition and available tourist behaviours. The image management plan acknowledges the distorted image and attempts to align it through, for example, noting the impact of inappropriate tourist behaviours on the reef.

The difficultly for the destination arises when attempting to generate a consistent response to distorted images. Some destinations may elect to promote distorted images, for example to maximise business opportunities (Lemelin et al., 2010). Generating a consistent destination response requires the destination community to have an agreed understanding of the needs and goals. This can be achieved through positioning. Within protected areas, positioning has largely focused on generating a shared understanding of the goals of the protected area agency (Crompton, 2009). Within this chapter's context, the focus is on positioning the protected area itself, i.e. the coral reefs.

Coral reefs need to be perceived as beneficial to all community stakeholders, not just users, businesses and researchers (Crompton, 2000). In a number of cases, there have been attempts to change peoples' perceptions of the importance of protecting reefs. For example, the Great Barrier Reef Foundation (2017) has released reports on the economic and employment benefits of the Great Barrier Reef to the local area, state and Australia. This is an example of an attempt to psychologically position the reef as a nationally significant contributor and hence emphasise the importance of investing in its protection (Crompton, 2000, 2009). However, the position needs to resonate with a range of stakeholders, their desired goals and the desired image. In the case of the Great Barrier Reef, it needs to be consistently perceived as the distinctive means to achieve the collective stakeholders' goals (Crompton, 2008). Positioning a reef's importance based on its economic contribution might, for example, prevent stakeholders from seeing its inherent value and conservation goals leading to a justification for economic exploitation. A slight positioning shift, for example, could provide the desired outcomes. A *quality* reef provides a significant economic and employment contribution; however, any reduction of reef quality significantly reduces the possible contributions. Two common challenges of positioning efforts, as highlighted earlier, are the perceived credibility and reach of the message.

Reefs in the media

For indications of reef images and how reefs are framed within a tourism context, we investigated a relatively credibly perceived tourism media. In particular, we focused on travel magazines, with their increasing international reach through online publication. Travel magazines are a popular media tool for tourism-related communication. Whilst the general media is mostly outside the control of destination managers, travel magazine are one means by which destination images can be reinforced or enhanced using positive image management strategies. Travel magazine as effective storytelling tools can encourage potential tourists to start dreaming about their next holiday and also create travel expectations. Although travel magazines may not always result in booking behaviours, they are still successful in raising awareness and educating potential tourists about certain destinations. In this research we examined coral reef images and how they are framed. The findings provide us with indications of reef image management and positioning needs.

Methods

This exploratory research reviewed online travel magazines to explore the image representations of coral reef destinations. Previous studies show that travel magazines are an influential source of information in forming tourists' destination image and in decisions regarding specific destinations, attractions and activities (Andereck, 2007; Tang et al., 2011; Hsu & Song, 2013). No previous studies have comprehensively identified how coral reefs are represented in online travel magazines and this research attempts to close this gap by addressing the aim of how coral reefs are represented in the media. To achieve this aim, the following research objectives are addressed. The first research objective seeks to identify the major reef-related themes that emerge from the online travel magazine articles. The second research objective explores the framing-sentiment of representations (positive, neutral or negative) driven by how the content about coral reefs is portrayed in the selected travel articles.

To identify the indicative media representation of coral reefs, five prominent international travel magazines were selected: *Condé Nast Traveler; Elite Traveler; Global Traveler; National Geographic Travel;* and *Travel and Leisure.* Each magazine's online version was searched using the keyword "coral reef" (no date parameters were set). The first two pages of search results for each were collected (ranked by order of relevance). A total of 109 online travel magazine articles were collected. Five were deleted because they did not include a single mention of coral reefs. The contents of the remaining 104 articles were analysed for emergent themes. Both authors completed this process and consensus was reached.

To investigate the framing of coral reefs, we classified the *sentiment* framing of representations driven by the article focus in relation to the coral reef portrayal in the travel magazines as positive, neutral, or negative (c.f. Kapuscinski & Richards, 2016). These three classifications were based on the following definitions:

- Positive frame: The content of the article was focused on enticing tourists to travel to a specific coral reef destination mentioned in the article.
- Neutral frame: The content of the article was focused on other destination attributes and travel-related information.
- Negative frame: The content of the article was focused on messages causing reasons for concern about coral reefs.

Findings and discussion

The coral reefs mentioned in the articles included a diverse range of destinations including the Great Barrier Reef, South East Asia, the Caribbean, the Red Sea and the South Pacific. The first objective of this research was to identify the major themes that emerged from the online travel magazine articles. Table 17.1 shows emergent theme representations of coral reefs in the travel magazines in the final column. Multiple themes were identified in some articles and, as a consequence, the percentage totals more than 100%. To address the second research objective, sentiment framing analysis of the coral reef themes was undertaken. The relationships between the framing and themes are presented in the middle columns. The numbers presented beneath each sentiment framing category represent the percentage of articles within that column. Specifically, 83% of the positively framed articles highlighted travel attractions and activities, whilst only 10% of the negatively framed articles did. Overall, 61% of all the articles highlighted travel attractions and activities.

TABLE 17.1 Coral reef themes and framing found in the travel magazine articles

Theme	Coral reef sentiment framing			Overall percentage
	Positive	Neutral	Negative	
Highlighting travel attractions and activities	83	41	10	61
Educating potential tourists about reef ecology	37	24	70	36
Highlighting the importance of reefs as habitat/marine life attraction	39	8	20	26
Highlighting the protection of natural reefs	23	16	20	20
Coral bleaching/death due to climate change	16	14	60	19
Reef damage due to human impact	4	5	40	8
Aquarium/underwater park or museum	5	11	0	7
Discovery of new reef systems	5	3	0	4
Cruising/cruise ship	4	5	0	4
Coral bleaching/death due to natural events	4	0	10	3
Reef mentioned as part of a name, i.e. hotel or restaurant	0	8	0	3
Dangers of snorkelling and diving on reefs	0	0	10	1
Older age specific dangers	0	0	10	1
Percentage of total articles	55	36	10	193

The travel magazines presented diverse coral reef destination images, focusing on aspects such as travel attractions and activities, educating potential visitors about reef ecology and highlighting the importance of reefs as habitat and marine life attractions. Unsurprisingly, the main themes highlighted reef-based travel attractions and activities (61% of all articles). There was also evidence of emerging themes including reef conservation and appropriate behaviours. These were complemented by themes highlighting human impacts and inappropriate behaviours. The frames within which these themes were presented were also interesting and will be discussed with examples.

Fifty-five percent of the articles were positively framed, 10% negatively and 36% were neutrally framed (focused on other destination attributes and behaviours). In the first instance, it is interesting that only 55% of the articles positively framed coral reefs. Given the travel magazine context, we assumed this percentage would be much higher. Interestingly, the positively and negatively framed representations might actually combine to have the same tourism consequences of enticing tourists to visit coral reefs. For the former, due to the positively framed themes, and for the latter, due to the climate change-related "last-chance tourism" (Piggott-McKellar & McNamara, 2016).

A number of specific patterns emerged from the relationships between the frames and themes. Unsurprisingly, the majority of positively framed representations highlighted travel attractions and activities themes (83%). Examples of these activities promoted coral reefs as premier snorkelling and scuba diving destinations. Another theme in the positive framing category was the importance of reefs as habitats for marine life (39%). Various articles attempted to promote reefs as an aquatic universe where encounters with exotic marine life (sharks, whale sharks, manta rays, turtles, etc.) are possible. A further positively framed theme highlighted the importance of protecting natural reefs (23%). In this case, the articles focused on coral reef conservation and restoration programmes that potential tourists would be able to partake whilst on holiday, specifically planting coral; removing destructive seaweeds; making a commitment to certain coral reef conservation initiatives; or taking an educational tour on a damaged coral reef system. Using travel magazine articles to showcase these and similar initiatives is important from a reef tourism management perspective, and reflects Croy's (2010) image management process.

In regards to neutrally framed representations of coral reefs, only smaller relationships with themes emerged. First, 41% of neutrally framed articles related to attractions and activities themes, and 24% related to educating potential tourists about reef ecology. Articles on attractions and activities encouraged potential tourists to snorkel or dive on specific artificial reefs (diverting pressures from natural reefs); to visit specific aquariums or underwater museums; or to promote newly built hotels/beach resorts close to reefs. In educating potential tourists about reef ecology, some articles indicated how impacts on coral reefs could be curbed by banning sunscreen or developing reef-safe sunscreens and encouraging participation in 360-degree virtual-reality underwater tours.

The negatively framed representations related predominately to three themes: educating potential tourists about reef ecology (70%); coral bleaching due to climate change (60%); and damage to reefs due to human impact (40%). The articles on educating potential tourists about reef ecology stated that many reefs were in danger due to coral bleaching reaching record levels because of continued high temperatures. Damage to reefs also occurred due to human impacts and some articles raised awareness about boat/ship damage to coral reefs.

To better manage recreational ecosystems, it is necessary to consider the perceptions of its users (Daily, 2000). The coral reef themes and frames appear to broadly reflect the situation many coral reefs are facing due to climate and tourism induced change. That is, there appears to be an alignment between the media representation and actual reef situations. This was not anticipated by the authors. This finding indicates that implementation of image management processes to reflect the possible reef experiences has not been seen as a priority by many DMOs. Collectively, it appears that many articles were framing appropriate expectations that were likely to be met and likely to influence potential tourists (Andereck, 2007; Tang et al., 2011; Hsu & Song, 2013).

Indeed, most negatively framed articles emphasised the challenges reefs are facing, and are educating potential tourists about impacts and appropriate behaviours to avoid exacerbating these impacts. A number of the positively framed and attraction-based themes also included references to challenges and impacts faced by coral reefs. We view these as an example of the use of the media to push a reef conservation image, aligning the influential portrayal to the destination reality. These magazine articles set out to raise awareness of the stresses experienced by coral reefs worldwide due to climate change and to manage on-site tourist behaviours through this kind of pre-travel media exposure (Moyle & Croy, 2009).

Whilst framing of this type indicates reduced environmental condition, which has been shown to affect tourists' behaviour in marine protected areas (Petrosillo et al., 2007), the collective framing and its reflection in the generated image assumes that the magazine readership engages with a range of articles about coral reefs that can also inform the perceived image (c.f. Goffman, 1974). Therefore, the subjectivity of these perceptions must be taken into account. For instance, Dwyer and Kim (2003) point out that it is not so much the real but the perceived environmental quality that influences the decisions of the potential visitors, reinforcing Gartner's (2001) distinction between the perceived and portrayed image. Environmental perceptions and related attitudes may vary based on people's demographic factors such as nationality, gender and experience (Petrosillo et al., 2007; Cini & Saayman, 2013; Jin et al., 2015).

Coral bleaching or coral death due to climate change (19%) and natural events (3%) was also mentioned in the travel magazine articles. This appears to reflect other media's representations of reefs, including documentaries and newspaper articles, on reporting the negative aspects of climate change and its consequences for coral reefs. Although this could be critiqued as "cherry-picking" the worst affected reef sites to expose global mass coral bleaching events (Michael, 2017), this is increasingly the reality of peoples' perceptions. Undeniably, exposure to information about environmental quality affects overall perceptions of people intending to engage in certain leisure activities (Petrosillo et al., 2007).

Overall, indications are that the themes and frames reflect the portrayal of reefs as being fragile places increasingly impacted by climate, natural and human activity, as well as places for holidays and leisure. We assume that this representation is created by the media as compared to DMOs. We argue that this presentation is tracking the broader media's representation of reefs as places under threat, and these travel magazine articles may be an attempt to align the themes with their audiences' broader image sources. This is an important point to reflect upon – just as coral reefs do not exist in an impact vacuum, nor do the tourists who visit them. Reef tourists do not isolate tourism media sources to solely create tourism images and other sources to create separate images. The overlap means that different media increasingly acknowledge and refer to the other image sources audiences are exposed to.

However, instead of engaging in collaborative reef positioning and image management, it appears that many destinations and reef businesses do not acknowledge these broader conversations and images. What these findings indicate is a positioning priority, where key stakeholders need to be more aligned to real media and audience images (Crompton, 2008). However, many DMOs and operators use selective tourists' experiences of unblemished pristine reefs captured on video and promoted through social media to inform future tourists' desires (Jenkins, 2003; Ponting, 2009). Unfortunately, many of these videos do not hint at the challenges reefs are facing, or how tourists could best limit their impact. Rather than align images to the reality of the reef, operators are increasingly attempting to combat the fragility image through boosterist promotions (Lemelin et al., 2010). From the perspective of the tourist, videos produced by DMOs or operators may not be seen as highly credible (Gartner, 1993), particularly because they might be perceived to be diverging from the dominant themes and frames across the audience's image sources (Croy, 2010). Unfortunately, the lower credibility and reach will have limited impact on potential tourists' perceptions, except to perceive the destination or business as well out-of-step with the perceived reality (whether it is true or not).

To reposition, DMOs need to engage their immediate stakeholders to an aligned and consistent image of their reef habitat (Crompton, 2000, 2009). This image communication needs to align with the desired reef conservation focus, the actual experiences available and complementary land-based experiences (Daily, 2000; Fitzsimmons, 2009; Pabel & Coghlan, 2011) that travel magazines are (collectively) presenting. Importantly, all destination stakeholders will need to buy-in and communicate the same image (Daily, 2000; Crompton, 2008). For example, consistent messages of stories highlighting challenges coral reefs are facing and appropriate behaviours could be displayed on websites. Strategies such as these could be strengthened to move the destination representation and align it with the potential tourists' widely informed perception of coral reefs (Kim & Kim, 2014). The themes portrayed help to develop potential tourists' perceptions, expectations and on-site behaviours.

Conclusion

In conclusion, the beauty and fragility of coral reefs make them popular tourist attractions; however, they are very susceptible to the impacts of tourism (Saphier & Hoffmann, 2005; Danovaro et al., 2008; Sarmento & Santos, 2012) and a range of other impacts (Webster et al., 2005; Hoegh-Guldberg et al., 2007). Image is a conduit between reefs and potential tourists, and these form the basis for tourists' motivations and expectations, which might be increasingly difficult to deliver on-site. Nonetheless, we have argued that the image can and should be managed to achieve reef goals, including conservation. The ability to manage image is problematic. Destination-based marketing messages appear to have lower levels of credibility and reach. In comparison, the media has relatively high levels of credibility and reach, having the ability to generate and change image on an international scale. Indeed, tourism studies in protected areas have long demonstrated the role of media and the influence it has on destination image. The media frames potential tourists' ways of seeing destinations and activities and influences the nature of and desirability of reef experiences.

Whilst it is acknowledged that the international media cannot be managed by a destination, two approaches are put forward to align represented images. The first is the image management approach of monitoring and familiarising media with the real destination image. Second is a collective and consistent reef-focused position based upon the distinctive reef benefits. Within this context, we investigated the coral reef image presented in travel magazines, particularly exploring themes and framing. Although the majority of articles highlighted tourist attractions and activities, we were surprised to find only a small majority of positively framed reefs. Indeed, we propose that travel magazine articles are reflecting the broader reef fragility conversation, which appears to be in stark contrast to many reef destinations and businesses. Overall, the image management approaches proposed might be used in two ways. First, image management allows destinations to better represent their coral reefs and to manage potential tourists' experiences. Second, image management enables destinations to engage with the wider media's conversations about reef fragility and under threat statuses.

Nevertheless, this exploration does have limitations and areas for future research. In an attempt to find indications of coral reef tourism representations we selected a small range of travel magazines and a small number of articles from each. Broadening the media and article search would provide more conclusive findings. Second, the media representations are not necessarily the same as their audiences' perceptions, and future research could investigate the influence that travel magazines have on audiences' perceptions. Third, our focus was on the media representations, and not destinations. There are numerous destinations undertaking huge efforts to enhance reef tourism and tourists' reef experiences and in making active use of the media. These destinations should also be researched for lessons learnt to enhance worldwide reef image management strategies.

References

Andereck, K.L. (2007). Use of travel magazines as a tourism information source. *Tourism Travel and Research Association: Advancing Tourism Research Globally*, 3–12.

Blose, J.E., Mack, R.W., & Pitts, R.E. (2015). The influence of message framing on hotel guests' linen-reuse intentions. *Cornell Hospitality Quarterly*, *56*(2), 145–154.

Cater, C.I. (2009). The life aquatic: Scuba diving and the experiential imperative. *Tourism in Marine Environments*, *5*(4), 233–244.

Chew, E.Y.T., & Jahari, S.A. (2014). Destination image as a mediator between perceived risk and revisit intention: A case of post-disaster Japan. *Tourism Management*, *40*, 382–393.

Cini, F., & Saayman, M. (2013). Understanding visitors' image of the oldest marine park in Africa. *Current Issues in Tourism*, *16*(7–8), 664–681.

Coghlan, A., McLennan, C., & Moyle, B. (2017). Contested images, place meaning and potential tourists' responses to an iconic nature-based attraction 'at risk': The case of the Great Barrier Reef. *Tourism Recreation Research*, *42*(3), 299–315.

Crompton, J. (2000). Repositioning leisure services. *Managing Leisure, 5*(2), 65–75.

Crompton, J. (2008). Evolution and implications of a paradigm shift in the marketing of leisure services in the USA. *Leisure Studies, 27*(2), 181–206.

Crompton, J. (2009). Strategies for implementing repositioning of leisure services. *Managing Leisure, 14*, 87–111.

Croy, W.G. (2010). Planning for film tourism: Active destination image management. *Tourism and Hospitality Planning and Development, 7*(1), 21–30.

Daily, G.C. (2000). Management objectives for protection of ecosystem services. *Environment Science Policy, 3*, 333–339.

Danovaro, R., Bongiorni, L., Corinaldesi, C., Giovannelli, D., Damiani, E., Astolfi, P., . . ., & Pusceddu, A. (2008). Sunscreens cause coral bleaching by promoting viral infections. *Environmental Health Perspectives, 116*(4), 441–447.

De'ath, G., Fabricius, K.E., Sweatman, H., & Putinen, M. (2012). The 27-year decline of coral cover on the Great Barrier Reef and its causes. *Proceedings of the National Academy of Sciences, 109*(44), 17995–17999.

Dwyer, L., & Kim, C. (2003). Destination competitiveness: Determinants and indicators. *Current Issues in Tourism, 6*(5), 369–414.

Fitzsimmons, C. (2009). Why dive? And why here? A study of recreational diver enjoyment at a Fijian eco-tourist resort. *Tourism in Marine Environments, 5*(2–3), 159–173.

Fridgen, J.D. (1984). Environmental psychology and tourism. *Annals of Tourism Research, 11*(1), 19–39.

Gartner, W. (1993). Image formation process. *Journal of Travel and Tourism Marketing, 2*(2–3), 191–215.

Gartner, W.C. (2001). Image, destination. In J. Jafari (Ed.), *Encyclopaedia of tourism* (p. 296). Abingdon, UK: Routledge.

Goffman, E. (1974). *Frame analysis: An essay on the organisation experience.* Cambridge, MA: Harvard University Press.

Great Barrier Reef Foundation. (2017). The value: The economic, social and icon value of the Great Barrier Reef. Accessed 12 June 2017 https://www.barrierreef.org/the-reef/the-value

Hoegh-Guldberg, O., Mumby, P.J., Hooten, A.J., Steneck, R.S., Greenfield, P., Gomez, E., . . ., & Hatziolos, M.E. (2007). Coral reefs under rapid climate change and ocean acidification. *Science, 318*(5857), 1737–1742.

Hsu, C.H.C., & Song, H. (2013). Destination image in travel magazines: A textual and pictorial analysis of Hong Kong and Macau. *Journal of Vacation Marketing, 19*(3), 253–268.

Hughes, T.P., Baird, A.H., Bellwood, D.R., Card, M., Connolly, S.R., Folke, C., . . ., & Lough, J.M. (2003). Climate change, human impacts, and the resilience of coral reefs. *Science, 301*(5635), 929–933.

Jenkins, O. (2003). Photography and travel brochures: The circle of representation. *Tourism Geographies, 5*(3), 305–328.

Jin, N., Lee, S., & Lee, H. (2015). The effect of experience quality on perceived value, satisfaction, image and behavioural intention of water park patrons: New versus repeat visitors. *International Journal of Tourism Research, 17*, 82–95.

Kapuscinski, G., & Richards, B. (2016). News framing effects on destination risk perception. *Tourism Management, 57*, 234–244.

Kihima, B. (2014). Unlocking the Kenyan tourism potential through park branding exercise. *Tourism Recreation Research, 39*(1), 51–64.

Kim, S.B., & Kim, D.Y. (2014). The effects of message framing and source credibility on green messages in hotels. *Cornell Hospitality Quarterly, 55*(1), 64–75.

Kruger, M., & Saayman, M. (2010). Travel motivation of tourists to Kruger and Tsitsikamma National Parks: A comparative study. *South African Journal of Wildlife Research, 40*(1), 93–102

Lecheler, S., & De Vreese, C.H. (2012). News framing and public opinion: A mediation analysis of framing effects on political attitudes. *Journalism & Mass Communication Quarterly, 89*(2), 185–204.

Lemelin, H., Dawson, J., Stewart, E.J., Maher, P., & Lueck, M. (2010). Last-chance tourism: The boom, doom, and gloom of visiting vanishing destinations. *Current Issues in Tourism, 13*(5), 477–493.

Lin, C., Morais, D., Kerstetter, D., & Hou, J. (2007). Examining the role of cognitive and affective image in predicting choice across natural, developed, and theme-park destinations. *Journal of Travel Research, 46*, 183–194.

Maingi, S., Ondigi, A., & Wadawi, J. (2016). Market profiling and positioning of park brands in Kenya (case of premium and under-utilized parks). *International Journal of Tourism Research, 18*, 91–104.

Michael, P. (2017). Great Barrier Reef tourism operators warn of 'exaggerated' reports over latest mass bleaching event. *The Courier-Mail*, 25 March 2017, http://www.couriermail.com.au/news/queensland/great-barrier-reef-tourism-operators-warn-of-exaggerated-reports-over-latest-mass-bleaching-event/news-story/efd5b1a2cf27826eb5f2bbd460ccc6d0

Moyle, B.D., & Croy, W.G. (2009). Media in the pre-visit stage of the tourist experience: Port Campbell National Park. *Tourism Analysis, 14*(2), 199–208.

Pabel, A., & Coghlan, A. (2011). Dive market segments and destination competitiveness: a case study of the Great Barrier Reef in view of changing reef ecosystem health. *Tourism in Marine Environments, 7*(2), 55–66.

Petrosillo, I., Zurlini, G., Corliano, M.E., Zaccarelli, N., & Dadamoa, M. (2007). Tourist perception of recreational environment and management in a marine protected area. *Landscape and Urban Planning, 79*, 29–37.

Piggott-McKellar, A.E., & McNamara, K.E. (2016). Last chance tourism and the Great Barrier Reef. *Journal of Sustainable Tourism, 25*(3), 397–415.

Ponting, J. (2009). Projecting paradise: The surf media and the hermeneutic circle in surfing tourism. *Tourism Analysis, 14*(2), 175–185.

Prideaux, B., Thompson, M., Pabel, A., & Anderson, A. (2017). A preliminary investigation into tourists' reaction to coral bleaching on the Great Barrier Reef. In C. Lee, S. Filep, J.N. Albrecht & J.L. Coetzee (Eds.), *CAUTHE 2017: Time for big ideas? Re-thinking the field for tomorrow.* Proceedings of the 27th CAUTHE conference, Council for Australasian Tourism and Hospitality Education (CAUTHE), 7–10 February 2017, University of Otago, Dunedin, New Zealand.

Ramis, M., & Prideaux, B. (2013). The importance of visitor perceptions in estimating how climate change will affect future tourists flows on the Great Barrier Reef. In M. Reddy & K. Wilkes (Eds.), *Tourism, climate change and sustainability* (pp. 173–188). Abingdon, UK: Routledge.

Saarinen, J. (2016). Wilderness use, conservation and tourism: What do we protect and for and from whom? *Tourism Geographies, 18*(1), 1–8.

Saphier, A.D., & Hoffmann, T.C. (2005). Forecasting models to quantify three anthropogenic stresses on coral reefs from marine recreation: Anchor damage, diver contact and copper emission from antifouling paint. *Marine Pollution Bulletin, 51*(5–7), 590–598.

Sarmento, V.C., & Santos, P.J.P. (2012). Trampling on coral reefs: Tourism effects on harpacticoid copepods. *Coral Reefs, 31*(1), 135–146.

Sparks, B., & Browning, V. (2011). The impact of online reviews on hotel booking intentions and perception of trust. *Tourism Management, 32*(6), 1310–1323.

Tang, L., Scherer, R., & Morrison, A.M. (2011). Web site-based destination images: A comparison of Macau and Hong Kong. *Journal of China Tourism Research, 7*(1), 2–19.

Walters, G., Mair, J., & Lim, J. (2016). Sensationalist media reporting of disastrous events: Implications for tourism. *Journal of Hospitality and Tourism Management, 28*, 3–10.

Webster, P.J., Holland, G.J., Curry, J.A., & Chang, H.R. (2005). Changes in tropical cyclone number, duration, and intensity in a warming environment. *Science, 309*(5742), 1844–1846.

Žugić, J., Gojović, I., & Perazić, M. (2015). Strategic approach to the promotion of protected areas. *Agriculture & Forestry, 61*(3), 87–100.

18

FINDINGS AND RESEARCH ISSUES – WHERE TO GO FROM HERE?

Bruce Prideaux and Anja Pabel

Introduction

It is tempting to conclude the discussion on coral reef tourism on a gloomy note given the declining state of reefs globally, ongoing concerns about climate change, the lack of a long-term vision about marine resources and the environment more generally, the greed of some firms in the business sector in continuing to support environmentally damaging development, and the continuing hesitancy of the global citizenry to take personal responsibility for environmental protection of the world's coral reefs and other environmental resources. But among the gloom are growing areas of hope. Many, though not all, governments have begun to realise the high economic and environmental cost of not protecting the environment. Citizens are also beginning to work together to support a growing number of programmes to protect coral reef ecosystems. More generous science-based funding is allowing the exploration of new techniques to rejuvenate coral reefs and to identify science-based strategies to reduce the impact of human actions. The 2015 Paris Climate Change Accord has produced a broad consensus on the need for all nations to work towards reducing global warming. Finally, there is growing recognition at citizen level that unless urgent action is taken to stop further global warming, many communities will face consequences ranging in extent from mild to catastrophic.

While there are a large number of factors that can be considered about the tourism use of coral reefs, this chapter will focus specifically on issues related to their long-term sustainability, use and management, water quality and climate change. These issues are arguably the most pressing problems faced by coral reefs and the destinations that depend on them. While there are numerous problems confronting coral reef destinations, there are also some grounds for optimism that many of these problems can be successfully addressed, at least at destination and national levels.

Table 18.1 suggests a ten-point policy framework that if adopted and enforced by governments and management agencies will greatly assist in the process of addressing the many issues that are associated with tourism use of coral reefs.

Sustainability

Sustainability was defined in the 1987 Brundtland Report as meeting the needs of the present without compromising the ability of future generations to meet their own needs. This approach to sustainability is based on the assumption of long-term ecosystem stability where the strength of relationships between species, the landscape and weather does not change over time. The triple bottom-line approach first suggested by Elkington (1994) provided a useful tool for extending the measurement of resource use from an economic perspective to include social and environmental aspects. The underlying assumptions of both the triple bottom-line approach and the carrying capacity approach are ecosystem stability. We now know that climate change has injected a significant element of uncertainty into long-term ecosystem relationships. Ecosystems can no longer be assumed to be stable but are more likely to be in a state of flux driven by changes in temperature and weather (Ramis & Prideaux, 2013). Rather than stability, the current situation appears to be moving towards a low order of entropy, or level of disorder, with ecosystem instability leading to increases in migration and outmigration by an increasing number of species. As Prideaux et al. (2010, p. 187) observed, "climate change will force a revision of the way we view the world, or in the language of Urry (1990), the way we gaze on the landscapes that constitute the tourism experience". From this perspective, the triple bottom-line and carrying capacity models will need reengineering to include a dynamic capability that recognises the previous relationships based on stability are being replaced by new and rapidly changing ecosystem relationships. As part of this shift in emphasis there is also an urgent need to extend the original triple bottom-line concept to a quadruple bottom-line with the addition of governance and associated compliance requirements to the original three elements of sustainability. As previous chapters in this book have highlighted, without effective governance backed up by a compliance capability, short cuts that damage the environment will continue.

Ecosystem instability creates problems when attempts are made to estimate the extent to which current use will impact on future use. Increasing temperatures lead to inward and outward migration of species which may change ecosystem resilience. In terms of managing tourism use, this may mean that the current patterns of coral reef use may transition from sustainable in the present to unsustainable in the future. Climate driven uncertainty has reduced the validity of the assumption of long-term ecosystem stability implicit in the 1987 Brundtland Report's definition of sustainability. For this reason, a new understanding of the meaning of long-term sustainability is required. Any new understanding of sustainability must be built around the ideas of future instability, uncertainty and unpredictable change. Because the upper limit to global temperature increase and the time when temperature plateauing will occur are both unknown and unknowable, it is becoming increasingly difficult to establish if current levels of use are sustainable in the future. This observation has significant implications for tourism use of all coral reef systems. Even without change in use intensity, current – apparently sustainable – levels of tourism use may become unsustainable in the future. From a planning perspective, the changing definition of sustainability creates significant problems for the management of contemporary reefs where managers are also conscious of the need to ensure future sustainability.

Drivers of coral reef use and management

Four factors appear to be driving the manner in which coral reefs are used and managed as tourism resources. The first factor is the juxtaposition of the demand for coral reef tourism based on the benefits derived by participants and the economic benefits this demand confers on firms, local communities and the public sector through profits, employment and tax revenue. The second factor is the way that coral reef ecosystems respond to their use by human agencies, including tourists, fishers and land-based industries such as farming. Unlike humans who can adapt to changes in heat and cold by modifying their external environment, reef ecosystems are unable to adapt and may die if the level of change exceeds their biological tolerance levels. The third factor is growing signs of the willingness of human agencies to recognise the limits to which coral reef ecosystems can be pushed before they degrade. The final factor is the willingness of communities and the tourism industry, to allow their use of coral reef ecosystems to be regulated to ensure that they do not degrade the resources that provide them with part of their livelihood. The following discussion focuses on these four factors.

Juxtaposition of demand and supply

Between 2010 and 2030 global tourism arrivals are forecast to grow by an average of 3.3% per annum to reach 1.8 billion (UN World Tourism Organization [UNWTO], 2017). As in the past, coral reef destinations can expect to benefit from this growth, provided their reef systems are not destroyed by climate change or other anthropogenic factors. As a number of chapters in this book have highlighted, numerous coral reef destinations have reported a decline in coral cover as demand for coral reef experiences has increased. Even in Marine National Parks (MNP) such as Brazil's Abrolhos MNP, significant damage has been caused by scuba divers touching coral and generating sediment (Giglio et al., 2016; see Chapter 14 in this volume).

In many destinations, the short-term prospects of an increase in export dollars and jobs through supporting intensification of coastal resort development has blinded governments to the long-term economic cost of the loss of tourism if the attractiveness of coral reefs are diminished. Failure by government to adequately protect the coral reefs under their control is demonstrated by Atzori and Fyall's discussion (Chapter 8 in this volume) on how the public sector has been slow to protect the Florida Keys coral reefs. While acknowledging the value of the Florida Keys as a tourism resource, failure to effectively protect the regions' coral reefs has led to a decline in coral cover to as little as 3% in some areas. *The Guardian* newspaper (Ellis-Petersen, 2018) reported on similar problems with unregulated development and mass tourism in Thailand and the Philippines. In Thailand, Maya Bay on the island of Koh Phi Phi Leh gained almost instant fame after the screening of the 2000 film *The Beach* staring Leonardo DiCaprio. However, intense and unregulated use has led to the loss of most of the islands' coral cover. Boracay Island in the Philippines provides a further example of almost unregulated development leading to significant loss of coral cover through mass tourism. The problems found in the Florida Keys, Koh Phi Phi Leh and Boracay illustrate the type of problems that occur when there is a failure to implement protective measures such as limits based on an assessment of carrying capacity and controls over the manner in which tourist operators are able to use coral reefs.

Coral and other reef organisms have very defined environmental parameters that if exceeded will lead to their decline and even death. If coral reefs are to be used as tourism experiences, the tourism industry must understand the importance of these environmental parameters and ensure that they are not exceeded. It is the tourism industry that must work towards accommodating the environmental needs of coral reefs; the industry cannot expect coral reefs to adapt to the needs of mass tourism. The question of how to balance the economic benefits of coral reef tourism with the cost of protection is important and an issue that needs the close attention of governments.

The response by coral reef ecosystems to human use

A number of chapters in this book have discussed the responses of coral reef ecosystems to human use and how use can be managed to neutralise adverse impacts. Tourism impacts can be measured on a number of scales beginning with the on-site, in-water interaction between tourists and coral. The main impacts at this level are the use of sunscreen lotions (see Chapter 15 in this volume; Botta et al., 2011; Danovaro et al., 2008), touching of coral, impacts from boat anchors on reefs and discharges such as sewage. The next scale occurs on a destination level and includes coastal development to accommodate the tourism industry, infrastructure such as ports, airfields and urban development to accommodate tourists, residents and services. Principle impacts include clearing of mangrove forests, sewerage and other waste disposal, sediment run-off, in-filling of coastal swamps, dredging for ports and reclamation. At the national scale, national environmental and planning policies can have a significant effect on tourism impacts. Declaration of national parks and regulation of use can determine the level of tourism use permitted. Other policies include the use and disposal of plastics and chemicals that find their way into coral reef systems and controls over the production of greenhouse gases that can also indirectly influence coral reef health. At the global level, the scale of tourism impact is indirectly related to the success, or failure, of global strategies to reduce the impact of climate change. If the 2015 Paris Climate Change Accord is successful in limiting global temperature increase to 1.5°C, many coral reef ecosystems may survive. However, if temperatures increase beyond this point, survival becomes problematic.

Managing tourism use of coral reefs

It is apparent that in many destinations the tourism industry has not been able or perhaps is not willing to accept responsibility for self-regulation of its impact on coral reefs. There are a number of explanations for this including a desire for quick profits, a failure to understand that coral reefs are relatively fragile and, in some jurisdictions, corruption. The position of firms within the supply chain in coral reef destinations may be another factor. For example, while the developers of a marina may understand that dredging generates sediment that can kill coral, the developer of a hillside resort may not be aware that sediment run-off can end up in the ocean and affect water quality.

Given that the private sector is a vast and loosely connected network comprised of firms that generally regard profit as the key objective, it is difficult for the private sector to self-regulate its management of the ecosystem. In the face of this failure, responsibility must be assumed by the public sector.

Regulation of the use of coral reefs

As discussed previously, there are number of scales that can be used for assessing how coral reefs may be managed, ranging from an individual scale to an international scale. Achieving reef friendly outcomes at any of these scales requires intervention by the public sector. At the individual level, use of coral reefs may require interventions such as prescription of the type of permittable sun cream and regulating behaviours relating to interactions with coral and fishes. At the destination scale, interventions may include planning controls over coastal development, sewerage discharge and pollution. At the national level, interventions may include designation of some areas as protected areas, licencing of coral reef tourism operators and the use of carrying capacity controls (Zakai & Chadwick-Furman, 2002; Hawkins et al., 2005) backed up with a compliance capability. While carrying capacity remains a widely debated issue, the use of this technique does provide a useful guide to how many people can undertake a particular reef related activity without causing serious damage to the ecosystem. According to Zakai and Chadwick-Furman (2002), carrying capacity in relation to scuba diving depends on three related factors: the vulnerability of organisms found at the dive site; the level of training and environmental awareness of divers; and the presence of anthropogenic stressors that can degrade reefs. Under this approach, calculation of carrying capacity requires detailed understanding of specific sites as well as the level of training that site visitors have.

One of the most popular methods for regulating the use of coral reefs, and other ecosystems, is to designate them as protected areas. The effectiveness of protection, however, relies on the level and form of protection given, allocation of adequate financial resources to achieve desired ecosystem protection outcomes, efficiency of governance organisations and enforcement to ensure compliance by all actors involved in coral reef tourism activities. Funding for research (both social science and scientific) is also required to ensure that sufficient knowledge is available to regulatory authorities to make informed, science-based decisions on issues related to reef protection and use.

Water quality

Water quality is one of the most important factors in coral reef health (De'ath & Fabricius, 2010). A decline in water quality places pressures on coral reef ecosystems that may lead to localised (and non-heat related) coral bleaching, coral diseases and coral death. Coral simply cannot thrive in conditions where water quality is poor. The causes of poor water quality are numerous and include sediment flow from dredging, farming use of pesticides and fertilisers, sewerage run-off, chemical pollution from industry, heavy metals in urban wastewater and even sun creams used by swimmers. Solving water quality issues is complicated because of the range of stakeholders involved and the cost of improving water quality and the cost of funding social science and scientific research.

To understand the range of factors that can directly and indirectly effect water quality it is useful to expand the discussion to include the impact that non-tourism related industries may have on water quality and, through this, the quality of coral experiences that are available for tourists. The following discussion based on Australia's Great Barrier Reef (GBR) expands this discussion to include the concerns of other industries that operate within the GBR hinterland and that may both affect and be affected by the protection of the GBR. Specifically, the discussion focuses on the conflicting demands of three major industry sectors in the GBR hinterland – coal mining, sugar cane production and tourism. While this discussion focuses on the GBR, similar conflicts occur in other coral reef destinations (see also Chapter 7 in this volume). Conflicts occur when the promotion and/or protection of one sector is seen to be at the expense of one or more other sectors sharing the same general area.

Inland from the GBR, the central Queensland region has enormous reserves of coal and in 2014/15 the value of coal exports was AU\$23.5 billion (The State of Queensland, 2017). Coal provides a significant number of jobs and makes a substantial contribution to public sector revenue but is also a major source of global CO_2 emissions that are one of the key drivers of global warming. It matters little if the coal is burnt in Australia or elsewhere, the effect on the atmosphere is the same. Although indirect, coal mining does contribute to the destruction of the GBR through the release of CO_2, but many politicians feel unable to ban coal mining because of its contribution to the state and national economies – an example of focusing on short-term profit rather than long-term environmental cost. Apart from CO_2 production, coal mining in Queensland necessitated the construction of new ports in the GBR region. Construction of these ports required reclamation works and dredging, which can affect water quality. One coal project in particular has become a flash point between conservationists concerned about the GBR and the public sector. A new and heavily criticised mega mine proposed by India's Adani group aims to mine 60 million tonnes of coal per annum for power generation in India (http://www.adaniaustralia.com/). Opponents argue that the net result of the mine will be to add to global CO_2 emission, which will in turn adversely affect the environment including the GBR. Initially, both the Queensland and federal governments supported the projects on economic grounds (jobs and tax revenue) but this support has weakened at state level. At the time this chapter was written, it appeared that the mine might not proceed even though it possessed the required federal governmental approvals. The refusal of the banking system to lend the capital required to establish the mine may lead to the project being abandoned (Topf, 2017).

Sugar cane production underpins the economies of many coastal towns in the GBR catchment. Sugar cane cultivation requires the use of pesticides and fertilisers, both of which may find their way into the GBR through run-off. Intensive agriculture may also increase the sediment load that is discharged into coastal rivers. While the public sector and the Great Barrier Reef Marine Park Authority (GBRMPA) in particular have been very active in promoting reduced fertiliser use and encouraging farming methods that reduce sediment, pesticide and fertiliser run-off continues to cause problems in the GBR lagoon (Furnas, 2003; Fabricius, 2005). The difficulty in reducing agricultural run-off stems from the need for farmers to use fertilisers and pesticides to produce yields that are profitable. If a reduction in pesticide and fertiliser use leads to yields that are uneconomic, farmers may abandon their farms leading to population loss in many of the single industry sugar cane towns in the GBR region. Achieving a balance between the needs of farmers and the damage that is caused to the GBR from farming activity is difficult.

Although a major beneficiary of coral reefs, the tourist industry may also be a contributor to poor water quality because of conflicting agendas within the tourism sector. In Cairns, Australia, for example, it is well understood that the GBR is one of the key pull factors that attracts international tourists. In recent years the global growth in cruise shipping has also led to an increasing number of port calls being made in Cairns by cruise lines. The opportunity to visit the GBR has been one of the key selling points for including the city in cruise schedules. To grow this sector of the tourism industry there has been significant support for deepening the harbour to allow large cruise lines to berth. However, even under the most controlled conditions, dredging has the potential to produce sediment that reduces water quality and, in turn, affects coral health, creating tensions between the actors promoting increased tourism and those advocating greater protection for the GBR.

The preceding discussion highlights the complexity of issues that are faced by destinations that are attempting to balance the ecological needs of coral reef systems with the demands for development not only by the tourism industry but also by other industries in the reef hinterland. Urbanisation of coastal areas, even for purposes other than tourism, adds another degree of complexity. Ultimately, the solution to water quality must lie in: the recognition that reefs are sensitive ecosystems that have little latitude for change beyond narrow tolerances; the findings of scientific monitoring systems that enables environmental stressors to be identified and that can provide information on what changes are able to be tolerated by coral reefs; a governance system that recognises the need to employ the precautionary principle when dealing with the natural environment; and a community that recognises that long-term sustainability of coral reefs yields greater economic benefits than short-term exploitation.

Climate change

One of the major frustrations in efforts to develop mitigation strategies to protect coral reef ecosystems while also determining the extent to which they can be safely used by tourists is that we simply do not know when efforts to mitigate climate change will result in a plateauing of the current annual increase in global temperatures. Four factors are at play in this situation. The principle factor is the continuing lack of an agreed global agenda of the type that enabled the ozone problem to be confronted and ultimately solved through the 1987 Montreal Protocol on Substances that Deplete the Ozone Layer (Guus et al., 2007). Although signs of the negative impacts of climate change are growing rapidly, the hesitancy of many governments to confront the issue in a serious way continues because of the fear of upsetting various special interest groups. A parallel can be drawn with the decades-long struggle to overcome the power of tobacco companies to have smoking accepted by the public as a serious health risk (Dearlove et al., 2002). Unfortunately, a parallel decades-long struggle to have climate change recognised as a major threat to human welfare poses an existential threat to human welfare. Until the global public recognises the threats posed by climate change and pushes governments to take affirmative action it is unlikely that global emissions will plateau at levels that minimise global wide threats to the environment.

The second factor is that the scientific community is unable to provide accurate forecasts of what changes can be expected at ecosystem level and over what time frames. Uncertainty about the extent to which global temperatures will rise to and over what time frames makes prediction difficult. We know, for example, that a rise in water temperature will eventually lead to coral death but we do not know how quickly temperatures will rise nor do we fully understand the factors that underpin the frequency and severity of the El Niño event that caused the 2014 to 2017 global bleaching events. This problem is compounded by the difficulty in developing climate models that accurately predict the extent of change. The problem is further compounded by sections of the international community continuing to resist the adoption of meaningful and enforceable mitigation strategies to halt atmospheric emissions of greenhouse gases.

The third factor relates to various tipping points that exist in global weather systems, as well as at ecosystem level and specific species level (Lenton et al., 2008). A tipping point can be described as the change that occurs when there is a change from one stable state to another stable state. The timing, extent and severity of the commencement of a tipping point is often unpredictable. Two examples demonstrate the type of ecosystem problems that can be caused by tipping points. The Younger Dryas event (approximately 12,900 to 11,700 BP) was the last of several interruptions to the gradual warming that occurred after the Last Glacial Maximum (27,000 to 24,000 BP) (Carlson, 2013). During the Younger Dryas event, temperatures in the northern hemisphere declined between 2°C and 6°C over a few decades reversing the melting of glaciers. Surprisingly, a different set of climatic conditions applied in the southern hemisphere where warming between 0.3°C and 1–9°C occurred (Carlson, 2013). At the end of the event, temperatures in the northern hemisphere had risen between 2°C and 6°C over a short period of time. Speculation over the cause of the event include a shift in the jet stream or a sudden shutdown of the Gulf Stream, which draws warm tropical water into the North Atlantic Ocean (Broecker, 2006). The Earth's geological history is full of unexpected events that have caused a radical shift in weather patterns and temperatures and demonstrates the importance of understanding the potential for tipping points to occur.

The second example points to the potential impact of a temperature related tipping point at ecosystem level. The 2016 and 2017 coral bleaching events can be described as a Mass Mortality Event (MME) where a single catastrophic event can kill large numbers of a particular species over a short time period. The 2015 MME of saiga antelopes in Kazakhstan over a 9-day period has also been linked to weather. The saiga MME was a result of increased humidity and elevated air temperatures in the days before the deaths that apparently triggered opportunistic bacterial invasion of the bloodstream causing septicaemia (blood poisoning) (Kock et al., 2018).

The fourth factor is predicting when the groundswell of public opinion will grow to the level where governments believe that they have a social licence (Moffat & Zhang, 2014) to enact mitigation strategies, even when these strategies requires sacrifices by individuals and communities. The idea of social licence emerged in the mining industry as a response to social risk but may be applied in a wider sense to include the communities' acceptance of the need for government to take action on behalf of the community to combat climate change events, even though the actions taken may bring discomfort to the community. Before government can exercise social licence of this type, society in general and industry in particular must be prepared to take personal responsibility for overcoming the problems they face. In relation to climate change, this has yet to occur. As research by Prideaux et al. (2018) highlights, the citizens of Cairns (a coral reef dependent destination) realise that greater action is required to protect the nearby GBR but they continue to see that it is a task for government, not a personal action. However, at a personal level actions may be as simple as moving away from using plastic, using only coral friendly sun cream, substituting non-renewable grid electricity for renewable power (rooftop solar systems with battery) and public advocacy for climate change mitigation.

Factors that appear to affect the point at which governments believe that they have a social licence to combat climate change include positive public opinion, the acceptance of industry bodies and the views of public opinion makers. It is not unusual for various conservation lobby groups to push a certain environmental view that is strongly opposed by other lobby groups (coastal developers, for example) that stand to lose from greater environmental protection. In the case of sensitive environments, such as coral reefs, the choice is relatively clear cut. If the environment is seen to be valuable, any proposed use must fall within the capacity of the ecosystem to sustain that use without degrading. The problem for regulatory authorities, as highlighted earlier in this chapter, is that it is becoming increasingly difficult to determine the future sustainability of ecosystems that are in a state of climate-driven change.

Another factor related to public opinion, and how the government reacts to public opinion, is how events such as coral bleaching are reported in the media. In a report on the 2015 and 2016 coral bleaching events on the GBR, Prideaux et al. (2018) noted that governments may reduce funding for research and management of the GBR if it forms the opinion that the GBR is on an unstoppable downward spiral and redirect those funds to other natural areas that may provide a better environmental return on funds invested. A similar situation can be noted with adverse publicity of coral bleaching and similar events leading to a decline in tourism (see Chapter 6 in this volume).

Collectively, these factors create a dilemma, best described as a wicked problem for individuals, the community, business and the public sector. The key questions that need to be addressed include when to start mitigation, what form mitigation should take, how are losers compensated and what form enforcement policies should take.

Many reasons for hope

Increasing recognition by communities and governments that coral reefs need protection has generated support for greater protection and in some cases restoration of coral reefs. This concern has assisted in the development of global level mitigation strategies such as the 2015 Paris Climate Change Accord. To complement global wide strategies and ensure their long-term success there continues to be a need for greater public education and participation in coral reef protection programmes. As Scott (2014) noted, there is an increasing number of successful reef restoration projects undertaken by small groups and communities. One example of a project of this type in Thailand is the Save Koh Tao Marine, which was initiated through a group chosen by community members and funded through private donations, merchandise sales and fundraising events (see Chapter 10 in this volume). Another example is the Reef Restoration Foundation, which has begun a small regeneration project at Fitzroy Island off Cairns, Australia. Its pilot project began in December 2017 with six coral growing frames and further expansions planned throughout 2018. This reef restoration project is a not-for-profit social enterprise that aims to create hope and optimism through undertaking practical, tangible and breakthrough solutions that will make positive improvements to the health of the GBR (http://www.fitzroyisland.com/explore-fitzroy/reef-restoration/). Fitzroy Island also hosts the Cairns Turtle Rehabilitation Centre, a non-profit organisation dedicated to rehabilitating sick and injured turtles and releasing them back into the ocean. Six of the word's seven species of marine turtle live in the waters of the Great Barrier Reef World Heritage Area and are exposed to a number of natural and predominantly human-induced threats (http://www.fitzroyisland.com/tours-day-trips/turtle-rehabilitation-tours/).

Coral Restoration Foundation Bonaire is another not-for-profit conservation organisation with a focus of restoring shallow water populations of elkhorn and stag-horn corals off Bonaire and Klein Bonaire located in the Caribbean. Healthy corals are grown in offshore nurseries and then transplanted onto degraded reefs, with the aim of building the resilience of the coral reefs along the coastline of Bonaire and Klein Bonaire. Since its inception in 2012, the organisation has transplanted over 14,000 corals back to Bonaire's reefs (http://crfbonaire.org/mission-history/).

In the Philippines, nearly half of the country's population live near the coast, with coral reefs representing an important source of food and livelihoods. Typhoon Haiyan in 2013 decimated many coral reefs in the east of the country. Concern Worldwide, an international humanitarian organisation, stepped in to help local communities rebuild the Philippines' reefs. In 2014, the organisation funded a coral restoration programme that targeted the reefs of Barangay Polopiña in the munici-pality of Concepcion. Healthy coral fragments were transplanted to artificial reef building blocks made out of concrete by the local communities (https://concern-worldwide.exposure.co/rebuilding-the-philippines-reefs). Programmes that aim to conserve and restore coral reef habitat have emerged in many global reef destinations, including the Seychelles (via the Reef Rescuers Project, http://natureseychelles.org/what-we-do/coral-reef-restoration), Mexico (via the Cozumel Coral Reef Restoration Program, https://reefbuilders.com/2018/02/03/cozumel-coral-reef-restoration-program/#), the Red Sea (via the Khaled bin Sultan Living Oceans Foundation, https://www.livingoceansfoundation.org/red-sea-coral-reef-biodiversity/) and the Caribbean (via the Reef Resilience Network, http://www.reefresilience.org/restoration/).

Other examples of initiatives that are directed towards building coral reef resil-ience include community-based programmes and citizen science initiatives such as the Reef Life Survey (https://reeflifesurvey.com/). The Reef Live Survey is a global project that trains volunteer scuba divers to monitor coral reefs. With support from recreational scuba diving organisations, training is provided to vol-unteers to monitor marine plants and animals. This has contributed to the building of a comprehensive database on marine life. Another example is the Citizens of the Great Barrier Reef foundation (https://citizensgbr.org/) in Australia, which aims to raise environmental awareness of the challenges the GBR is facing and to encourage the general public to learn about the reef.

Many marine tourism operators have also adopted proactive strategies to protect coral reefs and, by becoming coral reef stewards, have been able to take positive action through their business practices and by engaging guests with interpretation and targeted messages about coral reef management and future threats. Although it is apparent that marine tourism operators understand the gravity of the threat posed to coral reefs by climate change and other anthropogenic factors, many are unsure about how to position their education messages or even if they should tell their clients about climate change.

The way forward

For citizens of destinations that depend on coral reef tourism as a major tourism pull factor it is apparent that proactive rather than reactive strategies are required to ensure that coral reef ecosystems are not further damaged by individual and destination level use. This requires a radical change in the way that humans in general treat the environment and its ecosystems. It is now apparent that coral reefs have narrow tolerance levels to changes in water temperature, water quality and how the ecosystem is used for fishing, diving and so on. Preservation of coral reefs must therefore start with the question: "to what level of use can coral reefs be subjected to before the ecosystem begins to degrade?" Once this understanding is achieved it is possible to determine the limits of use and carrying capacity. Moving beyond these limits will cause reef decline, as demonstrated in many of the reef systems described in this book. To answer the question about ideal levels of use requires a comprehensive scientific evaluation and a governance system that is not inclined to weaken protection to accommodate the wishes of politicians and firms, some of which will operate within the sphere of the tourism industry.

The GBRMPA is an exemplar of how good governance backed by extensive science, a cooperative tourism industry and a supportive community is able to ensure that the level of use of the GBR is within scientifically determined parameters for both tourism and fishing use. However, even the GBRMPA has limits in its ability to protect the GBR because of coastal development, agriculture and mining in its land catchment area. The role of the public sector is therefore two-fold, to establish and empower organisations such as the GBRMPA to undertake ecosystem level management of coral reef resources, and to develop and administer planning controls that govern the development and use of resources such as land, urban development and agriculture that may affect coral reefs or other protected areas.

It is now apparent that there are a number of key actions that must scaffold all policies that are concerned with how coral reefs are used as tourism resources. These are illustrated in Table 18.1 and highlight the need to ensure that any use of coral reefs for tourism or other purposes must first recognise that coral reefs have a narrow level of tolerance in relation to water quality and ocean temperature. Any action that exceeds these tolerances will lead to coral degradation and must not be allowed to proceed. The framework outlined in Table 18.1 provides a succinct set of guidelines for policy makers, management authorities and the tourism industry. If adopted, these guidelines will assist in the minimisation of adverse impacts on coral reef ecosystems by the tourism industry.

TABLE 18.1 Key drivers that determine the use of coral reefs by the tourism industry

	Key drivers that must be used to determine tourism use of coral reefs
1	Coral reefs have a very narrow level of tolerance in relation to water quality and temperatures.
2	Exceeding the tolerance levels of coral reef systems will lead to their decline.
3	Industry and individuals need clear enforceable guidelines of use and cannot be trusted to self-regulate.
4	There will always be pressures by business and other special interest groups to make exceptions to planning guidelines.
5	The rate of change in global temperatures is such that it is difficult to predict what future levels of sustainable use will be.
6	Management of coral reefs requires a high level of scientific knowledge and active ecosystem monitoring.
7	Tourism use of coral reefs requires understanding of tourism demand and community resilience.
8	Protection of coral reefs will require negotiating with industries that may have development agendas that conflict with those designed to protect coral reefs.
9	Effective protection will require compliance with an enforcement capability.
10	Traditional Owners' rights in relation to coral reef resources need to be factored into any considerations about the use of reef resources.

Conclusion

This chapter has introduced a range of issues that are currently affecting the health of coral reef ecosystems and discussed how these issues are affecting tourism use of coral reefs. The chapter has specifically highlighted the impact of anthropogenic factors such as climate change, coastal land-use practices and use of coral reefs for food and income generation.

Many of the issues raised in this chapter are complex and include a large number of human actors and ecological interactions that range from ecosystem level to a global level. Given that tourism is best described as a system, it is obvious that any action in one part of the system will affect other parts of the system, often in unanticipated ways. Unanticipated tipping points such as those that caused the Younger Dryas event and Mass Mortality Events are likely to cause significant disruption as all ecosystems move from stability to instability. Global coral bleaching in 2015 and 2016 illustrates the effects that tipping points based on temperature can have on coral reef ecosystems.

It is also apparent that the protection of coral reefs and their long-term use as a tourism resource creates numerous problems for the community, government and the tourism industry. Successfully negotiating these issues is essential if coral reefs are to be protected. The key drivers illustrated in Table 18.1 provide a useful guide to coral reef protection and policy formation.

References

Botta, C., Labille, J., Auffan, M., Borschneck, D., Miche, H., Cabie, M., . . ., & Bottero, J. (2011). TiO$_2$-baesd nanoparticles released in water from commercialized sunscreens in a life-cycle perspective: Structures and quantities. *Environmental Pollution, 159*, 1534–1550.

Broecker, W.S. (2006). Was the Younger Dryas triggered by a flood? *Science, 312*(5777), 1146–1148. doi:10.1126/science.1123253. PMID 1672862

Brundtland, G., Khalid, M., Agnelli, S., Al-Athel, S.A., Chidzero, B., Fadika, L.M., et al. (1987). *Our common future: The world commission on environment and development.* Oxford: Oxford University Press.

Carlson, A. (2013). The Younger Dryas climate event. In S. Elias (Ed.), *The encyclopedia of quaternary science,* vol. 3. Amsterdam: Elsevier, pp. 126–134.

Danovaro, R., Bongiorni, L., Corinaldesi, C., Giovannelli, D., Damiani, E., Astolfi, P., . . ., & Pusceddu, A. (2008). Sunscreens cause coral bleaching by promoting viral infections. *Environmental Health Perspectives, 116*(4), 441–447.

De'ath, G., & Fabricius, K. (2010). Water quality as a regional driver of coral biodiversity and macroalgae on the Great Barrier Reef. *Ecological, 20*(3), 840–850.

Dearlove, J., Bialous, S., & Glantz, S. (2002). Tobacco industry manipulation of the hospitality industry to maintain smoking in public places. *Tobacco Control, 11*, 94–104.

Elkington, J. (1994). Towards the sustainable corporation: Win-win-win business strategies for sustainable development. *California Management Review, 36*(2), 90–100.

Ellis-Petersen, H. (2018). Can a tourist ban save DiCaprio's coral paradise from destruction? Retrieved from *The Guardian,* https://www.theguardian.com/environment/2018/feb/25/can-tourist-ban-save-dicaprios-coral-paradise-thailand-maya-bay-philippines-boracay?CMP=share_btn_link

Fabricius, K.E. (2005). Effects of terrestrial runoff on the ecology of corals and coral reefs: Review and synthesis. *Marine Pollution Bulletin, 50*(2), 125–146.

Furnas, M.J. (2003). *Catchments and corals: Terrestrial runoff to the Great Barrier Reef.* Townsville, Australia: Australian Institute of Marine Science, CRC Reef.

Giglio, V.J., Luiz, O.J., & Schiavetti, A. (2016). Recreational diver behavior and contacts with benthic organisms in the Abrolhos National Marine Park, Brazil. *Environmental Management, 57*(3), 637–648.

Guus J., Velders, S., Andersen, J., Daniel, D., Fahey, D., & McFarland, M.M. (2007). The importance of the Montreal Protocol in protecting climate. *PNAS, 104*(12), 4814–4819.

Hawkins, J.P., Roberts, C.M., Kooistra, D., Buchan, K., & White, S. (2005). Sustainability of scuba diving tourism on coral reef of Saba. *Coastal Management, 33*, 373–387.

Kock, R., Orynbayev, M., Robinson, S., Zuther, S., Singh, N.J., Beauvais, W., . . ., & Milner-Gulland, E.J. (2018). Saigas on the brink: Multidisciplinary analysis of the factors influencing mass mortality events. *Science Advances, 4*(1), eaao2314, doi: 10.1126/sciadv.aao2314

Lenton, T., Held, H., Kriegler, E., Hall, J., Lucht, W., Rahmstorf, S., & Schellnhuber, H. (2008). Tipping elements in the Earth's climate system. *PNAS, 105*(6), 1786–1793. doi: https://doi.org/10.1073/pnas.0705414105

Moffat, K. & Zhang, A. (2014). The paths to social licence to operate: An integrative model explaining community acceptance of mining. *Resources Policy, 39*, 61–70.

Prideaux, B., Carmody, J., & Pabel, A. (2018). *Impacts of the 2016 and 2017 mass coral bleaching events on the Great Barrier Reef tourism industry and tourism-dependent coastal communities of Queensland.* Report to the Reef and Rainforest Research Centre Limited, Cairns.

Prideaux, B., Coghlan, A., & McNamara, K. (2010). Assessing the impacts of climate change on mountain tourism destinations using a climate change impact model. *Tourism Recreation Research, 35*, 187–200.

Ramis, M., & Prideaux, B. (2013). The importance of visitor perceptions in estimating how climate change will affect future tourist flows to the Great Barrier Reef. In M. Reddy & K. Wilks (Eds.), *Tourism, climate change and sustainability*. Oxon, UK: Routledge, pp. 173–188.

Scott, C.M. (2014). *The Koh Tao ecological monitoring program* (2nd ed). Koh Tao, Thailand: Conservation Divers.

The State of Queensland. (2017). 2017 calendar year: Coal sales statistics. Retrieved from https://data.qld.gov.au/dataset/annual-coal-statistics/resource/c522fcaa-89d7-4c76-bd6e-064d39617d38

Topf, A. (2017). Who will fund Adani's Carmichael coal mine? Not China. *Mining.com*, 3 December. Retrieved from http://www.mining.com/will-fund-adanis-carmichael-coal-mine-not-china/

UN World Tourism Organization (UNWTO). (2017). *UNWTO highlights 2017 edition*. Retrieved from https://www.e-unwto.org/doi/pdf/10.18111/9789284419029

Zakai, D., & Chadwick-Furman, N. (2002). Impacts of intensive recreational diving on reef corals at Eilat, northern Red Sea. *Biological Conservation, 105*, 179–187.

INDEX